ALGEBRAIC ANALYSIS.

SOLUTIONS AND EXERCISES

ILLUSTRATING

THE FUNDAMENTAL THEOREMS AND THE
MOST IMPORTANT PROCESSES OF
PURE ALGEBRA.

BY

G. A. WENTWORTH, A.M.,
PROFESSOR OF MATHEMATICS IN PHILLIPS EXETER ACADEMY;

J. A. McLELLAN, LL.D.,
INSPECTOR OF NORMAL SCHOOLS, AND CONDUCTOR OF
TEACHERS' INSTITUTES, FOR ONTARIO, CANADA;

AND

J. C. GLASHAN,
INSPECTOR OF PUBLIC SCHOOLS, OTTAWA, CANADA.

PART I.

BOSTON, U.S.A.:
PUBLISHED BY GINN & COMPANY.
1889.

PREFACE.

THE work of which this volume forms the first or introductory part is intended to supply students of mathematics with a well-filled storehouse of solved examples and unsolved exercises in the application of the fundamental theorems and processes of pure Algebra, and to exhibit to them the highest and most important results of modern algebraic analysis. It may be used to follow and supplement the ordinary text-books, or it may be employed as a guide-book and work of reference, in a course of instruction under a teacher of mathematics.

The following are some of the special features of this volume :

It gives a large number of solutions in illustration of the best methods of algebraic resolution and reduction, some of which are not found in any text-book.

It gives, classified under proper heads and preceded by type-solutions, a great number of exercises, many of them illustrating methods and principles which are generally ignored in elementary Algebras; and it presents these solutions and exercises in such a way that the student not only sees how algebraic transformations are effected, but also perceives how to form for himself as many additional examples as he may desire.

It shows the student how simple principles with which he is quite familiar, may be applied to the solution of questions which he has thought beyond the reach of these principles; and gives complete explanations and illustrations of important topics which are omitted or are barely touched upon in the ordinary books, such as the Principle of Symmetry, Theory of Divisors, and its application to Factoring, and Applications of Horner's Division.

A few of the exercises are chiefly supplementary to those proposed in the text-books, but the intelligent student will find that even these examples have not been selected in an aimless fashion; he will recognize that they are really expressions of certain laws; they are in fact proposed with a view to lead him to investigate these laws for himself as soon as he has sufficiently advanced in his course. Nos. 8, 9, 10, and 11 of Ex. 1 afford instances of such exercises.

Others of the questions proposed are preparatory or interpretation exercises. These might well have been omitted were it not that they are generally omitted from the text-books and are too often neglected by teachers. Practice in the interpretation of a new notation, and in expression by means of it, should always precede its use as a symbolism itself subject to operations. Nos. 23 to 36 of Ex. 3, and nearly the whole of Ex. 15, may serve for instances.

By far the greater number of the exercises is intended for practice in the methods exhibited in the solved examples. As many as possible of these have been selected for their intrinsic value. They have been gathered from the works of the great masters of analysis, and the student who proceeds to the higher branches of mathematics will meet again with these examples and exercises, and will find his progress aided by his familiarity with them, and will not have to interrupt his advanced studies to learn theorems and processes properly belonging to elementary Algebra. In making this selection, it has been found that the most widely useful transformations are, at the same time, those that best exhibit the methods of reduction here explained, so that they have thus a double advantage.

The present volume ends with an extensive collection of exercises in Determinants. These present under new forms and from a different point of view the greater number of the theorems proposed, and many of the general results obtained, in the earlier chapters, and to these they add many important propositions in other subjects; as, for example, in the method of least squares, in linear, homographic, orthogonal, and homaloid transformations, and in the degeneracy and the tangency of quadrics.

The second volume will treat of factorials and the combinatory analysis; finite differences and derived functions, both direct and inverse, of explicit functions of a single variable; expansion, summation, reversion, transformation, and interpolation of series; the arithmetic, harmonic, and geometric series of integral orders, including the theta-functions; recurring series; binomial, logarithmic, and exponential series; hyperbolic and circular functions; trigonometric series, direct and inverse; Legendre's, Bessel's, Lamé's, and Heine's series and their associated functions; double series; infinite products; continued fractions; indeterminate equations; theory of numbers; inequalities; maxima and minima; binomial equations and cyclotomic functions; transformation of binary forms; theory of the quintic and of higher equations; theory of substitutions. The whole will close with a chapter on the fundamental postulates and the general laws of algebra, illustrated by examples and problems in matrices, polar algebras, and ideal arithmetic.

In this second part of the work the authors hope to be able to give numerous historical notes and bibliographical references for the use of students who desire to pursue the subject further, or to consult the original memoirs.

A companion volume to the present is in course of preparation for the use of private students and of all who have not the advantage of instruction by a specialist in mathematics. The companion will contain proofs of the theorems employed and solutions of the exercises proposed in this volume, the whole accompanied by hints on the best method of attacking problems, and on the selection of processes for their reduction.

Notwithstanding that the utmost care has been taken in revising the proof-sheets, there doubtless remain many errors both in the examples and in the exercises. The authors would feel grate-

ful to teachers and students for notification of all errors which
may be discovered, and also for suggestions in relation to the im-
provement of the work.

Messrs. J. S. Cushing & Co. deserve special mention for their
masterly skill in overcoming all the difficulties in the typography
of this work, and for their excellent taste and judgment exhibited
in the beauty and elegance of these pages.

G. A. WENTWORTH.
J. A. McLELLAN.
J. C. GLASHAN.

NOTE. It is due Mr. Glashan to state that the main part of the
work on this Algebra has been done by him.

G. A. WENTWORTH.
J. A. McLELLAN.

CONTENTS.

CHAPTER I.

CHAPTER II.

CHAPTER III.

CHAPTER IV.

CHAPTER V.

CHAPTER VI.

CHAPTER VII.

CHAPTER VIII.

CHAPTER IX.

CHAPTER X.

DETERMINANTS.

ALGEBRA.

—◆◇◆—

CHAPTER I. — SUBSTITUTION.

Ex. 1.

1. If $a=1$, $b=2$, $c=3$, $d=4$, $x=9$, $y=8$, find the values of the following expressions:

$$1 - \{1 - (1 - \overline{1-x})\};$$

$$a - (x-y) - (b-c)(d-a) - (y-b)(x+c);$$

$$x - y[y - (y-a)\{d + c(b-c)\}];$$

$$(x+d)(y+b+c) + (x-d)(a-b-d)$$
$$+ (y+d)(a-x-d);$$

$$(d-x)^3 + (c+y)^3;$$

$$(a-b)(c^3 - b^2x) - (c-d)(b^3 - a^2x)$$
$$+ (d-b-c)(d^3 - a^3);$$

$$\frac{d-a}{d+a} + \frac{d+c}{d-c} - 2\frac{d+b}{d-b}.$$

2. If $a=3$, $b=-4$, $c=-9$, and $2s=a+b+c$, find the values of the following expressions:

$$s(s-a)(s-b)(s-c);$$

$$s^2 + (s-a)^2 + (s-b)^2 + (s-c)^2;$$

$$s^2 - (s-a)(s-b) - (s-b)(s-c) - (s-c)(s-a);$$

$$2(s-a)(s-b)(s-c) + a(s-b)(s-c)$$
$$+ b(s-c)(s-a) + c(s-a)(s-by).$$

3. If $a = 2$, $b = -3$, $c = 1$, $x = 4\frac{1}{3}$, find the values of the following expressions:

$$\frac{a^2 - b^2}{a^3 + b^3}; \quad \frac{a^2 + b^2}{a^3 - b^3}; \quad \frac{(a-b)^2}{(a+b)^3}; \quad \frac{(a-b)^3}{(a+b)^2};$$

$$\frac{a^2 + ab + b^2}{a^2 - ab + b^2}; \quad \frac{a^2 - b^3}{a^3 - b^2}; \quad \frac{x}{2}\left\{\frac{2x-3}{3} - \frac{3x-1}{4}\right\}\frac{x-1}{2};$$

$$\frac{(a+b)\{(a+b)^2 - c^2\}}{4b^2c^2 - (a^2 - b^2 - c^2)^2}; \quad \frac{a^2(b-c) + b^2(c-a) + c^2(a-b)}{(a-b)(b-c)(c-a)}$$

4. If $a = 6$, $b = 5$, $c = -4$, $d = -3$, find the values of the following expressions:

$$\sqrt{(b^2 + ac)} + \sqrt{(c^2 - 2ac)}; \quad \sqrt{\{b^2 + ac + \sqrt{(c^2 - 2ac)}\}};$$

$$\frac{a^2 - \sqrt{(b^2 + ac)}}{2a - \sqrt{(b^2 - ac)}}; \quad \frac{c + \sqrt{(d^2 + c^2)}}{c^3 + 2d(d^2 - c^2)}$$

5. If $x = 3$, $y = 4$, $z = 0$, find the values of

$$\{3x - \sqrt{(x^2 + y^2)}\}^2 \{2x + \sqrt{(x^2 + y^2 + z^2)}\};$$

$$x^y + y^z + z^x; \quad (x - y)^{x-y} + (y - z)^{y-z} + (z - x)^{z-x};$$

$$(x^3 - y^3) \div \sqrt[3]{\{8x^3 + 8(3x^2 + 3xy + y^2)y\}}.$$

6. Find the values of $\dfrac{(x + y + z)^3 - 3(x^3 + y^3 + z^3)}{xyz}$

when

(i.) $x = 1$, $y = 2$, $z = 3$;

(ii.) $x = 2$, $y = 3$, $z = 4$;

(iii.) $x = 3$, $y = 4$, $z = 5$;

(iv.) $x = 10$, $y = 11$, $z = 12$.

7. Given $x = 3$, $y = 4$, $z = -5$, find the values of

$$(x + y + x)^3 - 3(x + y + z)(xy + yz + zx);$$
$$x^2(y + z) + y^2(z + x) + z^2(x + y) + 2xyz;$$
$$x^2(y - z) + y^2(z - x) + z^2(x - y);$$
$$(5x - 4z)^2 + 9(4x - z)^2 - (13x - 5z)^2;$$
$$(3x + 4y + 5z)^2 + (4x + 3y + 12z)^2 - (5x + 5y + 13z)^2.$$

8. If $s = a + b + c$, find the values of
$$(2s - a)^2 + (2s - b)^2 - (2s + c)^2$$
when

  (i.) $a = 3$, $b = 4$, $c = 5$;

  (ii.) $a = 21$, $b = 20$, $c = 29$;

  (iii.) $a = 119$, $b = 120$, $c = 169$;

  (iv.) $a = 3$, $b = -4$, $c = 5$;

  (v.) $a = 5$, $b = 12$, $c = -13$.

9. If $a = 1$, $b = 3$, $c = 5$, $d = 7$, $e = 9$, $f = 11$, show that

$$a + b + c + d + e + f = \left(\frac{a + f}{2}\right)^2;$$

$$\frac{1}{ab} + \frac{1}{bc} + \frac{1}{cd} + \frac{1}{de} + \frac{1}{ef} = \frac{1}{2}\left(\frac{1}{a} - \frac{1}{f}\right);$$

$$\frac{1}{abc} + \frac{1}{bcd} + \frac{1}{cde} + \frac{1}{def} = \frac{1}{4}\left(\frac{1}{ab} - \frac{1}{ef}\right);$$

$$\frac{1}{abcd} + \frac{1}{bcde} + \frac{1}{cdef} = \frac{1}{6}\left(\frac{1}{abc} - \frac{1}{def}\right);$$

$$a^2 + b^2 + c^2 - ab - bc - ca = b^2 + c^2 + d^2 - bc - cd - db$$
$$= c^2 + d^2 + e^2 - cd - de - ec$$
$$= d^2 + e^2 + f^2 - de - ef - fd.$$

10. If $a=1$, $b=2$, $c=3$, $d=4$, $e=5$, $f=6$, $g=7$, show that

$$a+b+c=\tfrac{1}{2}cd; \quad a+b+c+d=\tfrac{1}{2}de;$$

$$a+b+c+d+e=\tfrac{1}{2}ef; \quad a+b+c+d+e+f=\tfrac{1}{2}fg;$$

$$a^2+b^2+c^2=\frac{cd(c+d)}{ab(a+b)};$$

$$a^2+b^2+c^2+d^2=\frac{de(d+e)}{ab(a+b)};$$

$$a^2+b^2+c^2+d^2+e^2=\frac{ef(e+f)}{ab(a+b)};$$

$$a^2+b^2+c^2+d^2+e^2+f^2=\frac{fg(f+g)}{ab(a+b)};$$

$$a^3+b^3+c^3=(a+b+c)^2;$$

$$a^3+b^3+c^3+d^3=(a+b+c+d)^2;$$

$$a^3+b^3+c^3+d^3+e^3=(a+b+c+d+e)^2;$$

$$a^3+b^3+c^3+d^3+e^3+f^3=(a+b+c+d+e+f)^2;$$

$$a^4+b^4+c^4=\frac{cd(c+d)(c^2d-1)}{bc(b+c)};$$

$$a^4+b^4+c^4+d^4=\frac{de(d+e)(cde-1)}{bc(b+c)};$$

$$a^4+b^4+c^4+d^4+e^4=\frac{ef(e+f)(cef-1)}{bc(b+c)};$$

$$a^4+b^4+c^4+d^4+e^4+f^4=\frac{fg(f+g)(cfg-1)}{bc(b+c)};$$

$$c^2+d^2=e^2; \quad c^3+d^3+e^3=f^3.$$

11. Assume *any* numerical values for x, y, and z, and find the values of the following expressions:

$$(x^5 - 10x^3 + 5x)^2 + (5x^4 - 10x^2 + 1)^2 - (x^2 + 1)^5;$$

$$(x+1)^3 - 2(x+5)^3 - (x+9)^3 + 2(x+11)^3$$
$$+ (x+12)^3 - (x+16)^3;$$

$$(x^2 - y^2)^2 + (2xy)^2 - (x^2 + y^2)^2;$$

$$(x^3 - 3xy^2)^2 + (3x^2 y - y^3)^2 - (x^2 + y^2)^3;$$

$$(3x^2 + 4xy + y^2)^2 + (4x^2 + 2xy)^2 - (5x^2 + 4xy + y^2)^2;$$

$$(x-y)^3 + (y-z)^3 + (z-x)^3 - 3(x-y)(y-z)(z-x).$$

§ 1. If $x =$ any number (as, for example, 3), then $x^2 (= x \times x) = 3x$; $x^3 (= x \times x^2) = 3x^2$; $x^4 (= x \times x^3) = 3x^3$; etc. Or, $3 = x$; $3x = x^2$; $3x^3 = x^4$; $3x^4 = x^5$; etc. Hence, problems like the following may be solved like ordinary arithmetical problems in "Reduction Descending."

EXAMPLES.

1. Find the value of $x^2 - 2x - 9$, when $x = 5$.

$$x^2 - 2x - 9$$
$$\underline{5}$$
$$5x$$
$$\underline{-2x}$$
$$3x$$
$$\underline{5}$$
$$15$$
$$\underline{-9}$$
$$6 \ \textit{Ans.}$$

2. Find the value of $x^4 - x^3 - 4x^2 - 3x - 5$ when $x = 3$.

$$x^4 - x^3 - 4x^2 - 3x - 5$$
$$3$$

p_1 . . . . . . $\overline{3\,x^3}$

$ \quad\quad\quad -x^3$

r_1 . . . . . . $\overline{2\,x^3}$

$ \quad\quad\quad 3$

p_2 . . . . . . $\overline{6\,x^2}$

$ \quad\quad\quad -4\,x^2$

r_2 . . . . . . $\overline{2\,x^2}$

$ \quad\quad\quad 3$

p_3 . . . . . . $\overline{6\,x}$

$ \quad\quad\quad -3\,x$

r_3 . . . . . . $\overline{3\,x}$

$ \quad\quad\quad 3$

p_4 . . . . . . $\overline{9}$

$ \quad\quad\quad -5$

r_4 . . . . . . $\overline{4}$ *Ans.*

3. Find the value of $2x^4 + 12x^3 + 6x^2 - 12x + 10$ when $x = -5$. Using coefficients only, we have

$$2 + 12 + 6 - 12 + 10$$
$$-5$$

p_1 . . . . . $\overline{-10}$

$ \quad\quad\quad +12$

r_1 . . . . . $\overline{+\,2}$

$ \quad\quad\quad -\,5$

p_2 . . . . . $\overline{-10}$

$ \quad\quad\quad +\,6$

r_2 . . . . . $\overline{-\,4}$

$ \quad\quad\quad -\,5$

p_3 . . . . . $\overline{20}$

$ \quad\quad\quad -12$

r_3 . . . . . $\overline{8}$

$ \quad\quad\quad -\,5$

p_4 . . . . . $\overline{-40}$

$ \quad\quad\quad +10$

r_4 . . . . . $\overline{-30}$ *Ans.*

§ **2.** If the coefficients, and also the values of x, are small numbers, much of the above may be done mentally, and the work will then be very compact. Thus, performing mentally the multiplications and additions (or subtractions) of the coefficients, and merely recording the partial reductions r_1, r_2, r_3, and the result r_4, the last example will appear as follows:

$$-5)2 \quad +12 \quad +6 \quad -12 \quad +10$$
$$2$$
$$-4$$
$$8$$
$$-30$$

§ **3.** In the above examples, the coefficients are "brought down" and written below the products p_1, p_2, p_3, p_4, and are added or subtracted, as the case may require, to get the partial reductions r_1, r_2, r_3, and the result r_4. Instead of thus "bringing down" the coefficients, we may "carry up" the products p_1, p_2, p_3, p_4, writing them beneath their corresponding coefficients, and thus get r_1, r_2, r_3, r_4, in a third (horizontal) line. Arranged in this way, Exam. 2 will appear

$$3 \begin{array}{|ccccc} 1 & -1 & -4 & -3 & -5 \\ & +3 & +6 & +6 & +9 \\ \hline 1 & +2 & +2 & +3; & 4 \end{array}$$

and Exam. 3 will appear

$$-5 \begin{array}{|ccccc} 2 & +12 & +6 & -12 & +10 \\ & -10 & -10 & +20 & -40 \\ \hline 2 & +2 & -4 & +8; & -30 \end{array}$$

Comparing these arrangements with those first given (Exams. 2 and 3), it will be seen that they are, figure for figure, the same, except that the multiplier is not repeated.

§ 4. When there are several figures in the value of x, they may be arranged in a column, and each figure used separately, as in common multiplication. When only approximate values are required, "contracted multiplication" may be used.

4. Find the value of $3x^5 - 160x^4 + 344x^3 + 700x^2 - 1910x + 1200$, given $x = 51$.

	3	-160	$+344$	$+700$	-1910	$+1200$
1		3	-7	-13	37	-23
50		150	-350	-650	1850	-1150
	3	-7	-13	$+37$	-23;	$+27$

∴ result is 27.

5. Given $x = 1.183$, find the value of $64x^4 - 144x + 45$ correct to three decimal places.

	64	0	0	-144	$+45$
1	64	75.712	89.5673	-38.0419
1	6.4	7.5712	8.9567	-3.8042
8	5.12	6.0570	7.1654	-3.0434
3	0.192	0.2271	0.2687	-0.1141
	64	75.712	89.5673	-38.0419	-0.0036

∴ result is -0.004.

Ex. 2.

Find the value of

1. $x^4 - 11x^3 - 11x^2 - 13x + 11$, for $x = 12$.

2. $x^4 + 50x^3 - 16x^2 - 16x - 61$, for $x = -17$.

3. $2x^4 + 249x^3 - 125x^2 + 100$, for $x = -125$.

4. $2x^3 - 473x^2 - 234x - 711$, for $x = 200$.

5. $x^5 - 3x^2 - 8$, for $x = 4$.

6. $x^6 - 515x^5 - 3127x^4 + 525x^3 - 2090x^2 + 3156x - 15792$, for $x = 521$.

7. $2x^5 + 401x^4 - 199x^3 + 399x^2 - 602x + 211$, for $x = -201$.

8. $1000x^4 - 81x$, for $x = 0.1$.

9. $99x^4 + 117x^3 - 257x^2 - 325x - 50$, for $x = 1\frac{2}{3}$.

10. $5x^5 + 497x^4 + 200x^3 + 196x^2 - 218x - 2000$, for $x = -99$.

11. $5x^5 - 620x^4 - 1030x^3 + 1045x^2 - 4120x + 9000$, for $x = 205$.

Calculate, correct to three places of decimals:

12. $x^3 + 3x^2 - 13x - 38$, for $x = 3.58443$, for $x = -3.77931$, and for $x = -2.80512$.

13. $y^4 - 14y^2 + y + 38$, for $y = 3.13131$, for $y = -1.84813$, and for $y = -3.28319$.

Ex. 3.

What do the following expressions become (i.) when $x = a$; (ii.) when $x = -a$?

1. $x^4 - 4ax^3 + 6a^2x^2 - 4a^3x + a^4$.

2. $\sqrt{(x^2 - ax + a^2)}$.　　　　3. $\sqrt{(x^2 + 2ax + a^2)}$.

4. $(x^2 + ax + a^2)^3 - (x^2 - ax + a^2)^3$.

If $x = y = z = a$, find the value of the following expressions:

5. $(x - y)(y - z)(z - x)$.

6. $(x + y)^2(y + z - a)(x + z - a)$.

7. $x(y+z)(y^2+z^2-x^2)+y(z+x)(z^2+x^2-y^2)$
   $+z(x+y)(x^2+y^2-z^2)$.

8. $\dfrac{x}{y+z}+\dfrac{y}{x+z}+\dfrac{z}{x+y}$.

Find the value of

9. $\dfrac{x}{a}+\dfrac{x}{b}$, when $x=\dfrac{abc}{a+b}$.

10. $\dfrac{1}{a(b-x)}+\dfrac{1}{b(c-x)}+\dfrac{1}{a(x-c)}$, when $x=\dfrac{b}{a}(a-b+c)$.

11. $\dfrac{x}{a}+\dfrac{x}{b-a}$, when $x=\dfrac{a^2(b-a)}{b(b+a)}$.

12. $(a+x)(b+x)-a(b+c)+x^2$, when $x=\dfrac{ac}{b}$.

13. $bx+cy+az$, when $x=b+c-a$, $y=c+a-b$,
    $z=a+b-c$.

14. $\dfrac{a(1+b)+bx}{a(1+b)-bx}-\dfrac{a}{a-2bx}$, when $x=-a$.

15. $\left(\dfrac{x+a}{x+b}\right)^3-\dfrac{x+2a+b}{x-a-2b}$, when $x=\frac{1}{2}(b-a)$.

16. $(p-q)(x+2r)+(r-x)(p+q)$, when $x=\dfrac{r(3p-q)}{2q}$.

17. $a^2(b-c)+b^2(c-a)+c^2(a-b)$, when $a-b=0$.

18. $(a+b+c)(bc+ca+ab)-(a+b)(b+c)(c+a)$,
    when $a=-b$.

19. $(a+b+c)^3-(a^3+b^3+c^3)$, when $a+b=0$.

20. $(x+y+z)^4-(x+y)^4-(y+z)^4-(z+x)^4+x^4+y^4+z^4$,
    when $x+y+z=0$.

21. $a^3(c-b^2)+b^3(a-c^2)+c^3(b-a^2)+abc(abc-1)$,
    when $b-a^2=0$.

22. $a^5 \left(\dfrac{a^5 + 5\,b^5}{a^5 - b^5}\right)^5 + b^5 \left(\dfrac{5\,a^5 + b^5}{b^5 - a^5}\right)^5$, when $a^5 + b^5 = 0$.

23. Express in words the fact that $(a - b)^2 = a^2 - 2\,ab + b^2$.

24. Express algebraically the fact that "the sum of two numbers multiplied by their difference is equal to the difference of the squares of the numbers."

25. The area of the walls of a room is equal to the height multiplied by twice the sum of the length and breadth. What are the areas of the walls in the following cases: (i.) length l, height h, breadth b; (ii.) height x, length b feet more than the height, and breadth b feet less than the height.

26. Express in words the statement that
$$(x + a)\,(x + b) = x^2 + (a + b)\,x + ab.$$

27. Express in symbols the statement that "the square of the sum of two numbers exceeds the sum of their squares by twice their product."

28. Express in words the algebraic statement,
$$(x + y)^3 = x^3 + y^3 + 3\,xy\,(x + y).$$

29. Express algebraically the fact that "the cube of the difference of two numbers is equal to the difference of the cubes of the numbers diminished by three times the product of the numbers multiplied by their difference."

30. If the sum of the cubes of two numbers be divided by the sum of the numbers, the quotient is equal to the square of their difference increased by their product. Express this algebraically.

31. Express in words the following algebraic statement:
$$\dfrac{x^3 - y^3}{x - y} = (x + y)^2 - xy.$$

32. The square on the diagonal of a cube is equal to three times the square on the edge. Express this in symbols, using l for length of the edge, and d for length of the diagonal.

33. Express in symbols that "the length of the edge of the greatest cube that can be cut from a sphere is equal to the square root of one-third the square of the diameter."

34. Express in symbols that any "rectangle is half the rectangle contained by the diagonals of the squares upon two adjacent sides."

The square on the diagonal of a square is double the square on a side.

35. The area of a circle is equal to π times the square of the radius. Express this in symbols. Also express in symbols the area of the ring between two concentric circles.

36. The volume of a cylinder is equal to the product of its height into the area of the base; that of a cone is one-third of this; and that of a sphere is two-thirds of the volume of the circumscribing cylinder. Express these facts in symbols, using h for the height of the cylinder, and r for the radius of its base.

Ex. 4.

Perform the additions in the following cases:

1. $(b-a)x + (c-b)y$ and $(a+b)x + (b+c)y$.

2. $ax - by$, $(a-b)x - (a+b)y$, and $(a+b)x - (b-a)y$.

3. $(y-z)a^2 + (z-x)ab + (x-y)b^2$
and $(x-y)a^2 - (z-y)ab - (x-z)b^2$.

4. $ax + by + cz,\ bx + cy + az,$ and $cx + ay + bz.$

5. $(a + b) x^2 + (b + c) y^2 + (a + c) z^2,\ (b + c) x^2 + (a + c) y^2$
   $+ (a + b) z^2,\ (a + c) x^2 + (a + b) y^2 + (b + c) z^2,$ and
   $-(a + b + c)(x^2 + y^2 + z^2).$

6. $x (a - b)^2 + y (b - c)^2 + z (c - a)^2,\ y(a-b)^2 + z(b-c)^2$
   $+ x(c - a)^2,$ and $z(a - b)^2 + x(b - c)^2 + y(c - a)^2.$

7. $(a - b) x^2 + (b - c) y^2 + (c - a) z^2,\ (b - c) x^2 + (c - a) y^2$
   $+ (a - b) z^2,$ and $(c - a) x^2 + (a - b) y^2 + (b - c) z^2.$

8. $(a + b) x + (b + c) y - (c + a) z,\ (b + c) z + (c + a) x$
   $-(a + b) y,$ and $(a + c) y + (a + b) z - (b + c) x.$

9. $a^2 - 3ab - \frac{14}{21} b^2,\ 2b^2 - \frac{2}{3} b^3 + c^2,\ ab - \frac{1}{3} b^2 + b^3,$ and
   $2ab - \frac{1}{3} b^3.$

10. $ax^n - 3bx^n,\ -9ax^n + 7bx^n,$ and $-8bx^n + 10ax^n.$

11. What will $(ax - by + cz) + (bx + cy - az) - (cx + ay + bz)$
    become when $x - y - z = 1$?

Fundamental Formulas and their Application.

By Multiplication we obtain

$$(x + r)(x + s) = x^2 + (r + s) x + rs \qquad [A]$$

$$(x + r)(x + s)(x + t)$$
$$= x^3 + (r + s + t) x^2 + (rs + st + tr) x + rst \qquad [B]$$

From [A] we obtain immediately

$$(x \pm y)^2 = x^2 \pm 2xy + y^2 \qquad [1]$$

$$(x + y + z)^2 = x^2 + 2xy + 2xz + y^2 + 2yz + z^2 \qquad [2]$$

$$(\Sigma a)^2 = \Sigma a^2 + 2 \Sigma ab \qquad [3]$$

$$(x + y)(x - y) = x^2 - y^2 \qquad [4]$$

The symbol Σ means "the sum of all such terms as."

From [B] we derive

$$(x \pm y)^3 = x^3 \pm 3x^2 y + 3xy^2 \pm y^3 \qquad\qquad [5]$$

$$= x^3 \pm y^3 \pm 3xy (x \pm y) \qquad\qquad [6]$$

$$(x+y+z)^3 = x^3 + y^3 + z^3$$
$$+ 3x^2 (y+z) + 3y^2 (z+x) + 3z^2 (x+y)$$
$$+ 6xyz \qquad\qquad [7]$$

$$= x^3 + y^3 + z^3 + 3(x+y)(y+z)(x+z) \qquad [8]$$

$$= x^3 + y^3 + z^3 + 3(x+y+z)(xy+yz+xz) - 3xyz \quad [9]$$

$$(\Sigma a)^3 = \Sigma a^3 + 3\Sigma a^2 b + 6\Sigma abc \qquad\qquad [10]$$

FORMULA [1]. EXAMPLES.

1. We have at once $(x+y)^2 + (x-y)^2 = 2(x^2+y^2)$ and $(x+y)^2 - (x-y)^2 = 4xy$.

2. $(a+b+c+d)^2 + (a-b-c+d)^2$ may be written $[(a+d)+(b+c)]^2 + [(a+d)-(b+c)]^2$,

which (Exam. 1)

$$= 2[(a+d)^2 + (b+c)^2];$$

similarly,

$$(a-b+c-d)^2 + (a+b-c-d)^2$$
$$= [(a-d)-(b-c)]^2 + [(a-d)+(b-c)]^2$$
$$= 2[(a-d)^2 + (b-c)^2].$$

$$\therefore (a+b+c+d)^2 + (a-b-c+d)^2 + (a-b+c-d)^2$$
$$+ (a+b-c-d)^2$$
$$= 2[(a+d)^2 + (b+c)^2 + (a-d)^2 + (b-c)^2],$$

by Exam. 1,

$$= 4(a^2 + b^2 + c^2 + d^2).$$

3. Simplify $(a+b+c)^2 - 2(a+b+c)c + c^2$.

This is the square of a binomial of which the first term is $(a+b+c)$, and the second $-c$. Hence it equals

$$[(a+b+c)-c]^2 = (a+b)^2.$$

4. Simplify $(a+b)^4 - 2(a^2+b^2)(a+b)^2 + 2(a^4+b^4)$.

By Exam. 1, $2(a^4+b^4) = (a^2+b^2)^2 + (a^2-b^2)^2$.

Hence, the given expression equals

$$(a+b)^4 - 2(a^2+b^2)(a+b)^2 + (a^2+b^2)^2 + (a^2-b^2)^2$$
$$= [(a+b)^2 - (a^2+b^2)]^2 + (a^2-b^2)^2$$
$$= a^4 + 2a^2b^2 + b^4 = (a^2+b^2)^2.$$

Ex. 5.

Simplify :

1. $(x+3y^2)^2 + (x-3y^2)^2$; $(\tfrac{1}{3}a^2+3b^2)^2 - (\tfrac{1}{3}a^2-3b^2)^2$.

Show that :

2. $(mx+ny)^2 + (nx-my)^2 = (m^2+n^2)(x^2+y^2)$.

3. $(mx-ny)^2 - (nx-my)^2 = (m^2-n^2)(x^2-y^2)$.

Simplify :

4. $[(a+3b)^2 + 2(a+3b)(a-b) + (a-b)^2](a-b)^2$

5. $(x+3)^2 + (x+4)^2 - (x+5)^2$,
  and $(\tfrac{1}{2}x^2 - 2y^2)^2 - (\tfrac{1}{2}y^2 + 2x^2)^2$.

6. $(a+b+c)^2 + (b+c)^2 - 2(b+c)(a+b+c)$.

Show that :

7. $(ax+by)^2 + (cx+dy)^2 + (ay-bx)^2 + (cy-dx)^2$
  $= (a^2+b^2+c^2+d^2)(x^2+y^2)$.

Simplify :

8. $(x-3y^2)^2 + (3x^2-y)^2 - 2(3x^2-y)(x-3y^2)$.

9. $(x^2 + xy - y^2)^2 - (x^2 - xy - y^2)^2$,
$(1 + 2x + 4x^2)^2 + (1 - 2x + 4x^2)^2$.

10. If $a + b = -\frac{3}{4}c$, show that
$(2a-b)^2 + (2b-c)^2 + (2c-a)^2 + 2(2a-b)(2b-c)$
$+ 2(2b-c)(2c-a) + 2(2c-a)(2a-b) = \frac{1}{16}c^2$.

Simplify:

11. $2(a-b)^2 - (a-2b)^2$; $(a^2 + 4ab + b^2)^2 - (a^2 + b^2)^2$.

12. $(a+b)^2 - (b+c)^2 + (c+d)^2 - (d+a)^2$.

13. $(\frac{1}{2}x - y)^2 + (\frac{1}{2}y - z)^2 + (\frac{1}{2}z - x)^2 + 2(\frac{1}{2}x - y)(\frac{1}{2}z - x)$
$+ 2(\frac{1}{2}y - z)(\frac{1}{2}z - x) + 2(\frac{1}{2}x - y)(\frac{1}{2}y - z)$.

Show that:

14. $(x-y)^2 + (y-z)^2 + (z-x)^2$
$= 2(x-y)(z-y) + 2(y-x)(z-x) + 2(z-y)(z-x)$.

Simplify:

15. $(1+x)^4 - 2(1+x^2)(1+x)^2 + 2(1+x^4)$.

16. $(x+y+z)^2 - (x+y-z)^2 - (y+z-x)^2 - (z+x-y)^2$.

17. $(x-2y+3z)^2 + (3z-2y)^2 + 2(x-2y+3z)(2y-3z)$.

18. $(a^2+b^2-c^2)^2 + (c^2-b^2)^2 + 2(b^2-c^2)(a^2+b^2-c^2)$.

19. $(x+y)^4 + (x-y)^4 - 2(x-y)^2(x+y)^2$.

20. $(5a+3b)^2 + 16(3a+b)^2 - (13a+5b)^2$.

Show that:

21. $(3a-b)^2 + (3b-c)^2 + (3c-a)^2 - 2(b-3a)(3b-c)$
$+ 2(3b-c)(3c-a) - 2(a-3c)(3a-b)$
$- 4(a+b+c)^2 = 0$.

22. If $z^2 = 2xy$, show that $(2x^2 - y^2)^2 + (z^2 - 2y^2)^2$
$+ (x^2 - 2z^2)^2 - 2(2x^2 - y^2)(z^2 - 2y^2)$
$+ 2(x^2 - 2z^2)(z^2 - 2y^2) - 2(x^2 - 2z^2)(2x^2 - y^2)$
$= (x+y)^4$.

Simplify

23. $(1 + x + x^2 + x^3)^2 + (1 - x - x^2 + x^3)^2$
$+ (1 - x + x^2 - x^3)^2 + (1 + x - x^2 - x^3)^2.$

24. $(ax + by)^4 - 2(a^2 x^2 + b^2 y^2)(ax + by)^2 + 2(a^4 x^4 + b^4 y^4).$

FORMULAS [2] AND [3]. EXAMPLES.

1. $(1 - 2x + 3x^2)^2 = 1 - 4x + 6x^2$
$+ 4x^2 - 12x^3$
$+ 9x^4$

$$= 1 - 4x + 10x^2 - 12x^3 + 9x^4$$

2. $(ab + bc + ca)^2$
$= a^2 b^2 + 2ab^2 c + 2a^2 bc + b^2 c^2 + 2abc^2 + c^2 a^2$
$= a^2 b^2 + b^2 c^2 + c^2 a^2 + 2abc(a + b + c).$

3. $[(x + y)^2 + x^2 + y^2]^2$
$= (x + y)^4 + 2(x + y)^2(x^2 + y^2) + x^4 + 2x^2 y^2 + y^4$
$= (x + y)^4 + (x + y)^2[(x + y)^2 + (x - y)^2]$
$+ x^4 + 2x^2 y^2 + y^4$
$= 2(x + y)^4 + (x^2 - y^2)^2 + x^4 + 2x^2 y^2 + y^4$
$= 2[(x + y)^4 + x^4 + y^4].$

4. $(x^2 + xy + y^2)^2$
$= x^4 + 2x^3 y + 2x^2 y^2 + x^2 y^2 + 2xy^3 + y^4$
$= (x + y)^2 x^2 + x^2 y^2 + y^2(x + y)^2.$

5. In Exam. 3, substitute $b - c$ for x, $c - a$ for y, and consequently $b - a$ for $x + y$; then, since $(b - a)^2 = (a - b)^2$, Exam. 3 gives
$[(a - b)^2 + (b - c)^2 + (c - a)^2]^2$
$= 2[(a - b)^4 + (b - c)^4 + (c - a)^4].$

6. Making the same substitutions in Exam. 4, we have
$(a^2 + b^2 + c^2 - ab - bc - ca)^2$
$= (a - b)^2(b - c)^2 + (b - c)^2(c - a)^2 + (c - a)^2(a - b)^2.$

or, multiplying both sides by 4,

$$[(a-b)^2+(b-c)^2+(c-a)^2]^2$$
$$=4(a-b)^2(b-c)^2+4(b-c)^2(c-a)^2+4(c-a)^2(a-b)^2.$$

Hence, from Exam. 5,

$$(a-b)^4+(b-c)^4+(c-a)^4$$
$$=2(a-b)^2(b-c)^2+2(b-c)^2(c-a)^2+2(c-a)^2(a-b)^2.$$

Ex. 6.

Expand:

1. $(1-2x+3x^2-4x^3)^2$; $(1-x+x^2-x^3)^2$.

2. $(1-2x+2x^2-3x^3-x^4)^2$; $(1+3x+3x^2+x^3)^2$.

3. $(2a-b-c^2-1)^2$; $(1-x+y+z)^2$; $(\frac{1}{2}x-\frac{1}{3}y+6z)^2$.

4. $(x^3-x^2y+xy^2-y^3)^2$; $(ax+bx^2+cx^3+dx^4)^2$.

5. Show that $(a^2+b^2+c^2)(x^2+y^2+z^2)-(ax+by+cz)^2$
   $$=(ay-bx)^2+(cx-az)^2+(bz-cy)^2.$$

6. Show that $(a+b)x+(b+c)y+(c+a)z$ multiplied by $(a-b)x+(b-c)y+(c-a)z$ is equal to the difference of the squares of two trinomials.

7. Show that $(a-b)(a-c)+(b-c)(b-a)+(c-a)(c-b)$
   $$-\tfrac{1}{2}[(a-b)^2+(b-c)^2+(c-a)^2]=0.$$

8. Simplify $[a-(b-c)]^2+[b-(c-a)]^2+[c-(a-b)]^2$.

9. Show that $(a^2+b^2-x^2)^2+(a_1^2+b_1^2-x^2)^2+2(aa_1+bb_1)^2$
   $$=(a^2+a_1^2-x^2)^2+(b^2+b_1^2-x^2)^2+2(ab+a_1b_1)^2.$$

10. Show that $[(a-b)(b-c)+(b-c)(c-a)+(c-a)(a-b)]^2$
    $$=(a-b)^2(b-c)^2+(b-c)^2(c-a)^2+(c-a)^2(a-b)^2.$$

11. Square $2a-\tfrac{1}{2}bx-\tfrac{1}{4}cx+2dx$.

12. If $x+y+z=0$, show that
    $$x^4+y^4+z^4=(x^2-y^2)^2+(y^2-z^2)^2+(z^2-x^2)^2.$$

13. Show that $a^2(b+c)^2+b^2(c+a)^2+c^2(a+b)^2$
    $$+2abc(a+b+c)=2(ab+bc+ca)^2.$$

§ 5. To apply formula [4] to obtain the product of two factors which differ only in the signs of some of their terms, group together all the terms whose signs are the same in one factor as they are in the other, and then form into a second group all the other terms.

1. Multiply $a + b - c + d$ by $a - b - c - d$.

   Here the first group is $a - c$, the second $b + d$. Hence, we have
   $$[(a - c) + (b + d)] [(a - c) - (b + d)]$$
   $$= (a - c)^2 - (b + d)^2.$$

2. $(1 + 3x + 3x^2 + x^3)(1 - 3x + 3x^2 - x^3)$
   $$= [(1 + 3x^2) + (3x + x^3)][(1 + 3x^2) - (3x + x^3)]$$
   $$= (1 + 3x^2)^2 - (3x + x^3)^2$$
   $$= 1 - 3x^2 + 3x^4 - x^6.$$

3. Find the continued product of $a + b + c$, $b + c - a$, $c + a - b$, and $a + b - c$.

   The first pair of factors gives $[(b + c) + a][(b + c) - a]$
   $$= (b + c)^2 - a^2 = b^2 + 2bc + c^2 - a^2.$$

   The second pair gives $[a - (b - c)][a + (b - c)]$
   $$= a^2 - b^2 + 2bc - c^2.$$

   The only term whose sign is the same in both these results is $2bc$; hence, grouping the other terms, we have
   $$[2bc + (b^2 + c^2 - a^2)] [2bc - (b^2 + c^2 - a^2)]$$
   $$= (2bc)^2 - (b^2 + c^2 - a^2)^2$$
   $$= 2a^2b^2 + 2b^2c^2 + 2c^2a^2 - a^4 - b^4 - c^4.$$

4. Show that $(a^2 + ab + b^2)^2 - a^2b^2 = (a^2 + ab)^2 + (ab + b^2)^2$.

   The expression $= (a^2 + b^2)(a^2 + 2ab + b^2)$
   $$= (a^2 + b^2)(a + b)^2$$
   $$= a^2(a + b)^2 + b^2(a + b)^2$$
   $$= (a^2 + ab)^2 + (ab + b^2)^2.$$

Ex. 7.

1. $(a^2 + 2ab + b^2)(a^2 - 2ab + b^2)$.
2. $(\frac{1}{2}x^2 - xy + y^2)(\frac{1}{2}x^2 + y^2 + xy)$.
3. $(a^2 - ab + 2b^2)(a^2 + ab + 2b^2)$; $(x^4 + 4xy)(x^4 - 4xy)$.
4. $[(x+y)x - y(x-y)][(x-y)x - y(y-x)]$.
5. Simplify $(x+3)(x-3) + (x+4)(x-4) - (x+5)(x-5)$.
6. Simplify $(1+x)^4 + (1-x)^4 - 2(1-x^2)^2$.
7. $(x^2 + y^2)^2 - (2xy)^2 - (x^2 - y^2)^2$.
8. $(2a^2 - 3b^2 + 4c^2)(2a^2 + 3b^2 - 4c^2)$.
9. $(2a + b - 3c)(b + 3c - 2a)$; $(2a - b - 3c)(b - 3c - 2a)$.
10. $(x^4 + y^4)(x^2 + y^2)(x+y)(x-y)$.
11. $(x^2 + xy + y^2)(x^2 - xy + y^2)(x^4 - x^2y^2 + y^4)$.
12. $(a + b - ab - 1)(a + b + ab + 1)$.
13. If $a^4 = b^4 + c^4$, show that
$(a^2 + b^2 + c^2)(b^2 + c^2 - a^2)(c^2 + a^2 - b^2)(a^2 + b^2 - c^2)$
$= 4b^4c^4$.

Simplify:

14. $(x^2 + y^2 - \frac{5}{4}xy)(x^2 + y^2 + \frac{5}{4}xy)$.
15. $(x^4 - 2x^3 + 3x^2 - 2x + 1)(x^4 + 2x^3 + 3x^2 + 2x + 1)$.
16. Multiply $(2x - y)a^2 - (x+y)ax + x^3$ by $(2x - y)a^2$
    $+ (x+y)ax - x^3$.

Show that:

17. $(a^2 + b^2 + c^2 + ab + bc + ca)^2 - (ab + bc + ca)^2$
    $= (a + b + c)^2(a^2 + b^2 + c^2)$.
18. $(a^2 + b^2 + c^2 + ab + bc + ca)^2 - (a^2 + ab + ca - bc)^2$
    $= [(a + b)(b + c)]^2 + [(b + c)(c + a)]^2$.
19. $4(ab + cd)^2 - (a^2 + b^2 - c^2 - d^2)^2$
    $= (a+b+c-d)(a+b-c+d)(c+d+a-b)(c+d-a+b)$.

20. Find the product of: $x^2 + y^2 + z^2 - 2xy + 2xz - 2yz$
and $x^2 + y^2 + z^2 - 2xy - 2xz + 2yz$.

21. $(x^2 + y^2 + xy\sqrt{2})(x^2 - xy\sqrt{2} + y^2)(x^4 - y^4)$.

22. $(1 - 6a + 9a^2)(\frac{1}{3} + 2a + 3a^2)$.

23. $[(m + n) + (p + q)](m - q + p - n)$.

24. $1 + x + x^2$, $x^2 + x - 1$, $x^2 - x + 1$, and $1 + x - x^2$.

25. $(a - b^2)^2(a + b^2)^2(a^2 + b^4)^2(a^4 + b^8)^2$.

26. Show that $(x^2 + xy + y^2)^2(x^2 - xy + y^2)^2 - (x^2 y^2)^2$
$= (x^4 + x^2 y^2)^2 + (x^2 y^2 + y^4)^2$.

· FORMULA A. EXAMPLES.

1. Multiply $x^2 - x + 5$ by $x^2 - x - 7$.

Here the common term is $x^2 - x$; the other terms, $+5$
and -7. Hence, the product equals

$(x^2 - x)^2 + (-7 + 5)(x^2 - x) + (-7 \times 5)$
$= (x^2 - x)^2 - 2(x^2 - x) - 35$
$= x^4 - 2x^3 - x^2 + 2x - 35$.

2. $(x - a)(x - 3a)(x + 4a)(x + 6a)$.

Taking the first and third factors together, and the
second and fourth, we have the product equals

$(x^2 + 3ax - 4a^2)(x^2 + 3ax - 18a^2)$
$= (x^2 + 3ax)^2 - (4a^2 + 18a^2)(x^2 + 3ax) - 72a^4$, etc.

Ex. 8.

Find the products of:

1. $(x^2 + 2x + 3)(x^2 + 2x - 4)$; $(x - y + 3z)(x - y + 5z)$.

2. $(x + 1)(x + 5)(x + 2)(x + 4)$; $(x^3 + a - b)(x^3 + 2b - a)$.

3. $(a^2 - 3)(a^2 - 1)(a^2 + 5)(a^2 + 7)$; $(x^4 + x^2 + 1)(x^4 + x^2 - 2)$.

4. $[(x+y)^2 - 4xy]\,[(x+y)^2 + 5xy]$.

5. $(x^n + a + 7)(x^n - a - 9)$; $\left(\dfrac{x}{y} + \dfrac{y}{x} - 1\right)\left(\dfrac{x}{y} + \dfrac{y}{x} + 3\right)$.

6. $(nx + y + 3)(nx + y + 7)$.

7. $(x + a - y)(x + a + 3y)$.

8. $(x^{2n} + x^n - a)(x^{2n} + x^n - b)$.

9. $(\tfrac{1}{2}x^4 - y^2 + 2)(\tfrac{1}{2}x^4 - y^2 - 4)$.

10. $\left(\dfrac{1}{x} + \dfrac{1}{y} - \dfrac{1}{2}\right)\left(\dfrac{1}{x} + \dfrac{1}{y} + 2\tfrac{1}{2}\right)$.

11. $x - 2 + \sqrt{2},\ \ x - 2 + \sqrt{3},\ \ x - 2 - \sqrt{2},\ \ x - 2 - \sqrt{3}$.

12. $(x + a + b)(x + b - c)(x - a + b)(x + b + c)$.

13. $(a + b + c)(a + b + d) + (a + c + d)(b + c + d)$
    $- (a + b + c + d)^2$.

14. Show that $(2a + 2b - c)(2b + 2c - a)$
    $+ (2c + 2a - b)(2a + 2b - c) + (2b + 2c - a)(2c + 2a - b)$
    $= 9(ab + bc + ca)$.

FORMULAS [5] AND [6].   EXAMPLES.

1. We get at once
$$(x + y)^3 + (x - y)^3 = 2x(x^2 + 3y^2);$$
$$(x + y)^3 - (x - y)^3 = 2y(3x^2 + y^2).$$

2. Simplify $(a + b + c)^3 - 3(a + b + c)^2 c + 3(a + b + c)c^2 - c^3$.
This comes under formula [5], the first term being $a + b + c$; the second, $-c$. Hence, the expression is
$$[(a + b + c) - c]^3 = (a + b)^3.$$

3. Show that $(x^2 + xy + y^2)^3 + (xy - x^2 - y^2)^3$
    $- 6xy(x^4 + x^2y^2 + y^4) = 8x^3y^3$.
This comes under formula [6], the first term being $(x^2 + xy + y^2)$, and the second $-(x^2 - xy + y^2)$; we have, therefore,
$$[(x^2 + xy + y^2) - (x^2 - xy + y^2)]^3 = (2xy)^3 = 8x^3y^3.$$

Simplify:

Ex. 9.

1. $(1-x^2)^3+(1+x^2)^3$; $(x^2+xy)^3-(x^2-xy)^3$.

2. $(a+2b)^3-(a-b)^3$; $(3a-b)^3-(3a-2b)^3$.

3. $(x+y-z)^3+3(x+y-z)^2 z+z^3+3(x+y-z)z^2$.

4. $(a-b)^3+(a+b)^3+6a(a^2-b^2)$.

5. $(x-y)^3+(x+y)^3+3(x-y)^2(x+y)-3(y-x)(x+y)^2$.

6. $(1+x+x^2)^3-(1-x+x^2)^3-6x(1+x^2+x^4)$.

7. $(a-b-c)^3+(b+c)^3+3(b+c)^2(a-b-c)$
   $+3(a-b-c)^2(b+c)$.

8. $(3x-4y+5z)^3-(5z-4y)^3+3(5z-4y)^2(3x-4y+5z)$
   $-3(3x-4y+5z)^2(5z-4y)$.

9. $(1+x+x^2)^3+3(1-x^3)(2+x^2)+(1-x)^3$.

Show that:

10. $a(a-2b)^3-b(b-2a)^3=(a-b)(a+b)^3$.

11. $a^3(a^3-2b^3)^3+b^3(2a^3-b^3)^3=(a^3-b^3)(a^3+b^3)^3$.

Simplify:

12. $(x^2+xy+y^2)^3+6(x^2+y^2)(x^4+xy+y^4)+(x^2-xy+y^2)^3$.

Show that:

13. $a^3(a^3+2b^3)^3+b^3(2a^3+b^3)^3+(3a^2b^2)^3$
    $=(a^6+7a^3b^3+b^6)^2$.

Simplify:

14. $(ax+by)^3+a^3y^3+b^3x^3-3abxy(ax+by)$.

15. What will $a^3+b^3+c^3-3abc$ become when
    $a+b+c=0$.

16. Find the value of $x^6-y^6+z^6+3x^2y^2z^2$ when
    $x^2-y^2+z^2=0$.

FORMULAS [7], [8], AND [9]. EXAMPLES.

1. Simplify $(2x - 3y)^3 + (4y - 5x)^3 + (3x - y)^3$
$- 3(2x - 3y)(4y - 5x)(3x - y).$

By [9] this is seen to be

$[(2x - 3y) + (4y - 5x) + (3x - y)]^3 = (0)^3 = 0.$

2. Prove that $(a - b)^3 + (b - c)^3 + (c - a)^3$
$= 3(a - b)(b - c)(c - a).$

In [9] substitute $a - b$ for x, $b - c$ for y, and $c - a$ for z; for these values $x + y + z = 0$, and the identity appears at once.

3. Prove $(a + b + c)^3 - (b + c - a)^3 - (a + c - b)^3 - (a + b - c)^3$
$= 24 abc.$

In [8] let $x = b + c - a$, $y = c + a - b$, $z = a + b - c$; and therefore, $x + y = 2c$, $y + z = 2a$, $z + x = 2b$; and this identity at once appears.

Ex. 10.

1. Cube the following:
$1 - x + x^2$; $a - b - c$; $1 - 2x + 3x^2 - 4x^3.$

Simplify:

2. $(x^2 + 2x - 1)^3 + (2x - 1)(x^2 + 2x - 2) - (x^3 + 3x^2 - 1)^3.$

Prove that:

3. $xyz + (x + y)(y + z)(z + x) = (x + y + z)(xy + yz + zx).$*

4. $(ax - by)^3 + a^3 y^3 - b^3 x^3 + 3 abxy (ax - by)$
$= (a^3 - b^3)(x^3 + y^3).$

---

* Note that the right-hand member is formed from the left-hand one by changing *additions* into *multiplications*, and *multiplications* into *additions*; hence, in $(x + y + z) \times (x \times y + y \times z + z \times x)$ the signs $+$ and $\times$ may be interchanged throughout without altering the value of the expression.

Simplify :

5. $(x+y+z)^3 + (x-2y)^3 + (y-2z)^3 + (z-2x)^3$
$\quad + 3(x-y-2z)(y-z-2x)(z-x-2y).$

6. $(2x^2 - 3y^2 + 4z^2)^3 + (2y^2 - 3z^2 + 4x^2)^3$
$\quad + (2z^2 - 3x^2 + 4y^2)^3.$

7. $(2ax - by)^3 + (2by - cz)^3 + (2cz - ax)^3$
$\quad + 3(2ax + by - cz)(2by + cz - ax)(2cz + ax - by).$

Prove :

8. $(x^3 + 3x^2y - y^3)^3 + [3xy(x+y)]^3$
$\quad = [(x-y)^3 + 9x^2y](x^2 + xy + y^2)^3.$

9. $9(x^3 + y^3 + z^3) - (x+y+z)^3 = (4x+4y+z)(x-y)^2$
$\quad + (4y+4z+x)(y-z)^2 + (4z+4x+y)(z-x)^2.$

10. If $x+y+z=0$, show that $x^3 + y^3 + z^3 = 3xyz.$

11. If $x=2y+3z$, show that $x^3 - 8y^3 - 27z^3 - 18xyz = 0.$

Show that :

12. $(x^2 + xy + y^2)^3 + (x^2 - xy + y^2)^3 + 8z^6 - 6z^2(x^4 + x^2y^2 + y^4)$
$\quad = 0$, if $x^2 + y^2 + z^2 = 0.$

Prove that :

13. $8(a+b+c)^3 - (a+b)^3 - (b+c)^3 - (c+a)^3$
$\quad = 3(2a+b+c)(a+2b+c)(a+b+2c).$

Prove the following :

14. $(ax - by)^3 + b^3y^3 = a^3x^3 + 3abxy(by - ax).$

15. $a^3 + b^3 + c^3 - 3abc$
$\quad = \tfrac{1}{2}[(a-b)^2 + (b-c)^2 + (c-a)^2](a+b+c).$

16. $(a+b+c)[(a+b-c)(b+c-a) + (b+c-a)(c+a-b)$
$\quad + (c+a-b)(a+b-c)]$
$\quad = (a+b-c)(b+c-a)(c+a-b) + 8abc.$

17. $(a+b+c)^3 - 3[a(b-c)^2 + b(c-a)^2 + c(a-b)^2]$
$\quad = a^3 + b^3 + c^3 + 24abc.$

18. $(a+b+7c)(a-b)^2+(b+c+7a)(b-c)^2$
$\qquad +(c+a+7b)(c-a)^2=2(a+b+c)^3-54abc.$

19. $(a+b+c)[(2a-b)(2b-c)+(2b-c)(2c-a)$
$\qquad +(2c-a)(2a-b)]=(2a-b)(2b-c)(2c-a)$
$\qquad +(2a+b-c)(2b+c-a)(2c+a-b).$

20. If $x^2(y+z)=a^3,\ \ y^2(z+x)=b^3,\ \ z^2(x+y)=c^3,$ and $xyz=abc$, show that
$$a^3+b^3+c^3+2abc=(x+y)(y+z)(z+x).$$

EXPANSION OF BINOMIALS.

We have, from formula [5],
$$(a+b)^3=a^3+3a^2b+3ab^2+b^3\ ;$$
multiplying by $a+b$, we obtain
$$(a+b)^4=a^4+4a^3b+6a^2b^2+4ab^3+b^4\ ;$$
multiplying this by $a+b$, we obtain
$$(a+b)^5=a^5+5a^4b+10a^3b^2+10a^2b^3+5ab^4+b^5.$$

From these examples we derive the following law for the formation of the terms in the expansion of $a+b$ to any required power:

I. The exponent of a, in the *first* term, is that of the given power, and *decreases* by unity in each succeeding term; the exponent of b begins with unity in the *second* term, and *increases* by unity in each succeeding term.

II. The coefficient of the first term is unity, and the co-efficient of any other term is found by multiplying the coefficient of the preceding term by the exponent of a in that term, and dividing the product by the number of that preceding term.

It will be observed that the coefficients equally distant from the extremes of the expansion are equal.

Ex. 11.

1. Expand $(x+y)^6$; $(x+y)^7$; $(x+y)^8$; $(x+y)^{12}$.

2. What will be the law of *signs*, if $-y$ be written for y in [1]?

3. Expand $(a-b)^5$; $(a-2b)^4$; $(2b-a)^4$.

4. Expand $(1+m)^6$; $(m+1)^5$; $(2m+1)^6$.

5. What is the coefficient of the fourth term in $(a-b)^{10}$?

6. Expand $(x^2-y)^4$; $(a-2b^2)^5$; $(a^3-2b^3)^6$.

7. In the expansion of $(a-b)^{12}$ the third term is $66\,a^{10}b^2$; find the fifth and sixth terms.

8. Show that $(x+y)^5-x^5-y^5 = 5\,xy\,(x+y)\,(x^2+xy+y^2)$.

9. From [8] show that $2[(a-b)^5+(b-c)^5+(c-a)^5]$
$= 5(a-b)(b-c)(c-a)[(a-b)^2+(b-c)^2$
$+(c-a)^2]$.

Horner's Methods of Multiplication and Division.

Examples.

1. Find the product of kx^3+lx^2+mx+n and ax^2+bx+c.
   Write the multiplier in a column to the left of the multiplicand, placing each term in the same horizontal line with the partial product it gives:

$$
\begin{array}{l|llll l}
 & kx^3 & +lx^3 & +mx & +n & \quad\cdots\cdots Q \\
\hline
ax^2 & akx^5 & +alx^4 & +amx^3 & +anx^2 & \quad\cdots\cdots p_1 \\
+bx & & +bkx^4 & +blx^3 & +bmx^2+bnx & \quad\cdots\cdots p_2 \\
+c & & & +ckx^3 & +clx^2 +cmx+cn & \cdots\cdots p_3 \\
\hline
\multicolumn{6}{l}{akx^5+(al+bk)\,x^4+(am+bl+ck)\,x^3+(an+bm+cl)\,x^2} \\
\multicolumn{6}{r}{+(bn+cm)\,x+cn \;\cdots\cdots P}
\end{array}
$$

§ 6. The above example has been given in full, the powers of x being inserted; in the following example detached coefficients are used. It is evident that, if the coefficient of the first term of the multiplier be unity, the coefficients of the multiplicand will be the same as those of the first partial product, and may be used for them, thus saving the repetition of a line.

2. Multiply $3x^4 - 2x^3 - 2x + 3$ by $x^2 + 3x - 2$.

$$
\begin{array}{r|rrrrrr}
1 & 3 & -2 & +0 & -2 & +3 \\
+3 & & +9 & -6 & +0 & -6 & +9 \\
-2 & & & -6 & +4 & -0 & +4 & -6 \\
\hline
& \multicolumn{7}{l}{3x^6 + 7x^5 - 12x^4 + 2x^3 - 3x^2 + 13x - 6}
\end{array}
$$

3. Find the product of $(x-3)(x+4)(x-2)(x-5)$.

$$
\begin{array}{r|rrrr}
& 1 & -3 \\
+4 & & +4 & -12 \\
\hline
& 1 & +1 & -12 \\
-2 & & -2 & -2 & +24 \\
\hline
& 1 & -1 & -14 & +24 \\
-5 & & -5 & +5 & +70 & -120 \\
\hline
& \multicolumn{5}{l}{x^4 - 6x^3 - 9x^2 + 94x - 120}
\end{array}
$$

4. Multiply $x^3 - 4x^2 + 2x - 3$ by $2x^3 - 3$.

$$
\begin{array}{r|rrrrrr}
& 1 & -4 & +2 & -3 \\
\hline
2 & 2 & -8 & +4 & -6 & & & (x^3 \times x^3 = x^6) \\
0 & & 0 & 0 & 0 & 0 \\
0 & & & 0 & 0 & 0 & 0 \\
-3 & & & & -3 & +12 & -6 & +9 \\
\hline
& \multicolumn{7}{l}{2x^6 - 8x^5 + 4x^4 - 9x^3 + 12x^2 - 6x + 9}
\end{array}
$$

In this example the missing terms of the multiplier are

supplied by zeros; but, instead of writing the zeros as in the example, we may, as in ordinary arithmetical multiplication, "skip a line" for every missing term.

5. Multiply $x^4 - 2x^2 + 1$ by $x^4 - x^2 + 3$.

$$
\begin{array}{r|l}
1 & 1+0-2+0+1 \qquad\qquad (x^4 \times x^4 = x^8) \\
-1 & \quad\;\; -1-0+2-0-1 \\
+3 & \qquad\qquad +3+0-6+0+3 \\
\hline
& x^8 \quad\;\; -3x^6 \;\; +6x^4 \;\; -7x^2 \;\; +3
\end{array}
$$

6. Find the value of
$$(x+2)(x+3)(x+4)(x+5) - 9(x+2)(x+3)(x+4)$$
$$+3(x+2)(x+3)+77(x+2)-85.$$

$$
\begin{array}{r|l}
& 1 \;+5 \\
& \quad -9 \\
\hline
& 1 \;-4 \\
+4 & \quad +4 \;\; -16 \\
& \qquad\quad +\;3 \\
\hline
& 1 \;+0 \;\; -13 \\
+3 & \quad +3 \;\; +\;0 \;\; -39 \\
& \qquad\qquad\quad +77 \\
\hline
& 1 \;+3 \;\; -13 \;\; +38 \\
+2 & \quad +2 \;\; +\;6 \;\; -26 \;\; +76 \\
& \qquad\qquad\qquad\quad -85 \\
\hline
& x^4 + 5x^3 - 7x^2 + 12x - 9
\end{array}
$$

7. Find the coefficient of x^4 in the product of
$$x^4 - ax^3 + bx^2 - cx + d \text{ and } x^2 + px + q.$$

$$
\begin{array}{r|l}
& 1 \;\; -a \;\; +b \;\; -c \;\; +d \\
+p & \qquad\quad -ap \\
+q & \qquad\quad +q \\
\hline
& \qquad\quad +(b - ap + q)
\end{array}
$$

Ex. 12.

Find the product of :

1. $(1+x+x^2+x^3+x^4)(1-x^2+x^3-x^7+x^8-x^{12}+x^{13})$.
2. $(1+x^5)(1-x^5+x^6)(1+x+x^2+x^3+x^4)$.
3. $(x-5)(x+6)(x-7)(x+8)$; $(2x^5-x^2+1)(x^4-x+2)$.
4. $(x^3+5x^2-16x-1)(x^3-5x^2-16x+1)$.
5. $(6x^6-x^5+2x^4-2x^3+2x^2+19x+6)(3x^2+4x+1)$.

Obtain the coefficients of x^4 and lower powers in

6. $(1+\frac{1}{2}x-\frac{1}{8}x^2+\frac{1}{16}x^3-\frac{5}{128}x^4)(1-\frac{1}{2}x-\frac{1}{8}x^2-\frac{1}{16}x^3-\frac{5}{128}x^4)$.
7. Multiply $2x^7-x^3+3x-4$ by $3x^5-2x^2-x-1$.

Simplify the following :

8. $(x+1)(x+2)(x+3)+3(x+1)(x+2)-10(x+1)+9$.
9. $x(x+1)(x+2)(x+3)-3x(x+1)(x+2)$
   $-2x(x+1)+2x$.
10. $x(x-1)(x-2)(x-3)+3x(x-1)(x-2)$
    $-2x(x-1)-2x$.
11. $(x-1)(x+1)(x+3)(x+5)-14(x-1)(x+1)+1$.
12. Given that the sum of the four following factors is -1, find (i.) the product of the first pair ; (ii.) the product of the second pair ; and (iii.) the product of the sum of the first pair by the sum of the second pair :

    (i.) $x+x^4+x^{13}+x^{16}$,
    (ii.) $x^2+x^8+x^9+x^{15}$,
    (iii.) $x^3+x^5+x^{12}+x^{14}$,
    (iv.) $x^6+x^7+x^{10}+x^{11}$.

13. Given that the sum of the three following factors is equal to -1, find their product :

    (i.) $x+x^5+x^8+x^{12}$,
    (ii.) $x^2+x^3+x^{10}+x^{11}$,
    (iii.) $x^4+x^6+x^7+x^9$.

§ 7. Were it required to *divide* the product P in the first of the above examples by $ax^2 + bx + c$, it is evident that could we find and *subtract* from P the partial products p_2, p_3 (or, what would give the same result, could we *add* them with the sign of each term changed), there would remain the partial product p_1, which, divided by the monomial ax^2, would give the quotient Q. This is what Horner's method does, the change of sign being secured by changing the signs of b and c, which are factors in each term of p_2, p_3, respectively.

1.

$$
\begin{array}{l|lllll}
 & akx^5+(al+bk)x^4+(am+bl+ck)x^3+(an+bm+cl)x^2+(bn+cm)x+cn..P \\
-bx & \quad\quad -bkx^4 \quad\quad\quad -blx^3 \quad\quad\quad -bmx^2 \quad -bnx\ldots\ldots p_2 \\
-c & \quad\quad\quad\quad\quad\quad\quad -ckx^3 \quad\quad\quad -clx^2 \quad -cmx-cn..p_3 \\
\hline
ax^2\,|\,akx^5 \quad +alx^4 \quad\quad\quad +amx^3 \quad\quad +anx^2\ldots\ldots\ldots\ldots p_1 \\
\hline
\quad\ \ kx^3 \quad +lx^2 \quad\quad\quad\quad +mx \quad\quad\quad +n\ldots\ldots\ldots\ldots\ldots Q
\end{array}
$$

The dividend and divisor are arranged as in the example, the sign of every term in the divisor, except the first, being changed in order to turn the subtractions into additions. The first term of the dividend (akx^5) is brought down into the line of p_1; dividing this by ax^2, the first term of the divisor, we get kx^3, *the first term of the quotient.* Multiplying this term kx^3 by $-bx$ and $-c$, respectively, and writing the products in the proper *columns* and *rows*, makes all ready to give the *second* term of p_1, which is got by simply adding up the second column of the work, giving alx^4. Dividing this second term of p_1 by ax^2 gives lx^2, *the second term of the quotient.* Multiply lx^2 by $-bx$ and $-c$, respectively, and proceed in the same way as was done in getting the second term of the quotient, and the third will be obtained. Repeating the steps, the complete quotient and the remainder will finally be obtained.

Should the coefficient of the first term of the divisor be *unity*, the coefficients of the line Q will be the same as those of p_1, and the line Q need not be written down, since one line does for both.

2. Divide $3x^6+7x^5-12x^4+2x^3-3x^2+13x-6$ by x^2+3x-2.

$$
\begin{array}{r|rrrrrrr}
 & 3 & +7 & -12 & +2 & -3 & +13 & -6 \quad (x^6 \div x^2 = x^4) \\
-3 & & -9 & +6 & -0 & +6 & -9 \\
+2 & & & +6 & -4 & +0 & -4 & +6 \\
\hline
 & 3x^4 & -2x^3 & +0 & -2x & +3
\end{array}
$$

Compare this example with the second example of Horner's Multiplication, performing a step in multiplication, then the corresponding step in division; then another step in multiplication, and the second (corresponding) step in division; and so on.

3. Divide $x^7-3x^6+4x^4+18x^3-7x+12$ by x^3-3x^2+3x-1.

$$
\begin{array}{r|rrrrrrr}
 & 1 & -3 & +0 & -4 & +18 & +0 & -7 & +12 \\
+3 & & +3 & +0 & -9 & -36 & -27 \\
-3 & & & -3 & -0 & +9 & +36 & +27 \quad (x^7 \div x^3 = x^4) \\
+1 & & & & +1 & +0 & -3 & -12 & -9 \\
\hline
 & x^4 & +0 & -3x^2 & -12x & -9; & 6x^2 & +8x & +3
\end{array}
$$

The quotient is therefore $x^4-3x^2-12x-9$, and the remainder $6x^2+8x+3$.

4. Divide $x^8-3x^7-5x^5+2x^4+5x^3+4x^2+1$ by x^3+2x-1.

The zero coefficient in the divisor may be inserted, or it may be omitted and allowance made for it in the $2x$-line. See Exams. 4 and 5 in Multiplication.

$$
\begin{array}{r|rrrrrrrrr}
 & 1 & -3 & +0 & -5 & +2 & +5 & +4 & +0 & +1 \\
-2 & & & -2 & +6 & +4 & -4 & -6 & +2 \\
+1 & & & & 1 & -3 & -2 & +2 & +3 & -1 \\
\hline
 & 1 & -3 & -2 & +2 & +3 & -1; & 0 & +5 & +0
\end{array}
$$

$(x^8 \div x^3 = x^5)$. The quotient is therefore $x^5-3x^4-2x^3+2x^2+3x-1$, and the remainder $5x$.

5. Divide $10\,x^6 - 11\,x^5 - 3\,x^4 + 20\,x^3 + 10\,x^2 + 2$
   by $5\,x^3 - 3\,x^2 + 2\,x - 2.$

Arranging as in the ordinary method, we have

$$
\begin{array}{r|rrrr|rrrr}
 & 10 & -11 & -3 & +20 & +10 & + & 0 & + & 2 \\
+3 & & 6 & -3 & -6 & +12 & & & \\
-2 & & & -4 & +2 & +4 & -8 & & \\
+2 & & & & +4 & -2 & -4 & +8 & \\
\hline
5 & 2 & -1 & -2 & +4 & 24 & -12 & +10 &
\end{array}
$$

$$\text{Quotient} = 2\,x^3 - x^2 - 2\,x + 4 + \frac{24\,x^2 - 12\,x + 10}{5\,x^3 - 3\,x^2 + 2\,x - 2}.$$

We first draw a vertical line with as many vertical columns to the right as are less by unity than the number of terms in the divisor. This will mark the point at which the remainder begins to be formed. We then divide 10 by 5, and thus obtain the first coefficient of the dividend. We next multiply the remaining terms of the divisor by the 2 thus obtained. Adding the second vertical column and dividing by 5, we obtain -1; we multiply by the -1, add the next column, and divide the sum by 5, and so on for the others.

This method is not, however, always convenient. If the first term of the dividend be not divisible by the first term of the divisor, the work would be embarrassed with fractions. We may then proceed as in the following examples:

6. Divide $x^5 - 3\,x^4 + x^3 + 3\,x^2 - x + 3$ by $2\,x^3 + x^2 - 3\,x + 1.$

Let $2\,x = y,$ or $x = \dfrac{y}{2}.$

Substitute $\dfrac{y}{2}$ for x in the dividend and divisor, and we
   have

$$\frac{y^5}{2^5} - \frac{3y^4}{2^4} + \frac{y^3}{2^3} + \frac{3y^2}{2^2} - \frac{y}{2} + 3 \div \frac{2y^3}{2^3} + \frac{y^2}{2^2} - \frac{3y}{2} + 1$$

$$= \frac{y^5 - 2 \times 3y^4 + 2^2 y^3 + 2^3 \times 3y^2 - 2^4 y + 2^5 \times 3}{2^5}$$

$$\div \frac{y^3 + y^2 - 2 \times 3y + 2^2}{2^2}$$

$$= \frac{y^5 - 6y^4 + 4y^3 + 24y^2 - 16y + 96}{2^3} \div y^3 + y^2 - 6y + 4 \ldots A$$

Dividing $y^5 - 6y^4 + 4y^3 + 24y^2 - 16y + 96$ by $y^3 + y^2 - 6y + 4$, by the ordinary method, and the quotient by 2^3, we have

$$\frac{y^2 - 7y + 17}{2^3} - \frac{1}{2^3} \cdot \frac{39y^2 - 114y - 28}{y^3 + y^2 - 6y + 4}.$$

Substituting for y its value $2x$, and simplifying, we get

$$\frac{x^2}{2} - \frac{7x}{4} + \frac{17}{8} - \frac{1}{8} \cdot \frac{39x^2 - 57x - 7}{2x^2 + x^2 - 3x + 1} \quad \ldots \ldots \ldots B$$

By comparing the dividend of A with the original question, we find that we have multiplied the successive coefficients of the dividend by 2^0, 2^1, 2^2, etc., and we have multiplied the successive coefficients of the divisor, *omitting the first term*, by the same numbers. Dividing then by Horner's division, we get the coefficients 1, -7, 17; and for coefficients of remainder, -39, 114, and 28. The first three of these divided by 2, 2^2, 2^3, are the coefficients of x^2, etc.; and -39, etc., are divided by 1, 2, 2^2. Hence, the work will stand as follows:

	x^5	$-3x^4$	$+x^3$	$+3x^2$	$-x$	$+3$	$\div 2x^3$	$+x^2$	$-3x$	$+1$
	1	2	4	8	16	32	1	2	4
	1 -6	$+4$	$+24$	$-16 +96$			1 -6 $+4$
-1	-1 $+7$	-17					
$+6$	$+6$	-42	$+102$				
-4		-4	$+28 -68$				
	1 -7 $+17$	-39	$+114 +28$				

$$\text{Quotient}* = \frac{x^2}{2} - \frac{7x}{4} + \frac{17}{8} - \frac{1}{8} \cdot \frac{39\,x^2 - \dfrac{114\,x}{2} - \dfrac{28}{4}}{2x^3 + x^2 - 3x + 1}.$$

$$= \frac{x^2}{2} - \frac{7x}{4} + \frac{17}{8} - \frac{1}{8} \cdot \frac{39\,x^2 - 57\,x - 7}{2x^3 + x^2 - 3x + 1}.$$

7. Divide $5\,x^5 + 2$ by $3\,x^2 - 2x + 3$.

	$5x^5$	0	0	0	$0+$	$2 \div 3x^2 - 2x + 3$		
	1	3	9	27	81	243	1	3
	5	0	0	0	$0+$ 486		-2	$+9$
$+2$		$10+$	$20-$	50	-280			
-9			$-45-$	90	$+225+1260$			
	$5+$	$10-$	$25-$	140	$-55+1746$			

$$\text{Coeffs. of Quotient} = \frac{5}{3} + \frac{10}{3^2} - \frac{25}{3^3} - \frac{140}{3^4} - \frac{1}{3^4} \cdot \frac{55 - \dfrac{1746}{3}}{3 - 2 + 3}.$$

$$\text{Quotient} = \frac{5\,x^3}{3} + \frac{10\,x^2}{9} - \frac{25\,x}{27} - \frac{140}{81} - \frac{1}{81} \cdot \frac{55\,x - 582}{3\,x^2 - 2x + 3}.$$

Ex. 13.

Divide :

1. $6x^5 + 5x^4 - 17x^3 - 6x^2 + 10x - 2$ by $2x^2 + 3x - 1$.

2. $5x^6 + 6x^5 + 1$ by $x^2 + 2x + 1$.

3. $a^6 - 6a + 5$ by $a^2 - 2a + 1$.

4. $x^5 - 4x^3y^2 - 8x^2y^3 - 17xy^4 - 12y^5$ by $x^2 - 2xy - 3y^2$.

5. $a^6 - 3a^4x^2 + 3a^2x^4 - x^6$ by $a^3 - 3a^2x + 3ax^2 - x^3$.

6. $4x^4 + 3x^2 - 3x + 1$ by $x^2 - 2x + 3$.

---

* It will, in general, be more convenient to multiply the dividend by such a number as will make its first term exactly divisible by the first term of the divisor, and afterwards divide the quotient by this multiplier.

7. $10x^5 + 5x^4 - 90x^3 - 44x^2 + 10x + 1$ by $x^2 - 9$.

8. $x^5 - x^4y + x^3y^2 - x^2y^3 + xy^4 - y^5$ by $x^3 - y^3$.

9. Multiply $x^4 - 4x^3a + 6x^2a^2 - 4xa^3 + a^4$ by $x^2 + 2xa + a^2$, and divide the product by $x^4 - 2x^3a + 2xa^3 - a^4$.

Divide:

10. $x^5 - ax^4 + bx^3 - bx^2 + ax - 1$ by $x - 1$.

11. $6x^5 + 7x^4 + 7x^3 + 6x^2 + 6x + 5$ by $2x^2 + x + 1$.

12. $60(x^4 + y^4) + 91xy(x^2 - y^2)$ by $12x^2 - 13xy + 5y^2$.

13. $6x^6 - 481x^5 + 79x^4 + 81x^3 - 81x^2 + 86x - 481$
    by $x - 80$.

14. $6x^6 - x^5 + 2x^4 - 2x^3 + 2x^2 + 19x + 6$ by $3x^2 + 4x + 1$.

15. $a(a + 2b)^3 - b(2a + b)^3$ by $(a - b)^3$.

16. $(x + y)^3 + 3(x + y)^2z + 3(x + y)z^2 + z^3$
    by $(x + y)^2 + 2(x + y)z + z^2$.

17. $10x^{10} + 10x^6 + 10x^3 - 200$ by $x^7 + x^3 - x + 1$.

18. $bmx^4 + (bn + cm)x^3 + cnx^2 + abx + ac$ by $bx + c$.

19. Multiply $1 + 1\frac{3}{2}x - 18x^3$ by $1 - 1\frac{3}{4}x^2 + \frac{3}{2}x^3$, and divide the product by $1 + 1\frac{1}{2}x - 3x^2$.

Find the remainders in the following cases:

20. $(x^3 + 3x^2 + 4x + 5) \div (x - 2)$.

21. $(x^4 - 3x^2 + x - 3) \div (x - 1)$.

22. $(x^4 + 4x^3 + 6x + 8) \div (x + 2)$.

23. $(27x^4 - y^4) \div (3x - 2y)$.

24. $(3x^5 + 5x^4 - 3x^3 + 7x^2 - 5x + 8) \div (x^2 - 2x)$.

25. $(5x^4 + 90x^3 + 80x^2 - 100x + 500) \div (x + 17)$.

§ 8. The following are examples of an important use of rner's Division :

1. Arrange $x^3 - 6x^2 + 7x - 5$ in powers of $x - 2$.

$$
\begin{array}{r|rrrr}
 & 1 & -6 & +7 & -5 \\
2 & & 2 & -8 & -2 \\
\hline
 & 1 & -4 & -1; & -7 \\
2 & & 2 & -4 & \\
\hline
 & 1 & -2; & -5 & \\
2 & & +2 & & \\
\hline
 & 1; & 0 & &
\end{array}
$$

Hence, $x^3 - 6x^2 + 7x - 5 = (x - 2)^3 - 5(x - 2) - 7$;
or, as it is generally expressed,

$$x^3 - 6x^2 + 7x - 5 = y^3 - 5y - 7 \text{ if } y = x - 2.$$

2. Express $x^4 + 12x^3 + 47x^2 + 66x + 28$ in powers of $x + 3$.

$$
\begin{array}{r|rrrrr}
 & 1 & 12 & 47 & 66 & 28 \\
-3 & & -3 & -27 & -60 & -18 \\
\hline
 & 1 & 9 & 20 & 6; & 10 \\
-3 & & -3 & -18 & -6 & \\
\hline
 & 1 & 6 & 2; & 0 & \\
-3 & & -3 & -9 & & \\
\hline
 & 1 & 3; & -7 & & \\
-3 & & -3 & & & \\
\hline
 & 1; & 0 & & &
\end{array}
$$

Hence, $x^4 + 12x^3 + 47x^2 + 66x + 28 = y^4 - 7y^2 + 10$,
if $y = x + 3$.

After a few solutions have been written out in full, as in the above examples, the writing may be lessened by omitting the lines opposite the increments (-2 in Exam. 1, and

3 in Exam. 2), the multiplication and addition being performed mentally. The last example, written in this way,
would appear as follows :

$$\begin{array}{r|ccccc} & 1 & 12 & 47 & 66 & 28 \\ -3 & 1 & 9 & 20 & 6 & (10) \\ & 1 & 6 & 2 & (0) \\ & 1 & 3 & (-7) \\ & 1 & (0) \end{array}$$

Ex. 14.

Express :

1. $x^3 - 5x^2 + 3x - 8$ in powers of $x - 1$.

2. $x^3 + 3x^2 + 6x + 9$ in powers of $x + 1$.

3. $x^4 - 8x^3 + 24x^2 - 32x + 97$ in powers of $x - 2$.

4. $x^4 + 12x^3 + 5x^2 - 7$ in powers of $x + 2$.

5. $3x^5 - x^3 + 4x^2 + 5x - 8$ in powers of $x - 2$.

6. $x^4 - 7x^3 + 11x^2 - 7x + 10$ in powers of $x - 1\frac{3}{4}$.

7. $x^3 - 2x^2 - 4x + 9$ in powers of $x - \frac{2}{3}$.

8. $x^3 - 9x^2y + 6xy^2 - 8y^3$ in powers of $x - 2y$.

9. $x^5 - 5x^4y + 5xy^4 - y^5$ in powers of $x - y$.

10. $8x^3 + 12x^2y + 10xy^2 + 8y^3$ in powers of $2x + y$.

11. $x^3 - \frac{3}{2}x^2 + \frac{3}{8}x - \frac{5}{72}$ in powers of $\frac{1}{8}x - \frac{1}{16}$.

12. $x^4 + 8x^3 - 15x - 10$ in powers of $x + 2$.

CHAPTER II.

SYMMETRY.

§ 9. An expression is said to be symmetrical with respect to two of its letters when these can be interchanged by substituting each for the other without altering the expression.

Thus, $a^3 + a^2 x + ax^2 + x^3$ is symmetrical with respect to a and x, for on substituting x for a and a for x it becomes $x^3 + x^2 a + xa^2 + a^3$, which differs from the original expression merely in the order of its terms and of their factors. So, also, $x^2 + a^2 x + ab + b^2 x$ is symmetrical with respect to a and b, for on substituting b for a and a for b it becomes $x^2 + b^2 x + ba + a^2 x$, which is identical with the given expression. On interchanging x and a, $x^2 + a^2 x + ab + b^2 x$ becomes $a^2 + x^2 a + xb + b^2 a$; this is not the same as the given expression, which is therefore not symmetrical with respect to x and a. In like manner, it may be shown that this expression is not symmetrical with respect to x and b.

An expression is symmetrical with respect to three or more of its letters if it is symmetrical with respect to each and every pair of these that can be selected.

Thus, $x^3 + y^3 + z^3 - 3xyz$ is symmetrical with respect to x, y, and z, for it remains the same on interchanging x and y, or y and z, or z and x; and these are all the pairs that can be selected from x, y, and z. So, also,

$$a^2 b + b^2 a + a^2 c + c^2 a + b^2 c + c^2 b$$

and $(x-a)(b-c)^2 + (x-b)(c-a)^2 + (x-c)(a-b)^2$

are each symmetrical with respect to a, b, and c,

but $a^2 b + b^2 c + c^2 a$

and $(x-a)(a-b)^2 + (x-b)(b-c)^2 + (x-c)(c-a)^2$

are not so.

$$ab + ac + ad + bc + bd + cd$$

is symmetrical with respect to a, b, c, and d,

but $ab + bc + cd + da$

is not symmetrical with respect to these letters, for on inter-changing a and b it becomes

$$ab + ac + cd + bd.$$

So, also,

$$bcd + acd + abd + abc$$

is symmetrical with respect to a, b, c, and d, as may be seen at once by writing it in the form

$$abcd\left(\frac{1}{a} + \frac{1}{b} + \frac{1}{c} + \frac{1}{d}\right).$$

An expression is cyclo-symmetric with respect to three of its letters, a, b, and c, if it remains the same expression when a is changed into b, b into c, and c into a.

Thus, $a^2 b + b^2 c + c^2 a$ is cyclo-symmetric with respect to the cycle $(a\,b\,c)$, for on changing a into b, b into c, and c into a, it becomes $b^2 c + c^2 a + a^2 b$, which differs from the original expression only in the order of its terms.

$(a-b)^3 + (b-c)^3 + (c-a)^3$ is not symmetrical with respect to a, b, and c, for interchanging a and b changes it into $(b-a)^3 + (a-c)^3 + (c-b)^3$ which $= -(a-b)^3 - (b-c)^3 - (c-a)^3$, and this differs from the original expression in the signs of all its terms. But this expression is cyclo-symmetric with respect to $(a\,b\,c)$.

So, also, $(x-a)(a-b)^2+(x-b)(b-c)^2+(x-c)(c-a)^2$ is cyclo-symmetric with respect to $(a\,b\,c)$ but is not completely symmetric with respect to a, b, and c.

Generally, an expression is cyclo-symmetric with respect to any set of letters, a, b, c, ..., h, k, called the *cycle* $(a\,b\,c.....\,h\,k)$, if it remains the same expression when a is changed into b, b into c, ..., h into k, and k into a.

Thus, $ab+bc+cd+da$ and $ac+bd$ are each cyclo-symmetric with respect to the cycle $(a\,b\,c\,d)$, but are not completely symmetric with respect to a, b, c, and d.

Every expression which is completely symmetric with respect to a set of letters is necessarily cyclo-symmetric with respect to them; but, as is seen by the above examples, an expression may be cyclo-symmetric without being completely symmetric.

PRINCIPLE OF SYMMETRY. *An expression which in any one form is completely symmetric, or is cyclo-symmetric, with respect to any set of letters will in every other form be completely symmetric, or be cyclo-symmetric, as the case may be, with respect to these letters.*

Thus, $a^3+b^3+c^3-3abc$ is symmetrical with respect to a, b, and c; hence, it will be symmetrical when written in any other form, as, for example, in the form

$$\tfrac{1}{2}(a+b+c)\,[(b-c)^2+(c-a)^2+(a-b)^2].$$

Again, $(a-b)^3+(b-c)^3+(c-a)^3$ is cyclo-symmetric, but not completely symmetrical, with respect to $(a\,b\,c)$; it will therefore remain thus cyclo-symmetric, but not completely symmetrical, under every change of form which may be given it; for example, when it is reduced to

$$3(a-b)(b-c)(c-a).$$

A symmetric function of several letters is frequently represented by writing each *type-term* once, preceded by the letter Σ; thus, for $a+b+c+\cdots+l$ we write Σa, and for $ab+ac+ad+\cdots+bc+bd+\cdots$ (that is, the sum of the products of *every* pair of the letters considered) we write Σab.

Ex. 15.

Write the following in full:

1. $\Sigma a^2 b$; $\Sigma(a-b)^2$; $\Sigma a(b-c)$; $\Sigma ab(x-c)$; $\Sigma a^3 b^2 c$; $\Sigma(a+b)(c-a)(c-b)$; $\Sigma[(a+c)^2-b^2]$; and $\Sigma a(b+c)^2$, each for a, b, c.

2. Σabc; $\Sigma a^2 b$; $\Sigma a^2 bc$; $\Sigma(a-b)$; and $\Sigma a^2(a-b)$, each with respect to a, b, c, d.

Show that the following are symmetrical:

3. $(x+a)(a+b)(b+x)+abx$, with respect to a and b.

4. $(a+b)^2+(a-b)^2$ with respect to a and b, and also with respect to a and $-b$.

5. $(ab-xy)^2-(a+b-x-y)[ab(x+y)-xy(a+b)]$ with respect to a and b, and also with respect to x and y.

6. $a^2(b-c)^2+b^2(c-a)^2+c^2(a-b)^2$ with respect to a, b, c.

7. $(ac+bd)^2+(bc-ad)^2$ with respect to a^2 and b^2, and also with respect to c^2 and d^2.

8. $x^6+y^6+3xy(x^2+xy+y^2)$ with respect to x and y.

9. $[x^3-y^3+3xy(2x+y)]^3+[y^3-x^3+3xy(2y+x)]^3$ with respect to x and y.

10. $a(a+2b)^3+b(b+2a)^3$ with respect to a and b, and also with respect to a and $-b$.

11. $ab\{[(a+c)(b+c)+2c(a+b)]^2 - (a-c)^2(b-c)^2\}$
with respect to a, b, c.

12. $a^2b^2 + b^2c^2 + c^2a^2 + 2abc(a+b+c)$ with respect to ab, bc, ca.

With respect to what letters are the following symmetrical?

13. $xyz + 5xy + 2(x^2 + y^2)$.

14. $2(a^2x^2 + b^2y^2) - 2ab(xy + by + ax)$.

15. $(f^2 - h^2)^2 + 4g^2(f+h)^2 + (2fh - 2g^2)^2$.

16. $(x+y)(x-z)(y-z) - xyz$.

17. $a^2b^2 + b^2c^2 + c^2a^2 - 2abc(a+b-c)$.

18. $x^6 - y^6 + z^6 - 3(x^2 - y^2)(y^2 - z^2)(z^2 + x^2)$.

19. $(a+b)^2 + (a+c)^2 + (b-c)^2$.

20. $(a+b)^4 + (a-c)^4 + (b+c)^4 + (a+c)^4$.

21. $(a+b)^4 + (a-c)^4 + (b+c)^4 + (a+c)^4 + (c-b)^4$.

Select the type-terms in :

22. $a^2 + 2ab + b^2 + 2bc + c^2 + 2ca$.

23. $a(b^2 - c^2) + b(c^2 - a^2) + c(a^2 - b^2) + (a+b)(b+c)(c+a)$.

24. $a(b+c)^2 + b(c+a)^2 + c(a+b)^2 - 12abc$.

Write down the *type-terms* in :

25. $(x+y)^5$; $(x-y)^5$; $(x+y)^5 - x^5 - y^5$.

26. $(x+y)^7 + (x-y)^7$; $(x+y)^7 - (x-y)^7$.

27. $(x+y+z)^4$; $(x-y-z)^4$.

28. $(a+b+c+d)^4$; $(a^2+b^2+c^2+d^2)^2$.

29. $(a+b)^3 + (b+c)^3 + (c+a)^3$.

§ 10. In reducing an algebraic expression from one form to another, advantage may be taken of the principle of symmetry; for, it will be necessary to calculate only the *type-terms*, and the others may be written down from these.

EXAMPLES.

1. Find the expansion of $(a + b + c + d + e + \dots)^2$.

This expression is symmetrical with respect to a, b, c, ...; hence the expansion also must be symmetrical, and, as it is a product of *two* factors, it can contain only the squares a^2, b^2, c^2, ..., and the products in pairs, ab, ac, ad, ..., bc, bd, ...; so that a^2 and ab are *type-terms*.

Now $(a + b)^2 = a^2 + 2ab + b^2$; and the addition of terms involving c, d, e, ..., will not alter the terms $a^2 + 2ab$, but will merely give additional terms of the same type. Hence, from symmetry we obtain

$$(a+b+c+d+e+\dots)^2 = a^2 + 2ab + 2ac + 2ad + 2ae + \dots$$
$$+ b^2 \quad + 2bc + 2bd + 2be + \dots$$
$$+ c^2 \quad + 2cd + 2ce + \dots$$
$$+ d^2 \quad + 2de + \dots$$
$$+ e^2 \quad + \dots$$

This may be compactly written

$$(\Sigma a)^2 = \Sigma a^2 + 2\Sigma ab.$$

2. Expand $(a + b)^3$.

(i.) The expression is of *three* dimensions, and is symmetrical with respect to a and b.

(ii.) The type-terms are a^3, $a^2 b$.

Hence, $(a + b)^3 = a^3 + b^3 + n(a^2 b + b^2 a)$, where n is *numerical*.

To find the value of n, put $a = b = 1$, and we have
$$(1 + 1)^3 = 1 + 1 + n(1 + 1). \quad \therefore n = 3.$$

3. Expand $(x+y+z)^3$.

This is of three dimensions, and is symmetrical with respect to x, y, z. We have

$$(x+y+z)^3 = [(x+y)+z]^3 = (x+y)^3 + \cdots$$
$$= x^3 + 3x^2y + \cdots$$

which are type-terms, the only other possible type-term being xyz.

Now, since the expression contains $3x^2y$, it must also contain $3x^2z$; that is, it must contain $3x^2(y+z)$. Hence,

$$\begin{aligned}(x+y+z)^3 = {} & x^3 + 3x^2(y+z) \\ & + y^3 + 3y^2(z+x) \\ & + z^3 + 3z^2(x+y) \\ & + n(xyz),\end{aligned}$$

where n is numerical, and may be found by putting $x=y=z=1$ in the last result, giving

$$(1+1+1)^3 = 1+1+1+3(1+1)+3(1+1)$$
$$+3(1+1)+n.$$

Hence, $n=6$.

4. Similarly, we may show that

$$\begin{aligned}(a+b+c+d)^3 = {} & a^3 + 3a^2(b+c+d) + 6bcd \\ & + b^3 + 3b^2(c+d+a) + 6cda \\ & + c^3 + 3c^2(d+a+b) + 6dab \\ & + d^3 + 3d^2(a+b+c) + 6abc.\end{aligned}$$

5. Expand $(a+b+c+\cdots)^3$.

The type-terms are a^3, a^2b, abc.

Expanding $(a+b+c)^3$, we get $a^3 + 3a^2b + 6abc + \cdots$

Hence, by symmetry, we have

$$(\Sigma a)^3 = \Sigma a^3 + 3\Sigma a^2 b + 6\Sigma abc.$$

6. Simplify $(a+b-2c)^2+(b+c-2a)^2+(c+a-2b)^2$.

This expression is symmetrical, involving terms of the types a^2 and ab. Now, a^2 occurs with 1 as a coefficient in the first square, with 4 as a coefficient in the second square, and with 1 as a coefficient in the third square, and hence $6a^2$ is one type-term of the result; ab occurs with 2 as a coefficient in the first square, with -4 as a coefficient in the second square, and with -4 as a coefficient in the third square, and hence $-6ab$ is the second type-term in the result. Hence, the total result is $6(a^2+b^2+c^2-ab-bc-ca)$.

7. Simplify $(x+y+z)^3+(x-y-z)^3+(y-z-x)^3$ $+(z-x-y)^3$.

This is symmetrical with respect to x, y, z; and the type-terms are x^3, $3x^2y$, $6xyz$:

(i.) x^3 occurs in each of the first two cubes, and $-x^3$ in each of the second two cubes; therefore, there are no terms of the type x^3 in the result.

(ii.) $3x^2y$ occurs in the *first* and *third* cubes, and $-3x^2y$ in the second and fourth; therefore, there are no terms of this type in the result.

(iii.) $6xyz$ occurs in each of the four cubes; therefore, $24xyz$ is the total result.

8. Prove $(a^2+b^2+c^2+d^2)(w^2+x^2+y^2+z^2)-(aw+bx+cy+dz)^2$ $=(ax-bw)^2+(ay-cw)^2+(az-dw)^2+(by-cx)^2$ $+(bz-dx)^2+(cz-dy)^2$.

The left-hand member (considered as given) is symmetrical with respect to the pairs of letters, a and w, b and x, c and y, d and z; that is, any two pairs may be interchanged without affecting the expression. As the expression is only of the second degree in these pairs, no term can involve three pairs as factors; hence, the type-terms may be obtained by considering all the terms involving a, b, w, x; these are a^2w^2, a^2x^2, b^2w^2, b^2x^2, $-a^2w^2$, $-b^2x^2$, $-2abwx$, and are the terms of $(ax-bw)^2$, which is consequently a type-term. From $(ax-bw)^2$ we derive the five other terms of the second member by merely changing the letters.

9. Prove that $(x^2 - yz)^3 + (y^2 - zx)^3 + (z^2 - xy)^3$
$- 3(x^2 - yz)(y^2 - zx)(z^2 - xy)$ is a complete square.

The expression will remain symmetrical if $(x^2 - yz)$ $(y^2 - zx)(z^2 - xy)$, instead of being multiplied by -3, be subtracted from each of the preceding terms, thus giving

$$(x^2 - yz)[(x^2 - yz)^2 - (y^2 - zx)(z^2 - xy)]$$
$$+ (y^2 - zx)[(y^2 - zx)^2 - (z^2 - xy)(x^2 - yz)]$$
$$+ (z^2 - xy)[(z^2 - xy)^2 - (x^2 - yz)(y^2 - zx)]$$
$$= (x^2 - yz)x(x^3 + y^3 + z^3 - 3xyz) + \cdots$$
$$= (x^3 + y^3 + z^3 - 3xyz)(x^3 + y^3 + z^3 - 3xyz).$$

Ex. 16.

Simplify the following:

1. $(a+b+c)^2 + (a+b-c)^2 + (b+c-a)^2 + (c+a-b)^2$.

2. $(a-b-c)^2 + (b-a-c)^2 + (c-a-b)^2$.

3. $(a+b+c-d)^2 + (b+c+d-a)^2 + (c+d+a-b)^2$
$+ (d+a+b-c)^2$.

4. $(a+b+c)^2 - a(b+c-a) - b(a+c-b) - c(a+b-c)$.

5. $(x+y+z+n)^2 + (x-y-z+n)^2 + (x-y+z-n)^2$
$+ (x+y-z-n)^2$.

6. $(a+b+c)^3 + (a+b-c)^3 + (b+c-a)^3 + (c+a-b)^3$.

7. $(x-2y-3z)^2 + (y-2z-3x)^2 + (z-2x-3y)^2$.

8. $(ma + nb + rc)^3 - (ma + nb - rc)^3 - (nb + rc - ma)^3$
   $- (rc + ma - nb)^3.$

9. $a(b + c)(b^2 + c^2 - a^2) + b(c + a)(c^2 + a^2 - b^2)$
   $+ c(a + b)(a^2 + b^2 - c^2).$

10. $(ab + bc + ca)^2 - 2abc(a + b + c).$

Prove the following :

11. $(ax + by + cz)^2 + (bx + cy + az)^2 + (cx + ay + bz)^2$
   $+ (ax + cy + bz)^2 + (cx + by + az)^2 + (bx + ay + cz)^2$
   $= 2(a^2 + b^2 + c^2)(x^2 + y^2 + z^2)$
   $+ 4(ab + bc + ca)(xy + yz + zx).$

12. $(a + b + c)^4 + (b + c - a)^4 + (c + a - b)^4 + (a + b - c)^4$
   $= 4(a^4 + b^4 + c^4) + 24(a^2b^2 + b^2c^2 + c^2a^2).$

13. $(a + b + c)^4 = \Sigma a^4 + 4\Sigma a^3 b + 6\Sigma a^2 b^2 + 12\Sigma a^2 bc.$

14. $(\Sigma a)^4 = \Sigma a^4 + 4\Sigma a^3 b + 6\Sigma a^2 b^2 + 12\Sigma a^2 bc + 24\Sigma abcd.$

15. $(a^2 + b^2 + c^2)^3 + 2(ab + bc + ca)^3$
   $- 3(a^2 + b^2 + c^2)(ab + bc + ca)^2$
   $= (a^3 + b^3 + c^3 - 3abc)^2.$

16. $(a - b)^2(b - c)^2 + (b - c)^2(c - a)^2 + (c - a)^2(a - b)^2$
   $= (a^2 + b^2 + c^2 - ab - ac - bc)^2.$

17. $(2a - b - c)^2(2b - c - a)^2 + (2b - c - a)^2(2c - a - b)^2$
   $+ (2c - a - b)^2(2a - b - c)^2$
   $= 9(a^2 + b^2 + c^2 - ab - bc - ca)^2.$

18. $(ar^2 + 2brs + cs^2)(ax^2 + 2bxy + cy^2)$
   $- [arx + b(ry + sx) + csy]^2 = (ac - b^2)(ry - sx)^2.$

19. $(a^2 + ab + b^2)(c^2 + cd + d^2)$
   $= (ac + ad + bd)^2 + (ac + ad + bd)(bc - ad) + (bc - ad)^2.$

20. Show that there are two ways in which the given product in the last example can be expressed in the form $p^2 + pq + q^2$, and two ways in which it can be expressed in the form $p^2 - pq + q^2$.

21. $6(w^2+x^2+y^2+z^2)^2 = (w+x)^4+(w-x)^4+(w+y)^4$
$+(w-y)^4+(w+z)^4+(w-z)^4+(x+y)^4+(x-y)^4$
$+(x+z)^4+(x-z)^4+(y+z)^4+(y-z)^4.$

22. $\frac{1}{5}[(a+b+c)^5+(a-b-c)^5+(b-c-a)^5+(c-a-b)^5]$
$=\frac{1}{3}[(a+b+c)^3+(a-b-c)^3+(b-c-a)^3+(c-a-b)^3]$
$\times\frac{1}{2}[(a+b+c)^2+(a-b-c)^2+(b-c-a)^2+(c-a-b)^2].$

THEORY OF DIVISORS.

Any expression which can be reduced to the form

$$ax^n + bx^{n-1} + cx^{n-2} + \cdots + hx + k,$$

in which n is a positive integer, and a, b, c, ..., h, k, are independent of x, is called a **Polynome in x of degree n.**

The expressions $f(x)^n$, $F(x)^n$, $\phi(x)^m$, are used as general symbols for polynomes; the exponents n and m indicate the *degree* of the polynome.

THEOREM I. If the polynome $f(x)^n$ be divided by $x-a$, the remainder will be $f(a)^n$.

Cor. 1. $f(x)^n - f(a)^n$ is always exactly divisible by $x-a$.

Cor. 2. If $f(a)^n = 0$, $f(x)^n$ is exactly divisible by $x-a$; that is, $f(x)^n$ is an algebraic multiple of $x-a$.

Cor. 3. If the polynome $f(x)^n$, on division by the polynome $\phi(x)^n$, leave a remainder independent of x, such remainder will be the value of $f(x)^n$ when $\phi(x)^m = 0$.

EXAMPLES. THEOREM I.

1. Find the remainder when $x^5 - 7x^4 + 13x^3 - 16x^2 + 9x - 12$
is divided by $x-5$.

The remainder will be the value of the given polynome when 5 is substituted for x. (See § 3.)

$$\begin{array}{r|rrrrrr} & 1 & -7 & +13 & -16 & +9 & -12 \\ 5 & & 5 & -10 & 15 & -5 & 20 \\ \hline & 1 & -2 & 3 & -1 & 4\,; & 8 \end{array}$$

Hence, the remainder is 8.

2. Find the remainder when $(x-a)^3 + (x-b)^3 + (a+b)^3$ is divided by $x+a$.

For x substitute $-a$, then

$$(-2a)^3 + (-a-b)^3 + (a+b)^3 = -8a^3.$$

3. Find the remainder when
$$x^3 + a^3 + b^3 + (x+a)(x+b)(a+b)$$
is divided by $x+a+b$.

For x substitute $-(a+b)$, and we obtain

$$-(a+b)^3 + a^3 + b^3 + ab(a+b) = -2ab(a+b).$$

See Formula [6].

4. Find the remainder when
$$(x^2 + 2ax - 2a^2)^3 (x^2 - 2ax - 2a^2) + 32(x-a)^4(x+a)^4$$
is divided by $x^2 - 2a^2$.

$x^2 - 2a^2$ may be struck out wherever it appears.
This reduces the dividend to

$$(2ax)^3(-2ax) + 32(x-a)^4(x+a)^4$$
$$= -16a^4x^4 + 32(x^2 - a^2)^4.$$

In this substitute $2a^2$ for x^2, and it becomes

$$-64a^8 + 32a^8 = -32a^8,$$

which is the required remainder.

Ex. 17.

1. Find the remainder when $3x^4 + 60x^3 + 54x^2 - 60x + 58$ is divided by $x+19$.

2. Find the remainder when $px^3 - 3qx^2 + 3rx - s$ is divided by $x-a$.

3. What number added to
$$4x^5 + 34x^4 + 58x^3 + 21x^2 - 123x - 41$$
will give a sum exactly divisible by $2x + 13$.

4. What number taken from
$$10x^{10} - 20x^8 - 10x^6 - 0.89x^4 - 8.9x^2 + 20$$
will leave a remainder exactly divisible by $10x^2 - 11$?

Find the remainders from the following divisions:

5. $(x+1)^5 - x^5 + x + 1$; $(x+a+3)^3 - (x+a+1)^3 \div x + 2$.

6. $x^n + y^n \div x - y$; $x^{2n} + y^{2n} \div x + y$; $x^{2n+1} + y^{2n+1} \div x + y$.

7. $(x+1)^3 + x^3 + (x-1)^3 \div x - 2$.

8. $(x-a)^3 (x+a)^3 + (x^2 - 2b^2)^3 \div x^2 + b^2$.

9. $(x^2 + ax + a^2)(x^2 - ax + a^2)$
$- (x^2 - 3ax + 2a^2)(x^2 + 3ax + 2a^2) \div x^2 + 2a^2$.

10. $(9a^2 + 6ab + 4b^2)(9a^2 - 6ab + 4b^2)(81a^4 - 36a^2b^2 + 16b^4)$
$\div (3a - 2b)^2$.

11. $a^2(x-a)^3 + b^2(x-b)^3 \div x - a - b$.

12. $(ax + by)^3 + a^3 y^3 + b^3 x^3 - 3abxy(ax + by)$
$\div (a+b)(x+y)$.

13. $x^3 + a^3 + b^3 - 3abx \div x - a + b$; also, $\div x + a - b$;
also, $\div x - a - b$.

14. Any polynome divided by $x - 1$ gives for remainder the sum of the coefficients of the terms.

EXAMPLES. *Cor.* 1.

1. $x^5 + y^5$ is exactly divisible by $x + y$.
   In "$x^5 - a^5$ is exactly divisible by $x - a$," substitute $-y$ for a.

2. $mx^3 - px^2 + qx + m + p + q$ is exactly divisible by $x + 1$.
This may be written

$$(mx^3 - px^2 + qx) - [m(-1)^3 - p(-1)^2 + q(-1)]$$

is exactly divisible by $x - (-1)$.

3. $(x^2 + 6xy + 4y^2)^5 + (x^2 + 2xy + 4y^2)^5$ is exactly divisible by $(x + 2y)^2$.

For $(x^2 + 6xy + 4y^2)^5 - (-x^2 - 2xy - 4y^2)^5$ is exactly divisible by $(x^2 + 6xy + 4y^2) - (-x^2 - 2xy - 4y^2)$, which is $2(x^2 + 4xy + 4y^2) = 2(x + 2y)^2$.

Ex. 18.

Prove that the following are cases of exact division :

1. $x^{2n+1} + y^{2n+1} \div x + y$; $x^{2n} - y^{2n} \div x + y$.

2. $x^{12} + y^{12} \div x^4 + y^4$; $x^{30} + y^{30} \div x^6 + y^6$;   also, $\div x^{10} + y^{10}$; also, $\div x^2 + y^2$.

3. $(ax + by)^5 + (bx + ay)^5 \div (a + b)(x + y)$.

4. $(ax + by + cz)^3 - (bx + cy + az)^3$
     $\div (a - b)x + (b - c)y + (c - a)z$.

5. $(2y - x)^n - (2x - y)^n \div 3(y - x)$.

6. $(2y - x)^{2n+1} + (2x - y)^{2n+1} \div y + x$.

7. $(my - nx)^5 - (mx - ny)^5 \div (m + n)(y - x)$.

8. $(x + y)^6 + (x - y)^6 \div 2(x^2 + y^2)$.

9. $(x^2 + xy + y^2)^3 + (x^2 - xy + y^2)^3 \div 2(x^2 + y^2)$.

10. $(a + b)^9 - (a - b)^9 \div 2b(3a^2 + b^2)$.

11. $(x^2 + 5bx + b^2)^7 + (x^2 - bx + b^2)^7 \div 2(x + b)^2$.

12. $(a + b)^{4n+2} + (a - b)^{4n+2} \div 2(a^2 + b^2)$.

13. $[x^3 + 3xy(x - y) - y^3]^3 + [x^3 - 9xy(x - y) - y^3]^3$
     $\div 2(x - y)^3$.

14. $3x^3 - 5x^2 + 4x - 2 \div x - 1$.

15. Any polynome in x is divisible by $x - 1$ when the sum of the coefficients of the terms is zero.

16. Any polynome in x is divisible by $x + 1$ when the sum of the coefficients of the even powers of x is equal to the sum of the coefficients of the odd powers. (The constant term is included among the coefficients of the even powers.)

EXAMPLES. Cor. 2.

1. Show that $a(a+2b)^3 - b(2a+b)^3$ is exactly divisible by $a+b$.

By *Cor.* 2, the substitution of $-b$ for a must cause the polynome to vanish.

Substituting, $a(a-2a)^3 + a(2a-a)^3 = -a^4 + a^4 = 0$.

2. Show that

$$(ab - xy)^2 - (a+b-x-y)[ab(x+y) - xy(a+b)]$$

is exactly divisible by $(x-a)(y-a)$, also, by $(x-b)(y-b)$.

For x substitute a, and the expression becomes

$$(ab - ay)^2 - (b-y)[ab(a+y) - ay(a+b)]$$
$$= a^2(b-y)^2 - (b-y)[a^2(b-y)] = 0.$$

The expression is, therefore, exactly divisible by $x-a$. But it is symmetrical with respect to x and y, hence it is divisible by $y-a$; and, as $x-a$ and $y-a$ are independent factors, the expression is exactly divisible by $(x-a)(y-a)$. Again, the given expression is symmetrical with respect to a and b; hence, making the interchange of a and b, the expression is seen to be divisible by $(x-b)(y-b)$.

3. Show that $6(a^5+b^5+c^5)-5(a^3+b^3+c^3)(a^2+b^2+c^2)$ is exactly divisible by $a+b+c$.

For a substitute $-(b+c)$, and the result, which would be the remainder were the division actually performed, must vanish.

$$6[-(b+c)^5+b^5+c^5]$$
$$-5[-(b+c)^3+b^3+c^3][(b+c)^2+b^2+c^2]$$
$$=6[-(b+c)^5+b^5+c^5]+30bc(b+c)(b^2+bc+c^2).$$

See [1] and [6].

The expansion being of the fifth degree, and symmetrical in b and c, it will be sufficient to show that the coefficients of b^5, b^4c, b^3c^2 vanish, the coefficients of b^2c^3, bc^4, c^5 being the coefficients of the former terms in reverse order. Calculating the coefficients of these type-terms, we get

$$6[-5b^4c-10b^3c^2-\cdots]+30(b^4c+2b^3c^2+\cdots),$$

which evidently vanishes. Hence, the truth of the proposition.

4. If $a+b+c=0$,
$$\tfrac{1}{5}(a^5+b^5+c^5)=\tfrac{1}{2}(a^2+b^2+c^2)\times\tfrac{1}{3}(a^3+b^3+c^3).$$

In the last example it has been proved that the *difference* of the quantities, here declared to be equal, is a multiple of $a+b+c$, that is, in this case, a multiple of zero. Hence, under the given condition they are equal.

Ex. 19.

Prove that the following are cases of exact division :

1. $(ax-by)^3+(bx-ay)^3-(a^3+b^3)(x^3-y^3)\div a,\ b,\ x,\ y,$
$\quad a+b,\ x-y.$

2. $ax^3-(a^2+b)x^2+b^2\div ax-b.$ (Substitute ax for b.)

3. $\begin{cases} (ax+by)^2-(a-b)(x+z)(ax+by)+(a-b)^2xz \div x+y. \\ (ax-by)^2-(a+b)(x+z)(ax-by)+(a+b)^2xz \div x+y. \end{cases}$

4. $6a^3x^2-4ax^3-10axy-3a^2xy+2x^2y+5y^2 \div 2ax-y.$

5. $1.2a^4x-5.494a^3x^2+4.8a^2x^3+0.9ax^4-x^5$
   $\div 0.6ax-2x^2.$

6. $x^8+x^6y^2+x^2y+y^3 \div x^6+y.$

7. $(c-d)a^2+6(bc-bd)a+9(b^2c-b^2d) \div a+3b.$

8. $x(x-\frac{1}{12}y)^5+y(\frac{1}{12}x-y)^5 \div x-y.$

9. $a(a+2b)^3-b(b+2a)^3 \div a-b,$ also $\div a+b.$

10. $a^5-2a^4b+a^3b^2+a^2x^3-2abx^3+b^2x^3 \div (a-b)(x+a).$

11. $a(b-c)^3+b(c-a)^3+c(a-b)^3 \div (a-b), (b-c),$
    $(c-a).$

12. $a^3(b-c)+b^3(c-a)+c^3(a-b) \div (a-b), (b-c),$
    $(c-a).$

13. $a^4(b-c)+b^4(c-a)+c^4(a-b) \div (a-b), (b-c),$
    $(c-a).$

14. $(a-b)^2(c-d)^2+(b-c)^2(d-a)^2-(d-b)^2(a-c)^2$
    $\div (a-b), (b-c), (c-d), (d-a).$

15. $[(a-b)^2+(b-c)^2+(c-a)^2][(a-b)^2c^2+(b-c)^2a^2$
    $+(c-a)^2b^2]-[(a-b)^2c+(b-c)^2a+(c-a)^2b]^2$
    $\div (a-b), (b-c), (c-a).$

16. $(x+y)(y+z)(z+x)+xyz \div x+y+z.$

17. $ab(a^2-b^2)+bc(b^2-c^2)+ca(c^2-a^2) \div a+b+c.$

18. $(ab-bc-ca)^2-a^2b^2-b^2c^2-c^2a^2 \div a+b-c.$

19. $(a+2b)^3+(2b-3c)^3-(3c-a)^3+a^3+8b^3-27c^3$
    $\div a+2b-3c.$

20. $a^3b^3+b^3c^3+c^3a^3-3a^2b^2c^2 \div ab+bc+ca.$

EXAMPLES.   *Cors. 3 and 2.*

1. Find the value of $4x^5 + 9x^3 - 5x^2 + 23x + 6$ when $2x^2 = 3x - 4$.

   Since $2x^2 - 3x + 4 = 0$, we have simply to find the remainder on division by $2x^2 - 3x + 4$; and, if it is independent of x, it is the value sought, *Cor.* 3.

$$
\begin{array}{r|rrrrrr}
 & 4 & 0 & 9 & -5 & 23 & 6 \\
3 & & 6 & 9 & 15 & -3 & \\
-4 & & & -8 & -12 & -20 & 4 \\
\hline
2 & 2 & 3 & 5 & -1; & 0 & 10
\end{array}
$$

   Hence, the required value is 10.

2. What value of c will make $x^3 - 5x^2 + 7x - c$ exactly divisible by $x - 2$.

   If 2 be substituted for x, the remainder must vanish, *Cor.* 2.

$$
\begin{array}{r|rrrr}
 & 1 & -5 & 7 & -c \\
2 & & 2 & -6 & 2 \\
\hline
 & 1 & -3 & 1; & 2-c
\end{array}
$$

   Hence, $2 - c = 0$, or $c = 2$.

3. What value of c will make $6x^5 - 5x^4 + cx^3 - 20x^2 + 19x - 5$ vanish when $2x^2 = 3x - 1$.

   By *Cor.* 3 the remainder must vanish when the given polynome is divided by $2x^2 - 3x + 1$. We may divide at once and find, if possible, a value of c that will make both terms of the remainder vanish; or, we may first express cx^3 in lower terms in x, and then divide and find the required value of c from the remainder.

First Method (see page 31).

$$\begin{array}{c|cccccc} & 6 & -10 & 4c & -160 & 304 & -160 \\ 3 & & 18 & 24 & 12c+36 & 36c-420 & \\ -2 & & & -12 & -16 & -8c-24 & -24c+280 \\ \hline & 6 & 8, & 4c+12, & 12c-140\,; & 28c-140 & -24c+120 \end{array}$$

Hence, $28c = 140$ and $24c = 120$. Both of these are satisfied by $c = 5$.

Second Method. $x^3 = \frac{1}{2}x(3x-1) = \frac{3}{2}x^2 - \frac{1}{2}x$

$= \frac{3}{4}(3x-1) - \frac{1}{2}x = 2\frac{1}{4}x - \frac{3}{4} - \frac{1}{2}x = 1\frac{3}{4}x - \frac{3}{4}.$

$\therefore cx^3 = 1\frac{3}{4}cx - \frac{3}{4}c.$

Substituting for cx^3 in the given polynome, it becomes

$$6x^5 - 5x^4 - 20x^2 + (1\frac{3}{4}c + 19)x - \frac{3}{4}c - 5.$$

Divide and apply *Cor.* 3.

$$\begin{array}{c|cccccc} & 6 & -10 & 0 & -160 & 28c+304 & -24c-160 \\ 3 & & 18 & 24 & 36 & -420 & \\ -2 & & & -12 & -16 & -24 & 280 \\ \hline & 6 & 8 & 12 & -140\,; & 28c-140 & -24c+120 \end{array}$$

We thus obtain the same remainder as by the former method, and consequently the same result. A comparison of the two methods shows that they are but slightly different in form, but the second method shows rather more clearly that c need not be introduced into the dividend at all, but the proper multiples of it found by the preliminary reduction can be added to or taken from the numerical remainder, and the "true remainder" be thus found, and c determined from it.

Ex. 20.

Find the value of :

1. $x^4 - 3x^3 + 4x^2 - 3x + 4$, given $x^2 = x - 1$.
2. $x^5 - 2x^4 - 4x^3 + 13x^2 - 11x - 10$, given $(x-1)^2 = 2$.

3. $2x^5 - 7x^4 + 12x^3 - 11x^2 + 2x - 5$, given $(x-1)^2 + 2 = 0$.

4. $3x^6 + 11x^5 + 10x^3 + 7x^2 + 2x + 3$,
   given $x^3 + 3x^2 - 2x + 5 = 0$.

5. $6x^7 + 9x^6 - 16x^4 - 5x^3 - 12x^2 - 6x + 60$,
   given $3x^4 + x - 4 = 0$.

What values of c will make the following polynome vanish under the given conditions:

6. $x^4 + 13x^3 + 26x^2 + 52x + 8c$, given $x + 11 = 0$.

7. $x^4 - 2x^3 - 9x^2 + 2cx - 14$, given $3x + 7 = 0$.

8. $x^4 - 4x^3 - x^2 + 16x + 6c$, given $x^2 = x + 6$.

9. $2x^4 - 10x^2 + 4cx + 6$, given $x^2 + 3 = 3x$.

10. $2x^4 + x^3 - 7cx^2 + 11x + 10$, given $2x = 5$.

11. $4x^4 + cx^2 + 110x - 105$, given $2x^2 - 5x + 15 = 0$.

12. $3x^5 - 16x^4 + cx^3 - 5x^2 - 114x + 200$, given $x^2 = 3x - 4$.

13. What values of p and q will make
   $x^4 + 2x^3 - 10x^2 - px + q$ vanish, if $x^2 = 3(x-1)$?

14. What values of p and q will make
   $a^{12} - 5a^{10} + 10a^8 - 15a^6 + 29a^4 - pa^2 + q$ vanish, if
   $(a^2 - 2)^2 = a^2 - 3$?

THEOREM II. If the polynome $f(x)^n$ vanish on substituting for x each of the n (different) values $a_1, a_2, a_3, \ldots, a_n$,

$$f(x)^n = A(x - a_1)(x - a_2)(x - a_3) \ldots (x - a_n),$$

in which A is independent of x, and consequently is the coefficient of x^n in $f(x)^n$.

Cor. If $f(x)^n$ and $\phi(x)^m$ both vanish for the same m different values of x, $f(x)^n$ is algebraically divisible by $\phi(x)^m$.

EXAMPLES.

1. $x^3 + ax^2 + bx + c$ will vanish if 2, or 3, or -4 be substituted for x; determine a, b, c.

The coefficient of the highest power of x is 1;

$\therefore x^3 + ax^2 + bx + c = (x-2)(x-3)(x+4)$
$= x^3 - x^2 - 14x + 24$;

$\therefore a = -1$; $b = -14$; $c = 24$.

2. $x^3 + bx^2 + cx + d$ will vanish if -3, or 2, or 5 be substituted for x; determine its value if 3 be substituted for x.

The given polynome $= (x+3)(x-2)(x-5)$;

$\therefore$ the required value is $(3+3)(3-2)(3-5) = -12$.

3. $ax^3 + 3bx^2 + 3cx + d$ will vanish if for x be substituted -3, or $\frac{1}{2}$, or $1\frac{1}{2}$, but it becomes 45 if for x there be substituted 3; determine the values of a, b, c, d.

The coefficient of the highest power of x is a;

$\therefore ax^3 + 3bx^2 + 3cx + d = a(x+3)(x-\frac{1}{2})(x-1\frac{1}{2})$;

$\therefore a(3+3)(3-\frac{1}{2})(3-1\frac{1}{2}) = 45$;

$\therefore a = 2$.

$\therefore 2x^3 + 3bx^2 + 3cx + d = 2(x+3)(x-\frac{1}{2})(x-1\frac{1}{2})$;

$\therefore b = \frac{2}{3}$; $c = -3\frac{1}{2}$; $d = 4\frac{1}{2}$.

4. If $x^3 + px^2 + qx + r$ vanish for $x = a$, or b, or c, determine p, q, and r in terms of a, b, c.

$x^3 + px^2 + qx + r = (x-a)(x-b)(x-c)$
$= x^3 - (a+b+c)x^2 + (ab+bc+ca)x - abc$.

$\therefore p = -(a+b+c)$ or $-\Sigma a$,

$q = ab+bc+ca$ or Σab,

$r = -abc$ or $-\Sigma abc$.

5. If $x^3 + px^2 + qx + r$ vanish for $x = a$, or b, or c, determine the polynome that will vanish for $x = b + c$, or $c + a$, or $a + b$.

Since $x^3 + px^2 + qx + r$ vanishes for $x = a$, or b, or c,
$x^3 - px^2 + qx - r$ will vanish for $x = -a$, or $-b$, or $-c$, and $-p = a + b + c$.

But the required polynome will vanish for

$x = -p - a$, or $-p - b$, or $-p - c$;

that is, for $x + p = -a$, or $-b$, or $-c$.

Hence, it is $(x+p)^3 - p(x+p)^2 + q(x+p) - r$
$= x^3 + 2px^2 + (p^2 + q)x + pq - r$.

The following is the calculation in the last reduction. (See page 37.)

$$
\begin{array}{c|cccc}
  & 1 & -p & q & -r \\
\hline
p & 1 & 0 & q\,; & pq - r \\
p & 1 & p\,; & p^2 + q & \\
p & 1\,; & 2p & & \\
  & 1 & & &
\end{array}
$$

6. In any triangle, the square of the area expressed in terms of the lengths of the sides, is a polynome of four dimensions; and the area of the triangle, the lengths of whose sides are 3, 4, and 5, respectively, is 6. Find the polynome expressing the square of the area.

Let a, b, and c be the lengths of the sides, and A the required polynome.

1. The area vanishes if any two of the sides become together equal to the third side; hence, if $a + b = c$, $A = 0$, and consequently A is divisible by $a + b - c$. Similarly it is divisible by $b + c - a$ and by $c + a - b$.

2. The area vanishes if the three sides vanish together; hence, if $a+b+c=0$, $A=0$, and consequently A is divisible by $a+b+c$.

We have thus found four linear factors, but A is of only four dimensions.

$$\therefore A = m(a+b+c)(b+c-a)(c+a-b)(a+b-c),$$

in which m is a numerical constant.

But 6^2 or $36 = m(3+4+5)(4+5-3)(5+3-4)(3+4-5)$
$= 576\,m$; $\therefore m = \frac{1}{16}$.

The above includes all the ways in which the area of a triangle can vanish, for the vanishing of only one side involves the equality of the other two; or, if $a=0$, $b=c$, and therefore $a+b=c$, which is included in (1). If two sides vanish simultaneously, the three must vanish.

Examples on the Corollary.

7. Prove that $(x+1)^{12} - x^{12} - 2x - 1$ is divisible by $2x^3 + 3x^2 + x$.

Factoring $2x^3 + 3x^2 + x$, we find it vanishes for $x = 0$, or -1, or $-\frac{1}{2}$. Substituting these values in the first polynome, it also vanishes. But these are different values of x, hence the truth of the proposition.

8. $(x+y+z)^5 - x^5 - y^5 - z^5$ is divisible by $(x+y+z)^3 - x^3 - y^3 - z^3$.

The last expression vanishes if $x = -y$, so also does the first.

By symmetry they both vanish if $y = -z$ and if $z = -x$. Hence, they are both divisible by $(x+y)(y+z)(z+x)$. But this expression is of three dimensions, as also is the second of the given polynomes, hence it is a divisor of the former.

9. Prove that
$$[(a + b)^5 + (c + d)^5](a - b)(c - d)$$
$$+ [(b + c)^5 + (a + d)^5](b - c)(a - d)$$
$$+ [(b + d)^5 + (c + a)^5](b - d)(c - a)$$
is algebraically divisible by
$$(a-b)(c-d)(b-c)(a-d)(b-d)(c-a)(a+b+c+d),$$
and find the quotient.

Let $a = b$, and the first polynome reduces to
$$[(a + c)^5 + (a + d)^5](a - c)(a - d)$$
$$+ [(a + d)^5 + (c + a)^5](a - d)(c - a)$$
which vanishes, the second complex term differing from the first only in the sign of one factor, having $(c - a)$ instead of $(a - c)$.

Hence, the first polynome is divisible by $a - b$, and by symmetry it is also divisible by $a - c$, by $a - d$, by $b - c$, by $b - d$, by $c - d$.

Again, $(a+b)^5 + (c+d)^5$ is divisible by $(a+b)+(c+d)$; for, on putting $a + b = -(c + d)$, it becomes
$$[-(c + d)]^5 + (c + d)^5 = 0.$$

Similarly the other terms of the first of the given polynomes are each divisible by $a + b + c + d$, and consequently the whole is so divisible.

Now all these factors are different from each other, hence the first of the given polynomes is divisible by the product of these factors; that is, by the second of the given polynomes.

Both of these polynomes are of seven dimensions, hence their quotient must be a number the same for all values of a, b, c, d.

Put $a = 2$, $b = 1$, $c = 0$, $d = -1$, and divide. The quotient will be found to be -5.

$\therefore [(a + b)^5 + (c + d)^5](a - b)(c - d)$
$+ [(b + c)^5 + (a + d)^5](b - c)(a - d)$
$+ [(b + d)^5 + (c + a)^5](b - d)(c - a) =$
$- 5(a-b)(c-d)(b-c)(a-d)(b-d)(c-a)(a+b+c+d).$

NOTE. It is not always necessary to find the factors of the divisor, as the following examples show.

10. Prove that $x^2 + x + 1$ is a factor of $x^{14} + x^7 + 1$.

$x^2 + x + 1$ will be a factor of $x^{14} + x^7 + 1$, provided $x^{14} + x^7 + 1 = 0$, if $x^2 + x + 1 = 0$.

If $x^2 + x + 1 = 0$,

$\therefore x^3 + x^2 + x = 0$,

$\therefore x^3 + x^2 + x + 1 = 1$,

$\therefore x^3 = 1$,

$\therefore x^6 = 1$ and $x^{12} = 1$,

$\therefore x^7 = x$ and $x^{14} = x^2$,

$\therefore x^{14} + x^7 + 1 = x^2 + x + 1 = 0$,

$\therefore x^2 + x + 1$ is a factor of $x^{14} + x^7 + 1$.

Two other methods of proving this proposition are worthy of notice.

I. $x^2 + x + 1$ will be a factor of $x^{14} + x^7 + 1$, provided it is a factor of

$[(x^{14} + x^7 + 1) \pm \text{a multiple of} (x^2 + x + 1)]$.

$x^{14} + x^7 + 1$ differs by a multiple of $x^2 + x + 1$ from

$x^{14} + x^{11}(x^2 + x + 1) + x^8(x^2 + x + 1) + x^7$
$\quad + x^4(x^2 + x + 1) + x(x^2 + x + 1) + 1$

$= x^{12}(x^2 + x + 1) + x^9(x^2 + x + 1) + x^6(x^2 + x + 1)$
$\quad + x^3(x^2 + x + 1) + (x^2 + x + 1)$

$= (x^{12} + x^9 + x^6 + x^3 + 1)(x^2 + x + 1)$.

Hence, $x^2 + x + 1$ is a factor of $x^{14} + x^7 + 1$.

II. $\dfrac{x^{14} + x^7 + 1}{x^2 + x + 1} = \dfrac{x^{21} - 1}{x^7 - 1} \times \dfrac{x - 1}{x^3 - 1}$

$= \dfrac{(x^{21} - 1)[(x^{15} - 1) - x(x^{14} - 1)]}{(x^7 - 1)(x^3 - 1)}$

$= \dfrac{(x^{21} - 1)(x^{15} - 1)}{(x^7 - 1)(x^3 - 1)} - \dfrac{x(x^{21} - 1)(x^{14} - 1)}{(x^3 - 1)(x^7 - 1)}$.

But we see at once that on reduction both of these fractions give an integral quotient; hence,

$(x^{14} + x^7 + 1) \div x^2 + x + 1$ gives an integral quotient.

11. $x^2 + x + 1$ is a factor of $(x+1)^7 - x^7 - 1$.

If $x^2 + x + 1 = 0$, $(x+1)^7 - x^7 - 1$ will vanish also; for in such case $x + 1 = -x^2$.

$\therefore (x+1)^7 - x^7 - 1 = (-x^2)^7 - x^7 - 1 = -x^{14} - x^7 - 1$,

which by the last example vanishes if $x^2 + x + 1 = 0$.

$\therefore x^2 + x + 1$ is a factor of $(x+1)^7 - x^7 - 1$.

For x substitute $\dfrac{x}{y}$, and multiply by y^2 and y^7, respectively, and this example becomes

$x^2 + xy + y^2$ is a factor of $(x+y)^7 - x^7 - y^7$.

Ex. 21.

Determine the values of a, b, c, d, e, in the following cases:

1. $x^3 + 3bx^2 + 3cx + d$ vanishes for $x = 2$, or 3, or 4.

2. $x^4 + cx^2 + dx + e$ vanishes for $x = 1\frac{1}{2}$, or -3, or $4\frac{1}{2}$.

3. $x^3 + bx^2 + cx + 24$ vanishes for $x = 2$, or -3.

4. $ax^3 + bx^2 + cx + 90$ vanishes for $x = 3$, or -5, or 2.

5. $ax^4 + cx^2 - 30x + e$ vanishes for $x = 1\frac{1}{2}$, or -4, or $2\frac{1}{2}$.

6. $81x^4 + 6cx^2 + 4dx + e$ vanishes for $x = 1\frac{2}{3}$, or $-3\frac{1}{3}$, or $1\frac{1}{3}$.

7. $ax^4 + bx^3 + cx^2 - 81$ vanishes for $x = \frac{3}{5}$, or $\frac{3}{4}$, or 3.

8. $ax^4 + cx^2 + dx + e$ vanishes for $x = 2$, or $1\frac{1}{2}$, or -1, and becomes 14 for $x = 1$.

9. $ax^3 + cx + d$ vanishes for $x = 1\frac{1}{4}$, or $2\frac{3}{4}$, and becomes 49 for $x = 3$, determine its value for $x = -3$.

Given that $x^3 - px^2 + qx - r$ vanishes for $x = a$, or b, or c, determine the polynome that vanishes for :

10. $x = a + 1$, or $b + 1$, or $c + 1$.

11. $x = a - 1$, or $b - 1$, or $c - 1$.

12. $x = 1 - \dfrac{1}{a}$, or $1 - \dfrac{1}{b}$, or $1 - \dfrac{1}{c}$.

13. $x = ab$, or bc, or ca.

14. $x = a^2$, or b^2, or c^2.

15. $x = a(b + c)$, or $b(c + a)$, or $c(a + b)$.
$$\left[a(b + c) = q - \frac{r}{a} \right].$$

16. $x = \dfrac{a+b}{c}$, or $\dfrac{b+c}{a}$, or $\dfrac{c+a}{b}$. $\left(\dfrac{a+b}{c} = \dfrac{p}{c} - 1 \right).$

Prove that the following are cases of exact division :

17. $(x - 1)^{12} - x^6 + (x^2 - x + 1)^2 \div x^3 - 2x^2 + 2x - 1$.

18. $(x - 1)^{16} - x^8 + (x^2 - x + 1)^3 \div x^3 - 2x^2 + 2x - 1$

19. $(x - 2)^{10} (2x - 5)^{10} - x^{10} + 2^{10} (x^2 - 4x + 5)^5$
$\quad \div x^3 - 6x^2 + 13x - 10$.

20. $(x^2 + 4x + 3)^{18} - x^{18} - x^3 - 5x - 3 \div x^3 + 6x^2 + 8x + 3$.

21. $(9x - 4)^{21} (x - 1)^{21} - x^{21} - (9x^2 - 14x + 4)^{21}$
$\quad \div (x - 1)(9x - 4)(9x^2 - 14x + 4)$.

22. $[6(x - 1)]^{13} - (2x^2 + 3x - 4)^{13} + (2x^2 - 3x + 2)^{13}$
$\quad \div (2x^2 + 3x - 4)(2x^2 - 3x + 2)(x - 1)$.

23. $[2(x + 1)(x - 2)]^{17} + (x^2 - 3x + 3)^{17} - (3x^2 - 5x - 1)^{17}$
$\quad \div (x + 1)(x - 2)(x^2 - 3x + 3)(3x^2 - 5x - 1)$.

24. $[6(x - 1)]^{16} - (2x^2 + 3x - 4)^{16} - (2x^2 - 3x + 2)^{16}$
$\quad + 2(2x^2 + 3x - 4)^8 (2x^2 - 3x + 2)^4$
$\quad \div (x - 1)(2x^2 + 3x - 4)(2x^2 - 3x + 2)$.

25. $[2(x+1)(x-2)]^{20} - (x^2-3x+3)^{20} - (3x^2-5x-1)^{20}$
$+ 2(x^2-3x+3)^9 (3x^2-5x-1)^{11}$
$\div (x+1)(x-2)(x^2-3x+3)(3x^2-5x-1)$.

26. $1 + x^4 + x^8 \div 1 + x + x^2$.

27. $x^{10} + x^5 y^5 + y^{10} \div x^2 + xy + y^2$.

28. $1 + x^3 + x^6 + x^9 + x^{12} \div 1 + x + x^2 + x^3 + x^4$.

29. $1 + x^4 + x^8 + x^{12} + x^{16} \div 1 + x + x^2 + x^3 + x^4$.

30. $x^{15} + x^{10} y^5 + x^5 y^{10} + y^{15} \div x^3 + x^2 y + xy^2 + y^3$.

31. $x^{17} + x^4 + x^3 + x + 1 \div x^4 + x^3 + x^2 + x + 1$.

32. $1 + x + x^2 + x^3 + x^5 + x^6 + x^{53}$
$\div 1 + x + x^2 + x^3 + x^4 + x^5 + x^6$.

Find the quotient of the following divisions, in which D denotes the product:

$$(b-c)(c-a)(a-b)(a-d)(b-d)(c-d).$$

33. $(b^2 c^2 + a^2 d^2)(b-c)(a-d) + (c^2 a^2 + b^2 d^2)(c-a)(b-d)$
$+ (a^2 b^2 + c^2 d^2)(a-b)(c-d) \div D$.

34. $(bc+ad)(b^2-c^2)(a^2-d^2) + (ca+bd)(c^2-a^2)(b^2-d^2)$
$+ (ab+cd)(a^2-b^2)(c^2-d^2) \div D$.

35. $(b+c)(a+d)(b^2-c^2)(a^2-d^2) +$ the two similar terms
$\div D$.

36. $(b^2 + c^2)(a^2 + d^2)(b-c)(a-d) +$ the two similar
terms $\div D$.

37. $[bc(b+c)^2 + ad(a+d)^2](b-c)(a-d) +$ the two
similar terms $\div D$.

38. $[bc(b+c) + ad(a+d)](b^2 - c^2)(a^2 - d^2) +$ the two
similar terms $\div D$.

39. $[bc(b^3 + c^3) + ad(a^3 + d^3)](b-c)(a-d) +$ the two
similar terms $\div D$.

40. $(b+c-a-d)^4(b-c)(a-d)+$ the two similar terms $\div D.$

41. The sum of the fractions $\frac{1}{1}, \frac{1}{2}, \frac{1}{3}, \ldots, \frac{1}{n}$, increased by the sum of their products, two by two, increased by the sum of their products, three by three, ..., increased by their product is equal to n.

42. In any trapezium, the square of the area expressed in terms of the lengths of the parallel sides and the diagonals, is a polynome of four dimensions; determine that polynome.

43. In any quadrilateral inscribed in a circle, the square of the area, expressed in terms of the lengths of the sides, is a polynome of four dimensions; find that polynome.

THEOREM III. If the polynome $f(x)^n$ vanish for more than n different values of x, it vanishes identically, the coefficient of every term being zero.

Cor. If a rational integral expression of n dimensions be divisible by more than n linear factors, the expression is identically zero.

EXAMPLES.

1. $\dfrac{(x-a)(x-b)}{(c-a)(c-b)} + \dfrac{(x-b)(x-c)}{(a-b)(a-c)} + \dfrac{(x-c)(x-a)}{(b-c)(b-a)} - 1 = 0,$

if a, b, and c are unequal; for this is a polynome of *two* dimensions in x, but it vanishes for $x=a$, and, therefore, by symmetry, for $x=b$, and for $x=c$; that is, for *three* different values of x; hence, it vanishes identically.

2. $[(a+b)^2+(c+d)^2](a-b)(c-d)$
$+[(b+c)^2+(a+d)^2](b-c)(a-d)$
$+[(c+a)^2+(b+d)^2](c-a)(b-d) = 0.$

Substitute b for a, and the expression becomes

$$[(b+c)^2+(b+d)^2](b-c)(b-d)$$
$$+[(c+b)^2+(b+d)^2](c-b)(b-d),$$

which vanishes; hence, the given expression is divisible by $a-b$, and consequently, by symmetry, it is divisible by $(a-b)$, $(b-c)$, $(c-d)$, $(a-c)$, $(b-d)$, and $(a-d)$. But the given expression is of only *four* dimensions, while it appears to have *six* linear factors; hence, it vanishes identically.

Ex. 22.

Verify the following:

1. $\dfrac{x^2 y^2 z^2}{b^2 c^2} + \dfrac{(x^2-b^2)(y^2-b^2)(z^2-b^2)}{b^2(b^2-c^2)} + \dfrac{(x^2-c^2)(y^2-c^2)(z^2-c^2)}{c^2(c^2-b^2)}$
   $= x^2 + y^2 + z^2 - b^2 - c^2.$

2. $\dfrac{y^2 z^2}{b^2 c^2} + \dfrac{(y^2-b^2)(z^2-b^2)}{b^2(b^2-c^2)} + \dfrac{(y^2-c^2)(z^2-c^2)}{c^2(c^2-b^2)} = 1.$

3. $\dfrac{x^2 y^2}{a^2 b^2} + \dfrac{(x^2-a^2)(a^2-y^2)z^2}{(z^2-a^2)(b^2-a^2)a^2} + \dfrac{(x^2-b^2)(b^2-y^2)z^2}{(b^2-z^2)(b^2-a^2)b^2}$

   $+ \dfrac{(z^2-x^2)(z^2-y^2)}{(z^2-a^2)(b^2-z^2)} = 0.$

4. $\dfrac{1}{(x+a)(a-b)(a-c)} + \dfrac{1}{(x+b)(b-c)(b-a)}$

   $+ \dfrac{1}{(x+c)(c-a)(c-b)} = \dfrac{1}{(x+a)(x+b)(x+c)}.$

5. $bc(b^2-c^2) + ca(c^2-a^2) + ab(a^2-b^2)$
   $= (a+b+c)[a^2(b-c)+b^2(c-a)+c^2(a-b)].$

6. $\dfrac{a+x}{x(x-y)(x-z)} + \dfrac{a+y}{y(y-x)(y-z)} + \dfrac{a+z}{z(z-x)(z-y)}$
   $= \dfrac{a}{xyz}.$

7. $\dfrac{a^4(b^2-c^2)+b^4(c^2-a^2)+c^4(a^2-b^2)}{a^2(b-c)+b^2(c-a)+c^2(a-b)}$
$=\frac{1}{3}[(a+b+c)^3-a^3-b^3-c^3].$

8. $(adf+bcf+bed-ace)^2+(bce+aed+acf-bdf)^2$
$=(a^2+b^2)(c^2+d^2)(e^2+f^2).$

9. $\dfrac{(a-b)^5+(b-c)^5+(c-a)^5}{(a-b)(b-c)(c-a)}$
$=\frac{5}{2}[(a-b)^2+(b-c)^2+(c-a)^2].$

10. $(-x+y+z)(x-y+z)(x+y-z)$
$+x(x-y+z)(x+y-z)+y(x+y-z)(-x+y+z)$
$+z(-x+y+z)(x-y+z)=4xyz.$

11. $\dfrac{(a^2-b^2)^3+(b^2-c^2)^3+(c^2-a^2)^3}{(a+b)(b+c)(c+a)}$
$=(a-b)^3+(b-c)^3+(c-a)^3.$

12. $x^2(y+z)^2+y^2(z+x)^2+z^2(x+y)^2+2xyz(x+y+z)$
$=2(xy+yz+zx)^2.$

THEOREM IV. If the polynomes $f(x)^n$, $\phi(x)^m$ (n not less than m), are equal for more than n different values of x, they are equal for *all* values, and the coefficients of equal powers of x in each are equal to one another.

This is called the **Principle of Indeterminate Coefficients.**

EXAMPLES.

1. $\dfrac{a^2}{(a-b)(a-c)(a-d)}+\dfrac{b^2}{(b-a)(b-c)(b-d)}$
$+\dfrac{c^2}{(c-a)(c-b)(c-d)}+\dfrac{d^2}{(d-a)(d-b)(d-c)}=0.$

Assume $\dfrac{x^2}{(x-a)(x-b)(x-c)(x-d)}$

$=\dfrac{A}{x-a}+\dfrac{B}{x-b}+\dfrac{C}{x-c}+\dfrac{D}{x-d}$ \hspace{2em} (a)

in which A, B, C, D, are independent of x.

Multiply by $(x-a)(x-b)(x-c)(x-d)$.

$\therefore x^2 = (A+B+C+D)x^3 +$ terms in lower powers of x.

Now this equality holds for more than three values of x, holding in fact for *all* finite values of x.

$\therefore A+B+C+D=0.$ (β)

Again, multiply both sides of (α) by $x-a$,

$$\frac{x^2}{(x-b)(x-c)(x-d)} = A + \left(\frac{B}{x-b}+\frac{C}{x-c}+\frac{D}{x-d}\right)(x-a).$$

Put $x=a$,

$$\therefore \frac{a^2}{(a-b)(a-c)(a-d)} = A.$$

By symmetry, $\dfrac{b^2}{(b-a)(b-c)(b-d)} = B$, etc.

Adding,

$$\frac{a^2}{(a-b)(a-c)(a-d)} + \frac{b^2}{(b-a)(b-c)(b-d)}$$
$$+ \frac{c^2}{(c-a)(c-b)(c-d)} + \frac{d^2}{(d-a)(d-b)(d-c)}$$
$$= A+B+C+D=0, \text{ by } (\beta).$$

2. $\dfrac{a^2(a+b)(a+c)}{(a-b)(a-c)} + \dfrac{b^2(b+c)(b+a)}{(b-c)(b-a)} + \dfrac{c^2(c+a)(c+b)}{(c-a)(c-b)}$
$$= (a+b+c)^2.$$

Assume $x^3 - px^2 + qx - r = (x-a)(x-b)(x-c)$ (α)

$\therefore x^3 + px^2 + qx + r = (x+a)(x+b)(x+c)$ (β)

$$\frac{x^4 + px^3 + qx^2 + rx}{x^3 - px^2 + qx - r} = x + 2p + \frac{A}{x-a} + \frac{B}{x-b} + \frac{C}{x-c} \quad (\gamma)$$

Multiply by $x^3 - px^2 + qx - r$, and equate the coefficients of the terms in x^2. In multiplying the fractions in the right-hand member of (γ), use the factor side of (α).

$q = q - 2p^2 + A + B + C.$

$\therefore A + B + C = 2p^2.$

Multiply both members of (γ) by $x - a$.

$$\frac{x(x+a)(x+b)(x+c)}{(x-b)(x-c)}$$
$$= A + \left(x + 2p + \frac{B}{x-b} + \frac{C}{x-c}\right)(x-a).$$

Put $x = a$,

$$\frac{2a^2(a+b)(a+c)}{(a-b)(a-c)} = A.$$

By symmetry,

$$\frac{2b^2(b+c)(b+a)}{(b-c)(b-a)} = B, \text{ and } \frac{2c^2(c+a)(c+b)}{(c-a)(c-b)} = C.$$

$$\therefore \frac{a^2(a+b)(a+c)}{(a-b)(a-c)} + \frac{b^2(b+c)(b+a)}{(b-c)(b-a)}$$
$$+ \frac{c^2(c+a)(c+b)}{(c-a)(c-b)} = \tfrac{1}{2}(A+B+C) = p^2$$
$$= (a+b+c)^2.$$

3. Extract the square root of $1 + x + x^2 + x^3 + x^4 + \cdots$

Assume the square root to be

$1 + ax + bx^2 + cx^3 + dx^4 + \cdots$

$\therefore 1 + x + x^2 + x^3 + x^4 + \cdots$

$= (1 + ax + bx^2 + cx^3 + dx^4 + \cdots)^2$

$= 1 + 2ax + (a^2 + 2b)x^2 + 2(ab + c)x^3$

$+ (2d + 2ac + b^2)x^4 + \cdots$

$$\therefore 2a = 1, \qquad\qquad \therefore a = \tfrac{1}{2},$$
$$2b + a^2 = 1, \qquad b = \tfrac{1}{2}(1 - \tfrac{1}{4}) = \tfrac{3}{8},$$
$$2(c + ab) = 1, \qquad c = \tfrac{1}{2} - (\tfrac{1}{2} \times \tfrac{3}{8}) = \tfrac{5}{16},$$
$$2d + 2ac + b^2 = 1, \qquad d = \tfrac{1}{2}(1 - \tfrac{5}{16} - \tfrac{9}{64}) = \tfrac{35}{128}.$$
$$\therefore \sqrt{(1 + x + x^2 + \cdots)} = 1 + \tfrac{1}{2}x + \tfrac{3}{8}x^2 + \tfrac{5}{16}x^3$$
$$+ \tfrac{35}{128}x^4 + \cdots$$

NOTE. As it is frequently necessary to determine the coefficient of a particular power of x, a few preliminary exercises are given on this subject.

Ex. 23.

Determine the coefficient of:

1. x^4 in $(1 + ax)^3 + (1 + bx)^5 + (1 - cx)^6$.

2. x^5 in $(1 + x + 2x^2 + 3x^3)(1 - x + 3x^2 + x^3 - 5x^4)$.

3. x^4 in $(1 + x + 2x^2 + 3x^3 + 4x^4 + \cdots)$
    $(1 - x + x^2 - x^3 + x^4 - \cdots)$.

4. x^2 in $A(x - b)(x - c)(x - d) + B(x - a)(x - c)(x - d)$
    $\quad + C(x - a)(x - b)(x - d) + D(x - a)(x - b)(x - c)$

5. x^4 in $(1 - ax)^3(1 + ax)^5$.

6. x^4 in $(1 + ax)^3(1 - bx)^5$.

7. In the product
    $$(1 + ax + bx^2 + cx^3 + \cdots)(1 - ax + bx^2 - cx^3 + \cdots)$$
    prove that the coefficients of the odd powers of x must be all zeros.

Determine the value of the following expressions:

8. $\dfrac{1}{(a - b)(a - c)(a - d)} + \dfrac{1}{(b - a)(b - c)(b - d)}$
    $\quad + \dfrac{1}{(c - a)(c - b)(c - d)} + \dfrac{1}{(d - a)(d - b)(d - c)}.$

9. $\dfrac{a}{(a-b)(a-c)(a-d)} + \dfrac{b}{(b-a)(b-c)(b-d)} + \ldots$

10. $\dfrac{a^2}{(a-b)(a-c)(a-d)}$ + three similar terms.

11. $\dfrac{a^3}{(a-b)(a-c)(a-d)}$ + three similar terms.

12. $\dfrac{a^4}{(a-b)(a-c)(a-d)}$ + three similar terms.

13. $\dfrac{bcd}{(a-b)(a-c)(a-d)}$ + three similar terms.

14. $\dfrac{a(a+b)(a+c)}{(a-b)(a-c)}$ + two similar terms.

15. $\dfrac{a^3(a+b)(a+c)}{(a-b)(a-c)}$ + two similar terms.

16. $\dfrac{a^4(a+b)(a+c)}{(a-b)(a-c)}$ + two similar terms.

17. $\dfrac{a(a+b)(a+c)(a+d)}{(a-b)(a-c)(a-d)}$ + three similar terms.

18. $\dfrac{a^2(a+b)(a+c)(a+d)}{(a-b)(a-c)(a-d)}$ + three similar terms.

19. $\dfrac{a^3(a+b)(a+c)(a+d)}{(a-b)(a-c)(a-d)}$ + three similar terms.

20. $\dfrac{bc(b+c)}{(a-b)(a-c)}$ + two similar terms.

For numerator use $px^2 - p^2x + pq - r$.

21. $\dfrac{(2a+b)(2a+c)}{(a-b)(a-c)}$ + two similar terms.

For numerator use $3x^2 + px + q$.

22. $\dfrac{a(b+c)}{(a-b)(a-c)}$ + two similar terms.

  For numerator use $x(x-p)$.

23. $\dfrac{b+c+d}{(a-b)(a-c)(a-d)}$ + three similar terms.

24. $\dfrac{a^3(bc+cd+db)}{(a-b)(a-c)(a-d)}$ + three similar terms.

25. $\dfrac{bc+cd+db}{(a-b)(a-c)(a-d)}$ + three similar terms.

Extract the square root to four terms:

26. $1+x$. **27.** $1-x$. **28.** $1+2x+3x^2+4x^3+\cdots$

29. $1-4x+10x^2-20x^3+35x^4-56x^5+84x^6$.

30. Extract the cube root of $1+x$ to four terms.

§ 11. 1. Find the condition that $px^2+2qx+r$ and $p'x^2+2q'x+r'$ shall have a common factor.

Multiply the polynomials by p' and p respectively, and take the difference of the products; also, by r' and r, respectively, and divide the difference of the products by x.

$p'px^2+2p'qx+p'r$	$pr'x^2+2qr'x+rr'$
$pp'x^2+2pq'x+pr'$	$p'rx^2+2q'rx+r'r$
$2(pq'-p'q)x+(pr'-p'r)$	$(pr'-p'r)x+2(qr'-r'q)$

Multiply the former of these remainders by $(pr'-p'r)$, and the latter by $2(pq'-p'q)$, and the difference of the products is

$$(pr'-p'r)^2-4(pq'-p'q)(qr'-r'q).$$

But if the given polynomials have a linear factor, this remainder must vanish, or

$$(pr'-p'r)^2=4(pq'-p'q)(qr'-r'q).$$

If the given polynomes have a quadratic factor, the *linear* remainders must vanish identically, or (Th. III.),

$$pq' - p'q = 0, \; pr' - p'r = 0, \; \text{and} \; qr' - r'q = 0,$$

or, $\dfrac{p}{p'} = \dfrac{q}{q'} = \dfrac{r}{r'}.$

2. Find the condition that $px^3 + 3qx^2 + 3rx + s$ shall have a square factor.

Assume the square factor to be $(x - a)^2$. On division, the remainder must be zero for every finite value of x, and consequently (Th. III.) the coefficient of each term of the remainder must be zero. Divide by $(x - a)^2$, neglecting the first remainder.

	p	$3q$	$3r$	s
a		pa	$pa^2 + 3qa$	
	p	$pa + 3q$	$pa^2 + 3qa + 3r$; R	
a		pa	$2pa^2 + 3qa$	
	p	$2pa + 3q$;	$3(pa^2 + 2qa + r)$	

$\therefore pa^2 + 2qa + r = 0$;

$\therefore px^2 + 2qx + r$ is divisible by $x - a$;   (Th. I. *Cor.* 2.)
or, $px^3 + 3qx^2 + 3rx + s$ and $px^2 + 2qx + r$ have a common divisor. Multiply the latter polynome by x, and subtract the product from the former, and the proposition reduces to

If $px^3 + 3qx^2 + 3rx + s$ have a square factor, $px^2 + 2qx + r$ and $qx^2 + 2rx + s$ will have the square root of that factor for a common divisor.

Ex. 24.

1. Determine the condition necessary in order that
$$x^2 + px + q \; \text{and} \; x^3 + p'x + q' \; \text{may have a common}$$
divisor.

2. The expression $x^6 + 3a^2 x^5 + 3bx^4 + cx^3 + 3dx^2 + 3e^2 x + f^3$ will be a complete cube if

$$f = \frac{e}{a} = \frac{d}{b} = \frac{c - a^6}{6a^2} = b - a^4.$$

3. Prove that $ax^5 + bx + c$ and $a + bx^4 + cx^5$ will have a common quadratic factor if

$$b^2 c^2 = (c^2 - a^2 + b^2)(c^2 - a^2 + ab).$$

4. Prove that $ax^5 + bx^2 + c$ and $a + bx^3 + cx^5$ will have a common quadratic factor if

$$a^2 b^2 = (a^2 - c^2)(a^2 - c^2 + bc).$$

5. Prove that $ax^4 + bx^3 + cx + d$ and $a + bx + cx^3 + dx^4$ will have a common quadratic factor if

$$(a + d)(a - d)^2 = (b - c)(bd - ac).$$

6. $x^3 + px^2 + qx + r$ will be divisible by $x^2 + ax + b$ if

$$a^3 - 2pa^2 + (p^2 + q)a + r - pq = 0,$$
and $b^3 - qb^2 + rpb - r^2 = 0.$

7. $x^4 + px + q$ will be divisible by $x^2 + ax + b$ if

$$a^6 - 4qa^2 = p^2 \text{ and } (b^2 + q)(b^2 - q)^2 = p^2 b^3.$$

Determine the condition necessary in order that:

8. $x^4 + 4px^3 + 6qx^2 + 4rx + t$ may have a square factor.

9. $ax^4 + 4bx^3 + 6cx^2 + 4dx + e$ may have a complete cube as factor.

10. $x^5 + 10bx^3 + 10cx^2 + 5dx + bc$ may have a complete cube as factor.

CHAPTER III.

FORMULAS [1] AND [2]. $(x \pm y)^2 = x^2 \pm 2xy + y^2$, etc.

§ 12. From this it appears that a trinomial of which the extremes are squares, is itself a square if four times the product of the extremes is equal to the square of the mean, and that, to factor such a trinomial, we have simply to connect the square root of each of the squares by the sign of the other term, and write the result twice as a factor.

EXAMPLES.

1. $4x^4 - 80x^2y^2 + 400y^4 = (2x^2 - 20y^2)(2x^2 - 20y^2)$.

2. $1 - 12x^2y^2 + 36x^4y^4 = (1 - 6x^2y^2)(1 - 6x^2y^2)$.

3. $(a - b)^2 + (b - c)^2 + 2(a - b)(b - c)$.
   This equals
   $$(a - b + b - c)(a - b + b - c) = (a - c)(a - c).$$

4. $x^2 + y^2 + z^2 + 2xy - 2xz - 2yz$.

   Here the three squares and the three double products suggest that the expression is the square of a linear *trinomial* in x, y, z.

   An inspection of the signs of the double products enables us to determine the signs which are to connect x, y, z; we see that
   1. The signs of x and y must be *alike;*
   2. The signs of x and z must be *different;*
   3. The signs of y and z must be *different.*

   Hence, we have $x + y - z$, or $-x - y + z = -(x + y - z)$, and the factors are $(x + y - z)(x + y - z)$.

Ex. 25.

1. $9\,m^2 + 12\,m + 4$; $c^{2m} - 2\,c^m + 1$.

2. $y^6 - 2\,y^3 z^3 + z^6$; $16\,x^2 y^2 + 16\,xy^3 + 4\,y^4$.

3. $9\,a^2 b^2 + 12\,abc + 4\,c^2$; $36\,x^2 y^2 - 24\,xy^3 + 4\,y^4$.

4. $\frac{1}{4}x^4 + 16\,y^2 z^2 - 4\,x^2 yz$; $\frac{1}{4}a^4 - \frac{1}{3}a^2 b^2 c^2 + \frac{1}{9}b^4 c^4$.

5. $(a+b)^2 + c^2 - 2\,c(a+b)$; $9\,x^8 - \frac{3}{2}x^4 y^2 + \frac{1}{16}y^4$.

6. $z^2 + (x-y)^2 - 2\,z(x-y)$; $\left(\dfrac{a}{b}\right)^{2m} + \left(\dfrac{b}{a}\right)^{2m} - 2$.

7. $(x^2 - y)^2 + 2\,(x^2 - y)(y - z^2) + (y - z^2)^2$.

8. $(x^2 - xy)^2 - 2\,(x^2 - xy)(xy - y^2) + (xy - y^2)^2$.

9. $(a+b+c)^2 - 2\,c(a+b+c) + c^2$; $\frac{9}{16}p^6 - 2\,p^3 q^2 + \frac{16}{9}q^4$.

10. $(3\,x - 4\,y)^2 + (2\,x - 3\,y)^2 - 2\,(3\,x - 4\,y)(2\,x - 3\,y)$.

11. $(x^2 - xy + y^2)^2 + (x^2 + xy + y^2)^2 + 2\,(x^4 + x^2 y^2 + y^4)$.

12. $(5\,x^2 + 2\,xy + 7\,y^2)^2 + (4\,x^2 + 6\,y^2)^2$
    $- 2\,(4\,x^2 + 6\,y^2)(5\,x^2 + 2\,xy + 7\,y^2)$.

13. $\left(\dfrac{a}{b}\right)^{2m} + \left(\dfrac{b}{a}\right)^{2n} - 2\left(\dfrac{a}{b}\right)^{m-n}$.

14. $a^2 + b^2 + c^2 - 2\,ab - 2\,bc + 2\,ac$.

15. $a^4 + b^4 + c^4 - 2\,a^2 b^2 - 2\,a^2 c^2 + 2\,b^2 c^2$.

16. $(a-b)^2 + (b-c)^2 + (c-a)^2 + 2\,(a-b)(b-c)$
    $- 2\,(a-b)(c-a) + 2\,(b-c)(a-c)$.

17. $4\,a^4 - 12\,a^2 b + 9\,b^2 + 16\,a^2 c + 16\,c^2 - 24\,bc$.

FORMULA [4]. $x^2 - y^2 = (x+y)(x-y)$.

§ 13. In this case we have merely to take the square root of each of the squares, and connect the results with the sign $+$ for one of the factors, and with the sign $-$ for the other.

Examples.

1. $(a+b)^2 - (c+d)^2$.

This $= [(a+b) + (c+d)][(a+b) - (c+d)]$
$= (a+b+c+d)(a+b-c-d)$.

2. Factor $(x^2 + 5xy + y^2)^2 - (x^2 - xy + y^2)^2$.

Here we have $[(x^2 + 5xy + y^2) + (x^2 - xy + y^2)]$
$\times [(x^2 + 5xy + y^2) - (x^2 - xy + y^2)]$
$= 2(x^2 + xy + y^2)(6xy)$
$= 12xy(x+y)^2$.

3. $a^2 - b^2 - c^2 + 2bc$.

This $= a^2 - (b-c)^2 = (a+b-c)(a-b+c)$.

4. Resolve $(a^2 + b^2)^2 - (a^2 - b^2)^2 - (a^2 + b^2 - c^2)^2$.

This $= 4a^2b^2 - (a^2 + b^2 - c^2)^2$
$= (2ab + a^2 + b^2 - c^2)(2ab - a^2 - b^2 + c^2)$.

The former of these factors
$= (a+b)^2 - c^2 = (a+b+c)(a+b-c)$;

and the latter
$= c^2 - (a-b)^2 = (c+a-b)(c-a+b)$.

$\therefore$ the given expression
$= (a+b+c)(a+b-c)(c+a-b)(c-a+b)$.

Ex. 26.

1. $49a^2 - 4b^2$.

2. $9a^2 - \frac{1}{4}b^2$.

3. $81a^4 - 16b^4$.

4. $100x^2 - 36y^2$.

5. $5a^2b - 20bx^2y^2$.

6. $9x^6 - 16y^4$.

7. $\frac{9}{16}c^2 - 1$.

8. $4y^4 - \frac{4}{9}x^2z^2$.

9. $81a^4 - 1$.

10. $a^4 - 16b^4$.

11. $a^{16} - b^{16}$.

12. $a^2 - b^2 + 2bc - c^2$.

13. $(a+2b)^2 - (3x-4y)^2$.

14. $(x^2 + y^2)^2 - 4x^2y^2$.

15. $(x+y)^2 - 4z^2$.

16. $(3x+5)^2 - (5x+3)^2$.

17. $4x^2y^2 - (x^2 + y^2 - z^2)^2$.

18. $(x^2 + xy - y^2)^2 - (x^2 - xy - y^2)^2$.

19. $(x^2 - y^2 + z^2)^2 - 4x^2z^2$.

20. $(a + b + c + d)^2 - (a - b + c - d)^2$.

21. $(2 + 3x + 4x^2)^2 - (2 - 3x + 4x^2)^2$.

22. $(a^2 + b^2 + 4ab)^2 - (a^2 + b^2)^2$.

23. $(a^2 - b^2 + c^2 - d^2)^2 - (2ac - 2bd)^2$.

24. $(x^2 - y^2 - z^2)^2 - 4y^2z^2$.

25. $(a^6 - a^3b^3 + b^6)^2 - (a^6 - 5a^3b^3 + b^6)^2$.

26. $a^{12} - b^{12} + 6a^9b^3 - 6b^9a^3 + 8b^9a^3 - 8a^9b^3$.

27. $(x^2 + y^2 + z^2 - xy - yz - zx)^2 - (xy + yz + zx)^2$.

28. $(x^2 + y^2 + z^2 - 2xy + 2xz - 2yz) - (y + z)^2$.

29. $2a^2b^2 + 2b^2c^2 + 2c^2a^2 - a^4 - b^4 - c^4$.

30. $x^4 + y^4 + z^4 - 2x^2y^2 - 2y^2z^2 - 2z^2x^2$.

FORMULA [A]. $(x+r)(x+s) = x^2 + (r+s)x + rs$.

EXAMPLES.

1. $x^6 - 9x^3 + 20 = (x^3 - 5)(x^3 + 4)$.

2. $(x - y)^2 + x - y - 110 = (x - y + 11)(x - y - 10)$.

3. $(a^2 - ab + b^2)^2 + 6b(a^2 - ab + b^2) - 4a^2 + 9b^2$
   $= [(a^2 - ab + b^2) + (2a + 3b)]$
   $\times [(a^2 - ab + b^2) - (2a - 3b)]$.

4. $(x^2 - 5x)^2 - 6(x^2 - 5x) - 40$
   $= (x^2 - 5x + 4)(x^2 - 5x - 10)$.

5. $(ax + by + c)^2 - (m - n)(ax + by + c) - mn$
   $= (ax + by + c - m)(ax + by + c + n)$.

§ 14. It will be seen that the first (or *common*) term of the required factors is obtained by extracting the square root of the first term of the given expression, and that the other terms are determined by observing two conditions :

I. Their product must equal the *third* term of the given expression.

II. Their *algebraic* sum multiplied into the *common* term already found must equal the middle term of the given expression.

Hence, to make a systematic search for integral factors of an expression of the form $x^2 \pm bx \pm c$, we may proceed as follows:

1. Write down every pair of factors whose product is c.

2. If the sign before c is $+$, select the pair of factors whose *sum* is b, and write *both* factors $x+$, if the sign before b is $+$; $x-$, if the sign before b is $-$.

3. But if the sign before c is $-$, select the pair of factors whose *difference* is b, and write before the *larger* factor $x+$ or $x-$, and before the other factor $x-$ or $x+$, according as the sign before b is $+$ or $-$.

EXAMPLES.

1. $x^2 + 9x + 20$. The factors of 20 in pairs are 1 and 20, 2 and 10, 4 and 5. The sign before 20 is $+$; hence, select the factors whose *sum* is 9. These are 4 and 5. The sign before 9 is $+$; hence, the required factors are $(x+4)(x+5)$.

2. $x^2 - 8x + 12$. Pairs of factors of 12 are 1 and 12, 2 and 6, 3 and 4. Sign before 12 is $+$; therefore take the pair whose sum is 8. These are 2 and 6. Sign before 8 is $-$; hence, the factors are $(x-2)(x-6)$.

3. $x^2 - 21x - 100$.   Pairs of factors of 100 are 1 and 100,
2 and 50, 4 and 25, 5 and 20, 10 and 10.   Sign before 100 is $-$; therefore, take the pair whose difference is 21.   these are 4 and 25.   The sign before 21
is $-$; therefore, $x-$ goes before 25, the larger factor,
and the factors are $(x+4)(x-25)$.

4. $x^2 + 12x - 108$.   Pairs of factors of 108 are 1 and 108,
2 and 54, 3 and 36, 4 and 27, 6 and 18, 9 and 12.
Sign before 108 is $-$; therefore, take the pair whose
difference is 12.   These are 6 and 18.   Sign before
12 is $+$; therefore, $x+$ goes before 18, the larger
factor, and $x-$ before 6, the other factor.   Hence,
the factors are $(x-6)(x+18)$.

NOTE. It will be found convenient to write the factors in two
columns, separated by a short space.   Taking Exam. 2 above, proceed thus: Since the sign of the third term is $+$, write the sign of
the second term (in this case $-$) above both columns.

$$
\begin{array}{cc}
- & - \\
1 & 12 \\
(x-2) & (x-6)
\end{array}
$$

Exam. 3 above.   Since the sign of the third term is $-$, write the
sign of the second term (in this case $-$) above the column of larger
factors, and the other sign of the pair, $\pm$, above the other column.

$$
\begin{array}{cc}
+ & - \\
1 & 100 \\
2 & 50 \\
(x+4) & (x-25)
\end{array}
$$

5. $x^2 - 34x + 64$.

Here we have the factors

$$
\begin{array}{cc}
- & - \\
1 & 64 \\
x-2 & x-32 \\
4 & 16
\end{array}
$$

And since the last term has the sign $+$, and the
middle term has the sign $-$, we write $-$ over both
columns.

6. $x^2 + 12x - 64$.

$$
\begin{array}{cc}
- & + \\
1 & 64 \\
2 & 32 \\
x - 4 & x + 16
\end{array}
$$

Here, since the last term has the sign —, we write the sign (+) of the middle term over the column of larger factors, and the sign — over the other column.

7. $x^4 - 10x^2 - 144$.

Here we have the pairs of factors

$$
\begin{array}{cc}
+ & - \\
1 & 144 \\
2 & 72 \\
4 & 36 \\
x + 8 & x - 18
\end{array}
$$

And since the sign of the third term is —, we write the sign of the second term (in this case —) above the column of *larger* factors, and the other sign (of the pair, ±) above the other column.

Ex. 27.

1. $x^2 - 5x - 14$; $x^2 - 9x + 14$; $x^2 + 7x + 12$.

2. $x^2 - 8x + 15$; $x^2 - 19x + 84$; $x^2 - 7x - 60$.

3. $4x^2 - 2x - 20$; $9x^2 - 150x + 600$.

4. $\frac{1}{4}x^2 + 4\frac{1}{2}x - 36$; $25x^2 + 40x + 15$; $9x^6 - 27x^3 + 20$.

5. $\frac{1}{16}x^2 + 1\frac{3}{4}x + 12$; $16x^4 - 4x^2 - 20$.

6. $x^4 - (a^2 + b^2)x^2 + a^2b^2$; $4(x + y)^2 - 4(x + y) - 99$.

7. $(x^2 + y^2)^2 - (a^2 - b^2)(x^2 + y^2) - a^2b^2$.

8. $(a + b)^2 - 2c(a + b) - 3c^2$.

9. $(x + y)^2 + 2(x^2 + y^2)(x + y) + (x^2 - y^2)^2$.

10. $(a+b)^2 - 4ab(a+b) - (a^2 - b^2)^2$.

11. $(x^2 + xy + y^2)^2 + x^3 - y^3 - 5xy - 2y^2 - 2x^2$.

12. $a^2 - 2a(b-c) - 3(b-c)^2$.

13. $(x^3 + y^3)^2 + 2a^3(x^3 + y^3) + a^6 - b^6$.

14. $(x^2 - 10x)^2 - 4(x^2 - 10x) - 96$.

15. $(x^2 - 14x + 40)^2 - 25(x^2 - 14x + 40) - 150$.

16. $(x^2 - xy + y^2)^2 + 2xy(x^2 - xy + y^2) - 3x^2y^2$.

17. $z^4 - 3z^2 + 2$; $x^4 - 2x^2 - 3$; $9x^8 + 9x^4y^2 - 10y^4$.

18. $c^{2m} + c^m - 2$; $x^6 - x^3 - 2$; $x^{2m} - 2x^m y^n - 8y^{2n}$.

19. $x^{2m} - (a-b)x^m y^n - aby^{2n}$.

§ 15. Trinomials of the form $ax^2 + bx + c$ (a not a square) may sometimes be easily factored from the following considerations:

The product of two binomials consists of

1. The product of the *first* terms;
2. The product of the *second* terms;
3. The *algebraic* sum of the products of the terms taken diagonally.

These three conditions guide us in the converse process of resolving a trinomial into its binomial factors.

EXAMPLES.

1. Resolve $6x^2 - 13xy + 6y^2$.

Here the factors of the first term are x and $6x$, or $2x$ and $3x$; those of the third term are y and $6y$, or $2y$ and $3y$. These pairs of factors may be arranged:

(1)	(2)	(3)	(4)
x	$2x$	y	$2y$
$6x$	$3x$	$6y$	$3y$

Now, we may take (1) with (3) or (4), or (2) with (3)
or (4); but none of these combinations will satisfy
the third condition. If, however, in (4) we interchange the coefficients 2 and 3, then (2) and (4) give

$$2x \quad 3y$$
and $$\quad 3x \quad 2y,$$

where we can combine the "diagonal" products to
make 13, and the factors are

$$2x - 3y$$
and $$\quad 3x - 2y.$$

The coefficients of (2), instead of those of (4), might
have been interchanged, giving the same result.

2. $6x^2 - 15xy + 6y^2$.

Here, comparing (2) and (3), Exam. 1, we see that
their diagonal products may be combined to give
15, and the factors are $2x - y$ and $3x - 6y$.

3. $6x^2 - 20xy + 6y^2$.

Here, again referring to Exam. 1, we see at once that
it is useless to try *both* (2) *and* (4), since the diagonal products cannot be combined in any way to
give a higher result than $13xy$. But comparing (1)
and (4), we obtain, by interchanging the coefficients
in 4, $\quad\quad x - 3y$
and $\quad\quad 6x - 2y,$

which satisfy the third condition.

Or, we might interchange the coefficients of (3), and
take the resulting terms with (2), getting

$$2x - 6y$$
and $$\quad 3x - \ y.$$

4. $6x^2 + 35xy - 6y^2$.

Here the large coefficient of the middle term shows at once that we must take (1) and (3) together. Interchanging the coefficients of (1), we have

$$6x - y$$

and $$x + 6y.$$

The same result will be obtained by interchanging the coefficients of (3).

Ex. 28.

1. $6x^2 - 37xy + 6y^2$.		**11.** $6x^2 - 16xy - 6y^2$.
2. $6x^2 + 9xy - 6y^2$.		**12.** $6x^2 + 5xy - 6y^2$.
3. $56x^2 - 76xy + 20y^2$.		**13.** $56x^2 + 562xy + 20y^2$.
4. $56x^2 - 36xy - 20y^2$.		**14.** $56x^2 - 122xy + 20y^2$.
5. $56x^2 - 1121xy + 20y^2$.		**15.** $56x^2 - 102xy - 20y^2$.
6. $56x^2 - 68xy + 20y^2$.		**16.** $56x^2 - 229xy + 20y^2$.
7. $56x^2 - 558xy - 20y^2$.		**17.** $56x^2 - 94xy + 20y^2$.
8. $56x^2 + 36xy - 20y^2$.		**18.** $56x^2 - 276xy - 20y^2$.
9. $56x^2 - 67xy + 20y^2$.		**19.** $36x^2 - 33xy - 15y^2$.
10. $56x^2 + 3xy - 20y^2$.		**20.** $72x^2 - 19xy - 40y^2$.

§ 16. *Generally*, trinomials of the form $ax^2 + bx + c$ (a not a square) may be resolved by Formula [A]; thus,

Multiplying by a we get $a^2x^2 + bax + ac$. Writing z for ax, this becomes $z^2 + bz + ac$. Factor this trinomial, restore the value of z, and divide the result by a.

EXAMPLES.

1. $6x^2 + 5x - 4$.

Multiplying by 6, we get $(6x)^2 + 5(6x) - 24$ or $z^2 + 5z - 24$. Factoring, we get $(z - 3)(z + 8)$; hence, the required factors are $\frac{1}{6}(6x - 3)(6x + 8) = (2x - 1)(3x + 4)$.

2. $6x^2 - 13xy + 6y^2$.

Factoring $z^2 - 13zy + 36y^2$, we get $(z - 4y)(z - 9y)$; hence, the required factors are

$$\tfrac{1}{6}(6x - 4y)(6x - 9y) = (3x - 2y)(2x - 3y).$$

3. $33 - 14x - 40x^2$.

Factoring $1320 - 14z - z^2$, we get $(30 - z)(44 + z)$; hence, the required factors are

$$\tfrac{1}{40}(30 - 40x)(44 + 40x) = (3 - 4x)(11 + 10x).$$

NOTE. The factors may conveniently be arranged in two columns, each with its appropriate sign above it.

Exam. 1, above :

$-$	$+$
1	24
2	12

$$\tfrac{1}{6}(6x - 3)(6x + 8) = (2x - 1)(3x + 4).$$

Exam. 2, above :

$-$	$-$
1	36
2	18
3	12

$$\tfrac{1}{6}(6x - 4)(6x - 9) = (3x - 2)(2x - 3).$$

Another method of factoring trinomials of the form $ax^2 + bx + c$ is as follows:

Multiply by $4a$, thus obtaining $4a^2x^2 + 4abx + 4ac$. Add $b^2 - b^2$, which will not change the value, $4a^2x^2 + 4abx + b^2 - b^2 + 4ac$; by [1] this may be written $(2ax + b)^2 - (b^2 - 4ac)$. Factor this by [4], and divide the result by $4a$.

Example. Factor $56x^2 + 137x - 27,885$. Multiply by 4×56 or 2×112, $112^2x^2 + 2 \times 137 \times 112x - 6,246,240$. Add $137^2 - 137^2$; then,

$$112^2x^2 + 2 \times 137 \times 112x + 137^2 - (137^2 + 6,246,240)$$
$$= (112x + 137)^2 - 6,265,009$$
$$= [(112x + 137) + 2503][(112x + 137) - 2503]$$
$$= (112x + 2640)(112x - 2366).$$

We multiplied by 4×56; we must, therefore, now divide by that number. Doing so, we obtain as factors $(7x + 165)(8x - 169)$.

Ex. 29.

1. $10x^2 + x - 21$.
2. $10x^2 - 29x - 21$.
3. $10x^2 + 29x - 21$.
4. $6x^2 - 37x + 55$.
5. $12a^2 - 5a - 2$.
6. $12x^2 - 37x + 21$.
7. $12x^2 + 37x + 21$.

8. $15a^6 + 13a^3b^2 - 20b^4$.
9. $12x^2 - x - 1$.
10. $9x^2y^2 - 3xy^3 - 6y^4$.
11. $4x^2 + 8xy + 3y^2$.
12. $6b^2x^2 - 7bx^3 - 3x^4$.
13. $6x^4 - x^3y^2 - 35y^4$.
14. $2x^4 + x^2 - 45$.

15. $4x^4 - 37x^2y^2 + 9y^4$.
16. $4(x+2)^4 - 37x^2(x+2)^2 + 9x^4$.
17. $6(2x+3y)^2 + 5(6x^2+5xy-6y^2) - 6(3x-2y)^2$.
18. $6(2x+3y)^4 + 5(6x^2+5xy-6y^2)^2 - 6(3x-2y)^4$.
19. $6(x^2+xy+y^2)^2 + 13(x^4+x^2y^2+y^4) - 385(x^2-xy+y^2)^2$.
20. $21(x^2+2xy+2y^2)^2 - 6(x^2-2xy+2y^2)^2 - 5(x^4+4y^4)$.

EXTENDED APPLICATION OF THE FORMULAS.

§ 17. The methods of factoring just explained may be applied to find the rational factors, where such exist, of quadratic multinomials.

EXAMPLES.

1. Resolve $12x^2 - xy - 20y^2 + 8x + 41y - 20$.

In the first place we find the factors of the *first three* terms, which are

$$4x + 5y$$
and $$3x - 4y.$$

Now, to find the *remaining terms* of the required factors, we must observe the following conditions:

1. Their product must $= -20$.

2. The *algebraic* sum of the products, obtained by multiplying them diagonally into the y's, must $= 41\,y$.

3. The sum of the products, obtained by multiplying them diagonally into the x's, must $= 8\,x$.

We see at once, that -4, with the first pair already found, and $+5$, with the second pair, satisfy the required conditions; and hence the factors are

$$4x + 5y - 4$$
and $$3x - 4y + 5.$$

2. $p^2 + 2pr - 2q^2 + 7qr - 3r^2 + pq.$

Here the factors of $p^2 + pq - 2q^2$ are

$$p + 2q$$
and $$p - q.$$

Now find two factors which will give $-3r^2$, and which, multiplied diagonally into the p's and q's respectively, will give $2pr$ and $7qr$; these are found to be $-r$, taken with the *first* pair, and $+3r$, taken with the second pair. Hence, the required factors are

$$p + 2q - r$$
and $$p - q + 3r.$$

The work of seeking for the factors may be conveniently arranged as follows:

3. $x^2 + xy - 2y^2 + 2xz + 7yz - 3z^2.$

Reject:

1. The terms involving z;

2. The terms involving y;

3. The terms involving x;

and factor the expression that remains in each case.

   1. $x^2 + xy - 2y^2 = (x - y)(x + 2y)$.

   2. $x^2 + 2xz - 3z^2 = (x + 3z)(x - z)$.

   3. $-2y^2 + 7yz - 3z^2 = (-y + 3z)(2y - z)$.

Arrange these three pairs of factors in two sets of three factors each, by so selecting one factor from each pair that two of each set of three may have the same coefficient of x, two may have the same coefficient of y, and two the same coefficient of z (*coefficient* including *sign*). In this example there are

$$x - y, \quad x + 3z, \quad -y + 3z,$$
and $\quad x + 2y, \quad x - z, \quad 2y - z.$

From the first set select the common terms (including signs) and form therewith a trinomial, $x - y + 3z$.
Repeat with the second set, and we get $x + 2y - z$.

$\therefore x^2 + xy - 2y^2 + 2xz + 7yz - 3z^2$
$\quad = (x - y + 3z)(x + 2y - z)$.

4. $3x^2 - 8xy - 3y^2 + 30x + 27$.

   1. $3x^2 - 8xy - 3y^2 = (3x + y)(x - 3y)$.

   2. $3x^2 + 30x + 27 = (3x + 3)(x + 9)$.

   3. $-3y^2 + 27 = (y + 3)(-3y + 9)$.

   $\therefore$ the factors are $(3x + y + 3)(x - 3y + 9)$.

5. $6a^2 - 7ab + 2ac - 20b^2 + 64bc - 48c^2$.

   1. $6a^2 - 7ab - 20b^2 = (2a - 5b)(3a + 4b)$.

   2. $6a^2 + 2ac - 48c^2 = (2a + 6c)(3a - 8c)$.

   3. $-20b^2 + 64bc - 48c^2 = (-5b + 6c)(4b - 8c)$.

   $\therefore$ the factors are $(2a - 5b + 6c)(3a + 4b - 8c)$.

To find, where such exist, the factors of

$$ax^2 + \mathbf{b}xy + \mathbf{c}xz + ey^2 + gyz + hz^2.$$

Multiply by $4a$:

$$4a^2x^2 + 4abxy + 4acxz + 4aey^2 + 4agyz + 4ahz^2.$$

Select the terms containing x, and complete the square: thus,

$$4a^2x^2 + 4abxy + 4acxz + b^2y^2 + 2bcxz + c^2z^2$$
$$- (b^2 - 4ae)y^2 - 2(bc - 2ag)yz - (c^2 - 4ah)z^2$$
$$= (2ax + by + cz)^2$$
$$- [(b^2 - 4ae)y^2 + 2(bc - 2ag)yz + (c^2 - 4ah)z^2].$$

If the part within the double bracket is a square, say $(my + nz)^2$, the given expression can be written

$$(2ax + by + cz)^2 - (my + nz)^2,$$

which can be factored by [4]. Factor and divide the result by $4a$. If the part within the double bracket is not a square, the given expression cannot be factored. If b and c are *both* even, multiply by a instead of by $4a$, and the square can be completed without introducing fractions. If e be less than a, it will be easier to multiply by $4e$ instead of by $4a$, and select the terms containing y. A similar remark applies to h.

This method can evidently be extended to quadratic multinomials of any number of terms.

EXAMPLES.

1. Resolve $x^2 + xy + 2xz - 2y^2 + 7yz - 3z^2$ into factors.

Multiply by 4:

$$4x^2 + 4xy + 8xz - 8y^2 + 28yz - 12z^2.$$

Complete the square, selecting terms in x:

$$4x^2 + 4xy + 8xz + y^2 + 4yz + 4z^2 - 9y^2 + 24yz - 16z^2$$
$$= (2x + y + 2z)^2 - (3y - 4z)^2$$
$$= [(2x + y + 2z) + (3y - 4z)][(2x + y + 2z) - (3y - 4z)]$$
$$= (2x + 4y - 2z)(2x - 2y + 6z) = 4(x + 2y - z)(x - y + 3z).$$

$\therefore$ the factors are $(x + 2y - z)(x - y + 3z)$.

2. $6a^2 - 7ab + 2ac - 20b^2 + 64bc - 48c^2$.

Multiply by $4 \times 6 = 24$:

$\quad 144a^2 - 168ab + 48ac - 480b^2 + 1536bc - 1152c^2$
$\quad = (12a - 7b + 2c)^2 - 529b^2 + 1564bc - 1156c^2$
$\quad = (12a - 7b + 2c)^2 - (23b - 34c)^2$
$\quad = (12a + 16b - 32c)(12a - 30b + 36c)$
$\quad = 24(3a + 4b - 8c)(2a - 5b + 6c)$.

$\therefore$ the factors are $3a + 4b - 8c$ and $2a - 5b + 6c$.

3. $x^2 + 12xy + 2xz + 26y^2 - 8yz - 9z^2$
$\quad = (x^2 + 12xy + 2xz + 36y^2 + 12yz + z^2) - 10y^2 - 20yz - 10z^2$
$\quad = (x + 6y + z)^2 - [(y + z)\sqrt{10}]^2$
$\quad = [x + (6 + \sqrt{10})y + (\sqrt{10} + 1)z]$
$\quad \times [x + (6 - \sqrt{10})y - (\sqrt{10} - 1)z]$.

4. $3a^2 + 10ab - 14ac + 12ad - 8b^2 - 8bd + 8c^2 - 8cd$.

Multiply by 3, not 4×3, since the coefficients of the other terms in a are all even :

$\quad 9a^2 + 30ab - 42ac + 36ad - 24b^2 - 24bd + 24c^2 - 24cd$.

Select the terms containing a, and complete the square :

$\quad (3a + 5b - 7c + 6d)^2$
$\quad\quad - 49b^2 + 70bc - 84bd - 25c^2 + 60cd - 36d^2$
$\quad = (3a + 5b - 7c + 6d)^2 - (7b - 5c + 6d)^2$
$\quad = (3a + 12b - 12c + 12d)(3a - 2b - 2c)$
$\quad = 3(a + 4b - 4c + 4d)(3a - 2b - 2c)$.

$\therefore$ the factors are $a + 4b - 4c + 4d$ and $3a - 2b - 2c$.

Ex. 30.

1. $7x^2 - xy - 6y^2 - 6x - 20y - 16$.

2. $20x^2 - 15xy - 5y^2 - 68x - 42y - 88$.

3. $3x^4 + x^2y^2 - 4y^4 + 10x^2 - 17y^2 - 13$.

4. $20x^2 - 20y^2 + 9xy + 28x + 35y$.

5. $72\,x^2 - 8\,y^2 + 55\,xy + 12\,y - 169\,x + 20.$

6. $x^2 - xy - 12\,y^2 - 5\,x - 15\,y.$

7. $8\,x^2 + 18\,xy + 9\,y^2 + 2\,xz - z^2.$

8. $6\,x^2 + 6\,y^2 - 13\,xy - 8\,z^2 - 2\,yz + 8\,xz.$

9. $6\,x^4 - 10\,y^4 + 11\,x^2y^2 - 25\,z^2 + 10\,y^2 + 25\,y^2z^2 - 15\,x^2 + 10\,x^2z^2.$

10. $15\,x^4 - 16\,y^4 - 22\,x^2y^2 + 15\,z^4 + 14\,y^2z^2 + 50\,x^2z^2.$

11. $4\,a^2 - 15\,b^2 - 4\,ab - 21\,c^2 - 36\,bc - 8\,ac.$

12. $a^4 + b^4 + c^4 - 2\,a^2b^2 - 2\,b^2c^2 - 2\,c^2a^2.$

§ 18. Trinomials of the form $ax^4 + bx^2 + c$ can always be broken up into real factors.

If a and c have different signs, the expression may be factored by § 16.

If a and c are of the same sign, three cases have to be considered :

$$\text{(i.)} \quad b = 2\,\sqrt{(ac)}.$$
$$\text{(ii.)} \quad b > 2\,\sqrt{(ac)}.$$
$$\text{(iii.)} \quad b < 2\,\sqrt{(ac)}.$$

Case I. $b = 2\,\sqrt{(ac)}$. This case falls under § 12, Formula [1], where examples will be found.

Case II. $b > 2\,\sqrt{(ac)}$. This case falls under § 16, where examples will be found.

The following additional examples are resolved by the second method of that section :

Examples.

1. $4\,x^4 + 5\,x^2y^2 + y^4.$

Here we see that $(\frac{5}{4}y^2)^2$ will make, with the first two terms, a perfect square, and we therefore add to the given expression, $(\frac{5}{4}y^2)^2 - (\frac{5}{4}y^2)^2.$

The expression then becomes

$$4x^4 + 5x^2y^2 + (\tfrac{5}{4}y^2)^2 + y^4 - (\tfrac{5}{4}y^2)^2$$
$$= (2x^2 + \tfrac{5}{4}y^2)^2 - \tfrac{9}{16}y^4$$
$$= (2x^2 + \tfrac{5}{4}y^2 + \tfrac{3}{4}y^2)(2x^2 + \tfrac{5}{4}y^2 - \tfrac{3}{4}y^2)$$
$$= (2x^2 + 2y^2)(2x^2 + \tfrac{1}{2}y^2)$$
$$= (x^2 + y^2)(4x^2 + y^2).$$

2. $3x^4 + 6x^2 + 2.$

Here, multiplying by 4×3, and completing the square as in Exam. 1, we have

$$36x^4 + 72x^2 + 6^2 + 24 - 6^2$$
$$= (6x^2 + 6)^2 - 12$$
$$= (6x^2 + 6 - \sqrt{12})(6x^2 + 6 + \sqrt{12}),$$

which, divided by 4×3, gives the required factors.

3. $ax^4 + bx^2 + c.$

Proceeding as in Exam. 2, we have, by multiplying by $4a$,

$$ax^4 + bx^2 + c$$
$$= (4a^2x^4 + 4abx^2 + b^2 - b^2 + 4ac) \div 4a$$
$$= [2ax^2 + b + \sqrt{(b^2 - 4ac)}][2ax^2 + b - \sqrt{(b^2 - 4ac)}] \div 4a.$$

Ex. 31.

1. $x^4 + 7x^2 + 1$; $4x^4 + 14x^2 + 1.$

2. $x^4 + 7x^2y^2 + y^4$; $3x^4 + 5x^2y^2 + y^4.$

3. $4x^4 + 10x^2 + 3$; $3(x+y)^4 + 5z^2(x+y)^2 + z^4.$

4. $x^4 + 7x^2y^2 + 3\tfrac{1}{4}y^4$; $x^4 + 7x^2y^2 + 8\tfrac{1}{4}y^4.$

5. $4x^4 + 9x^2y^2 + \tfrac{17}{16}y^4$; $4(a+b)^4 + 10c^2(a+b)^2 + 3c^4.$

6. $3x^4 + 8x^2y^2 + 4\tfrac{7}{12}y^4$; $36x^4 + 96x^2 + 55.$

7. $5x^4 + 20x^2 + 2$; $4a^4 + 12a^2 + 1.$

8. $4(x+y)^4 + 12(x+y)^2z^2 + z^4$; $5x^4 + 20x^2y^2 + 2y^4.$

9. $9x^4 + 14x^2 + 4$; $2x^4 + 12x^2(y+z)^2 + 15(y+z)^4$.

10. $2x^4 + 12x^2 + 15$; $7x^4 + 40x^2 + 45$.

11. $8x^4 + 36x^2y^2 + 29y^4$; $7x^4 + 20x^2y^2 - 20y^4$.

12. $7(a-b)^4 + 16(a-b)^2c^2 + 5c^4$; $\frac{3}{2}a^4 + 3a^2b^2 + b^4$.

13. $3x^4 + 6x^2y^2 + 2y^4$; $3(a+b)^4 + 6(a^2-b^2)^2 + 2(a-b)^4$.

14. $49a^4 - 84a^2b^2 + 22b^4$; $25m^4 + 60m^2n^2 + 27n^4$.

15. $49(m+n)^4 - 84(m^2-n^2)^2 + 22(m-n)^4$.

CASE III. $b < 2\sqrt{(ac)}$. This case may be brought under § 13.

The following examples illustrate the process of reduction and resolution :

EXAMPLES.

1. $x^4 - 7x^2 + 1$.

We have to throw this into the form $a^2 - b^2$:
$$x^4 - 7x^2 + 1 = (x^2+1)^2 - 9x^2$$
$$= (x^2+1+3x)(x^2+1-3x).$$

2. $9x^4 + 3x^2y^2 + 4y^4 = (3x^2 + 2y^2)^2 - 9x^2y^2$
$$= (3x^2 + 2y^2 - 3xy)(3x^2 + 2y^2 + 3xy).$$

3. $x^4 + y^4 = (x^2 + y^2)^2 - 2x^2y^2$
$$= (x^2 + y^2 + xy\sqrt{2})(x^2 + y^2 - xy\sqrt{2}).$$

4. $x^4 - \frac{1}{4}x^2y^2 + y^4 = (x^2 + y^2)^2 - \frac{9}{4}x^2y^2$
$$= (x^2 + y^2 + \frac{3}{2}xy)(x^2 + y^2 - \frac{3}{2}xy).$$

5. $ax^4 + bx^2 + c = (x^2\sqrt{a} + \sqrt{c})^2 - (2\sqrt{ac} - b)x^2$
$$= \{x^2\sqrt{a} + \sqrt{c} - (\sqrt{2\sqrt{ac} - b})x\}$$
$$\times \{x^2\sqrt{a} + \sqrt{c} + (\sqrt{2\sqrt{ac} - b})x\}.$$

§ 19. It is seen from these examples that we have merely to add to the given expression what will make with the *first* and *last* terms (arranged as in Exam. 5) a perfect square, and to subtract the same quantity. In Exam. 2, for instance, the square root of $9x^4 = 3x^2$, the square root of $4y^4 = 2y^2$; hence, $3x^2 + 2y^2$ is the binomial whose square is required; we need, therefore, $12x^2y^2$; but the expression contains $3x^2y^2$; hence, we have to *add* and *subtract* $12x^2y^2 - 3x^2y^2 = 9x^2y^2$.

Hence, we derive a practical rule for factoring such expressions:

1. Take the square roots of the two extreme terms, and connect them by the proper sign; this gives the first two terms of the required factors.

2. Subtract the middle term of the given expression from twice the product of these two roots, and the square roots of the difference will be the third terms of the required factors.

6. $x^4 + \frac{7}{16}x^2y^2 + y^4$.

Here $\sqrt{x^4} = x^2$, $\sqrt{y^4} = y^2$, and the first two terms of the required factors are $x^2 + y^2$; twice the product of these is $+2x^2y^2$, from which, subtracting the middle term, $\frac{7}{16}x^2y^2$, we get $\frac{25}{16}x^2y^2$; the square roots of this are $\pm\frac{5}{4}xy$. Hence, the factors are

$$x^2 + y^2 \pm \frac{5}{4}xy.$$

Note that, since $\sqrt{y^4} = +y^2$, or $-y^2$, it may sometimes happen that while the former sign will give irrational factors, the latter will give rational factors, and conversely.

7. $x^4 - 11x^2y^2 + y^4$.

Here, taking $+y^2$, we have
$$x^2 + y^2 + xy\sqrt{13} \text{ and } x^2 + y^2 - xy\sqrt{13}.$$

But, taking $-y^2$, we have
$$x^2 - y^2 + 3xy \text{ and } x^2 - y^2 - 3xy.$$

Sometimes *both* signs will give rational factors.

8. $16x^4 - 17x^2y^2 + y^4$.

Here we have $(4x^2 + y^2 + 3xy)(4x^2 + y^2 - 3xy)$,

and also $\quad (4x^2 - y^2 + 5xy)(4x^2 - y^2 - 5xy)$.

Ex. 32.

1. $x^4 + 2x^2y^2 + 9y^4$; $x^4 - x^2y^2 + y^4$; $x^4 + x^2y^2 + y^4$.

2. $x^4 + 4y^4$; $16x^4 + y^4 - x^2y^2$; $\frac{1}{4}x^4 + y^4$.

3. $x^4 + 1$; $x^4 + 9y^4$; $1 - 12y^2 + 16y^4$.

4. $x^4 - 7x^2 + 1$; $x^4 + 9$; $\frac{1}{4}x^4 + y^4 - 3x^2y^2$.

5. $y^4 - x^4 + 11x^2y^2$; $x^8 + 4y^8$; $x^4 + 4x^2 + 16$.

6. $4x^4 + y^4 - 8\frac{1}{4}x^2y^2$; $x^4 + y^4 - \frac{7}{16}x^2y^2$; $4x^4 + 1$.

7. $x^{4m} + 64y^{4m}$; $x^{4m} + 4y^{4m}$; $\frac{1}{4}x^4 + \frac{9}{16}y^4 - 5\frac{3}{4}x^2y^2$.

8. $4x^4 - 8x^2 + 1$; $7x^2y^2 - \frac{1}{4}x^4 - 36y^4$; $x^4 + a^4y^4$.

9. $m^2x^4 + n^2y^4 - (2mn + p)x^2y^2$; $x^{4m} + 2^{4m-2}y^{4m}$.

10. $16x^4 - 25x^2 + 9$; $4x^4 - 16x^2 + 4$; $13x^2y^2 - 9x^4 - 4y^4$.

11. $4x^4 - 12\frac{16}{25}x^2y^2 + 9y^4$; $x^4 + 6x^2 + 25$.

12. $a^4 + b^4 + (a + b)^4$; $1 + a^4 + (1 + a)^4$.

13. $(x + y)^4 - 7z^2(x + y)^2 + z^4$.

14. $(a + b)^4 + 7c^2(a + b)^2 + c^4$.

15. $16a^4 + 4(b - c)^4 - 9a^2(b - c)^2$.

16. $4(a + b)^4 + 9(a - b)^4 - 21(a^2 - b^2)^2$.

17. $(x^2 + y^2 - xy)^4 - 7(x^3 + y^3)^2 + (x + y)^4$.

18. $(a^2 + ab + b^2)^4 + 7(a^3 - b^3)^2 + (a - b)^4$.

19. $16a^4 + 4a^2 + 1$; $x^4 - 41x^2 + 16$.

20. $(a^2 + 1)^4 + 4(a^2 + 1)^2a^2 + 16a^4$;

$\quad (x + 1)^4 + 2(x^2 - 1)^2 + 9(x - 1)^4$.

§ 20. We can apply [4], § 13, to factor expressions of the form $ax^4 + bx^3 + rbx - r^2a$. This may be written,

$$a(x^4 - r^2) + bx(x^2 + r) = [a(x^2 - r) + bx](x^2 + r).$$

EXAMPLES.

1. $6x^4 + 4x^3 + 12x - 54$
   $$= 6(x^4 - 9) + 4x(x^2 + 3)$$
   $$= (x^2 + 3)[6(x^2 - 3) + 4x]$$
   $$= (x^2 + 3)(6x^2 + 4x - 18).$$

2. $11x^4 + 10x^3 - 40x - 176$
   $$= 11(x^4 - 16) + 10x(x^2 - 4)$$
   $$= (x^2 - 4)[11(x^2 + 4) + 10x]$$
   $$= (x^2 - 4)(11x^2 + 10x + 44).$$

3. $40x^4 + 30x^3 + 60x - 160$
   $$= 10(4x^4 - 16) + 15x(2x^2 + 4)$$
   $$= (2x^2 + 4)[10(2x^2 - 4) + 15x]$$
   $$= (2x^2 + 4)(20x^2 + 15x - 40).$$

NOTE. To determine r, take the ratio of the coefficient of x^3 to the coefficient of x.

Ex. 33.

Resolve into factors:

1. $x^4 + 2x^3 + 6x - 9.$

2. $2x^4 + 2x^3 + 6x - 18.$

3. $x^4 + 3x^3 + 12x - 16.$

4. $3x^4 + x^3 - 4x - 48.$

5. $5x^4 + 4x^3 - 12x - 45.$

6. $10x^4 + 5x^3 + 30x - 360.$

7. $\frac{1}{4}x^4 + 20x^3 + 4x - \frac{1}{100}.$

8. $25x^4 - 40x^3 + 8x - 1.$

9. $37\frac{1}{2}x^4 - 30x^3 + 48x - 96.$

10. $63x^4 - 39x^3 + 52x - 112.$

11. $810x^4 + \frac{81}{4}x^3 + \frac{9}{8}x - 2\frac{1}{2}.$

12. $242x^4 - 33x^2 - 3x - 2.$

13. $\frac{1}{4}x^4 + \frac{1}{10}x^3 - \frac{2}{25}x - \frac{4}{25}$.

14. $80x^4 - 32x^3y + 64xy - 320y^4$.

15. $24x^4 - 12x^3y + 30xy^3 - 150y^4$.

16. $2x^4 + \frac{1}{2}x^3y - 8xy^3 - 512y^4$.

17. $11x^4 + 10x^3 - 12x - 15\frac{21}{25}$.

18. $40x^4 + 30x^3 + 60x - 160$.

19. $13x^4 - 12x^3y + 72xy^3 - 468y^4$.

20. $3x^4 + 3x^3y + 12xy^3 - 48y^4$.

21. $5x^4 + 4x^3y - 12xy^3 - 45y^4$.

22. $4x^4 - 14x^3y + 28xy^3 - 16y^4$.

23. $x^4 + 80x^3y + 16xy^3 - \frac{1}{25}y^4$.

24. $2x^4 - x^3y + 6xy^3 - 72y^4$.

§ 21. Formulas [1] and [4] may sometimes be applied to factor expressions of the form

$$ax^4 + bx^3 + cx^2 + rbx + r^2a.$$

This may be put under the form

$$a(x^4 + r^2) + bx(x^2 + r) + cx^2$$
$$= a(x^2 + r)^2 + bx(x^2 + r) + (c - 2ar)x^2,$$

which can sometimes be factored.

Examples.

1. $x^4 + 6x^3 + 27x^2 + 162x + 729$.

$\quad x^4 + 729 + 6x(x^2 + 27) + 27x^2$
$\quad = (x^2 + 27)^2 + 6x(x^2 + 27) + 9x^2 - 36x^2$
$\quad = (x^2 + 27 + 3x)^2 - 36x^2$, which gives the factors
$\quad x^2 - 3x + 27$ and $x^2 + 9x + 27$.

2. $x^4 + 4x^3 + 4x^2 + 20x + 25$
$= (x^2 + 5)^2 + 4x(x^2 + 5) - 6x^2$
$= (x^2 + 5)^2 + 4x(x^2 + 5) + 4x^2 - 10x^2$
$= (x^2 + 5 + 2x - x\sqrt{10})(x^2 + 5 + 2x + x\sqrt{10}).$

Ex. 34.

Resolve into factors:

1. $x^4 - 6x^3 + 27x^2 - 162x + 729.$

2. $x^4 + 2x^3 + 3x^2 + 8x + 16.$

3. $x^4 + x^3 + x^2 + x + 1.$

4. $x^4 - 4x^3 + x^2 - 4x + 1.$

5. $4x^4 - 12x^3 - 6x^2 - 12x + 4.$

6. $x^4 + 14x^3 - 25x^2 - 70x + 25.$

7. $16x^4 - 24x^3 - 16x^2 + 12x + 4.$

8. $x^4 + 5x^3 - 16x^2 + 20x + 16.$

9. $x^4 + 6x^3 - 11x^2 - 12x + 4.$

10. $x^4 + 4x^3 y + x^2 y^2 + 12xy^3 + 9y^4.$

11. $x^4 + 6x^3 - 9x^2 - 6x + 1.$

12. $x^4 + 4x^3 y - 19x^2 y^2 + 4xy^3 + y^4.$

13. $4x^4 + 4x^3 y - 65x^2 y^2 - 10xy^3 + 25y^4.$

14. $x^4 + 6x^3 y - 9x^2 y^2 - 6xy^3 + y^4.$

15. $x^4 + 6x^3 y + 10x^2 y^2 + 12xy^3 + 4y^4.$

16. $9x^4 + 18x^3 y - 52x^2 y^2 - 12xy^3 + 4y^4.$

17. $11x^4 + 10x^3 y + 39\frac{96}{121} x^2 y^2 + 20xy^3 + 44y^4.$

FACTORING BY PARTS.

§ 22. To factor an expression which can be reduced to the form $a \times F(x) + b \times f(x)$.

When the expression is thus arranged, any factor common to a and b, or to $F(x)$ and $f(x)$, will be a factor of the whole expression. The method about to be illustrated will be found useful in cases where only *one* power of some letter is found.

EXAMPLES.

1. Factor $acx^2 - abx - bc^2x + b^2c$.

   Here we see that only one power of a occurs, and we therefore group together the terms involving this letter, and those not involving it, getting

   $a(cx^2 - bx) - bc^2x + b^2c$
   $= ax(cx - b) - bc(cx - b) = (ax - bc)(cx - b)$.

2. Factor $m^2x^2 - mna^2x - mnx + n^2a^2$.

   Here we observe that a occurs in only one power (a^2). Therefore, we have

   $-a^2(mnx - n^2) + m^2x^2 - mnx$
   $= -na^2(mx - n) + mx(mx - n)$
   $= (mx - n)(mx - na^2)$.

3. Factor $2x^2 + 4ax + 3bx + 6ab$.

   Here we observe that the expression contains only one power of both a and b. We may, therefore, collect the coefficients in either of the following ways:

   $a(4x + 6b) + (2x^2 + 3bx)$,
   or $b(3x + 6a) + (2x^2 + 4ax)$.

Now the expressions in the brackets ought to have a common factor; and we see that this is the case. Hence,

$a(4x + 6b) + (2x^2 + 3bx)$
$= 2a(2x + 3b) + x(2x + 3b)$
$= (2x + 3b)(x + 2a)$.

4. $abxy + b^2y^2 + acx - c^2$

$$= a(bxy + cx) + b^2y^2 - c^2$$
$$= ax(by + c) + (by + c)(by - c)$$
$$= (by + c)(ax + by - c).$$

5. $y^3 - (2a + b)y^2 + (2ab + a^2)y - a^2b$

$$= -b(y^2 - 2ay + a^2) + y^3 - 2ay^2 + a^2y$$
$$= -b(y^2 - 2ay + a^2) + y(y^2 - 2ay + a^2)$$
$$= (y - b)(y - a)^2.$$

6. $2x^3y + 2bx^4 - bx^3y + 4abx^2y - x^2y^2 + 4axy^2$
$\quad\quad - 2abxy^2 - 2ay^3$

$$= b(2x^4 - x^3y + 4ax^2y - 2axy^2) + 2x^3y - x^2y^2$$
$$\quad + 4axy^2 - 2ay^3$$
$$= bx(2x^3 - x^2y + 4axy - 2ay^2)$$
$$\quad + y(2x^3 - x^2y + 4axy - 2ay^2)$$
$$= (y + bx)(2x^3 - x^2y + 4axy - 2ay^2).$$

And $2x^3 - x^2y + 4axy - 2ay^2$

$$= a(4xy - 2y^2) + 2x^3 - x^2y$$
$$= 2ay(2x - y) + x^2(2x - y)$$
$$= (2ay + x^2)(2x - y).$$

7. $x^3 + (2a - b)x^2 - (2ab - a^2)x - a^2b$

$$= b(-x^2 - 2ax - a^2) + x^3 + 2ax^2 + a^2x$$
$$= -b(x + a)^2 + x(x + a)^2$$
$$= (x - b)(x + a)^2.$$

8. $px^3 - (p - q)x^2 + (p - q)x + q$

$$= q(x^2 - x + 1) + px^3 - px^2 + px$$
$$= q(x^2 - x + 1) + px(x^2 - x + 1)$$
$$= (px + q)(x^2 - x + 1).$$

Ex. 35.

1. $x^2y - x^2z - y^2 + yz.$ **3.** $x^2z^2 + ax^2 - a^2z^2 - a^3.$

2. $abxy + b^2y^2 + acx - c^2.$ **4.** $2x^2 - ax - 4bx + 2ab.$

5. $x^2 + 2bx + 3ax + 6ab$.   8. $8x^2 + 12ax + 10bx + 15ab$.

6. $x^3 - b^2 x^2 - a^2 x + a^2 b^2$.   9. $a^2 + (ac - b^2) x^2 + bcx^3$.

7. $x^5 - a^3 x^2 - b^2 x^3 + a^3 b^2$.   10. $a^2 + (ac - b^2) x^2 - bcx^3$.

11. $abx^3 + (ac - bd) x^2 - (af + cd) x + df$.

12. $px^3 - (p + q) x^2 + (p + q) x - q$.

13. $a^2 + ab + 2ac - 2b^2 + 7bc - 3c^2$.

14. $x^3 + (a + 1) x^2 + (a + 1) x + a$.

15. $mpx^3 + (mq - np) x^2 - (mr + nq) x + nr$.

16. $x^3 - (a + b + c) x^2 + (ab + bc + ac) x - abc$.

17. $x^3 + (a - b - c) x^2 - (ab - bc + ca) x + abc$.

18. $x^3 + (a + b - c) x^2 - (bc - ca - ab) x - abc$.

19. $a^2 x^3 - a^3 x^2 y - a^2 xy + a^3 y^2 - ax^2 yz + x^3 z - xyz + ay^2 z$.

20. $a^2 bx^2 + ab^2 xy + acdxy + bcdy^2 - aefxz - befyz$.

21. $a^2 x^3 - a(b - c) x^2 + c(a - b) x + c^2$.

22. $mx^3 - nx^2 y + rx^2 z - mxy^2 + ny^3 - ry^2 z$.

23. $amx^2 + (mby - nay + mcz) x - nby^2 - ncyz$.

24. $(am - bcm) x^2 + (am - bcn) x + an + nax$.

25. $a^2 b^2 c^2 - b^2 c^2 xy - a^2 c^2 yz + c^2 xy^2 z - a^2 b^2 zx + b^2 x^2 yz$
    $+ a^2 z^2 xy - x^2 y^2 z^2$.

26. $x^5 - m^2 x^4 - (n - n^2) x^3 + (m^2 n - m^2 n^2) x^2 - a(x^2 + n^2 - n)$.

27. $1 - (a - 1) x - (a - b + 1) x^2 + (a + b - c) x^3$
    $- (b + c) x^4 + cx^5$.

28. $a^3 x^3 - a^2 (b - c + d) x^2 y - (abc - abd + acd) xy^2 + bcdy^3$.

29. $m^2 npx^3 - (n^2 p - m^2 n^2 - m^2 pq) x^2$
    $- (n^3 + npq - m^2 nq) x - n^2 q$.

30. $m^2 p^2 x^5 + m^2 p^2 x^4 - (p^2 n^2 - q^2 m^2) x^3 y^2$
    $- (p^2 n^2 - q^2 m^2) x^2 y^2 - (n^2 q^2 + n^3 q^2 x) y^4$.

§23. Sometimes an expression which does not come directly under the preceding form may be resolved by first finding the factors of its parts.

EXAMPLES.

1. $abx^2 + aby^2 - a^2xy - b^2xy$.

> Here, taking ax out of the first and third terms, and by out of the second and fourth terms, we have
> $$ax(bx - ay) - by(bx - ay),$$
> and hence $(ax - by)(bx - ay)$.

2. $x^4 - (a+b)x^3 + (a^2b + ab^2)x - a^2b^2$.

> Here, taking the first and last terms together, and the two middle terms together, we have
> $$(x^2 + ab)(x^2 - ab) - (a+b)x^3 + ab(a+b)x$$
> $$= (x^2 + ab)(x^2 - ab) - (a+b)x(x^2 - ab)$$
> $$= (x^2 - ab)[x^2 + ab - (a+b)x]$$
> $$= (x^2 - ab)(x - a)(x - b).$$

3. $x^{3m} - 4x^m + 3$
$$= x^{3m} - x^m - 3(x^m - 1)$$
$$= x^m(x^{2m} - 1) - 3(x^m - 1)$$
$$= x^m(x^m + 1)(x^m - 1) - 3(x^m - 1)$$
$$= (x^m - 1)[x^m(x^m + 1) - 3].$$

Ex. 36.

1. $a^2 - ab + ax - bx$.

2. $abx^2 + b^2xy - a^2xy - aby^2$.

3. $x^4 + ax^3 - a^3x - a^4$.

4. $a^3x + 2a^2x^2 + 2ax^3 + x^4$.

5. $acx^2 + (ad - bc)x - bd$.

6. $25x^4 - 5x^3 + x^2 - 1$.

7. $a^2 - b^2 + ax - ac - bx + bc$.

8. $a^3 + (1+a)ab + b^2$.

9. $x^4 + 2xy(x^2 - y^2) - y^4$.

10. $x^3 - y^3 + x^2 + xy + y^2$.

11. $2b + (b^2 - 4)x - 2bx^2$.

12. $x^3 + 3x^2 - 4$.

13. $p^3 - p^2 q - 2pq^2 + 2q^3$.　　**20.** $a^3 - 4ab^2 + 3b^3$.

14. $a^3 + a^2 - 2$.　　**21.** $a^{2m} - 3a^m c^n + 2c^{2n}$.

15. $3a^2 b^4 - 2ab^2 - 1$.　　**22.** $ax^3 - (a^2 + b)x^2 + b^2$.

16. $y^3 - 3y + 2$.　　**23.** $35x^{2n} - 6a^2 x^n - 9a^4$.

17. $2a^3 - a^2 b - ab^2 + 2b^3$.　　**24.** $a^2 b^2 + 2abc^2 - a^2 c^2 - b^2 c^2$.

18. $b^{3m} + b^{2m} - 2$.　　**25.** $am^2 - ab^2 + b^2 m - m^3$.

19. $y^{3n} - 2y^{2n} z^n - 2y^n z^{2n} + z^{3n}$.　　**26.** $\frac{1}{3} - 6a^2 + 27a^4$.

27. $(x - y)^3 + (1 - x + y)(x - y)z - z^2$.

28. $24m^3 - 28m^2 n + 6mn^2 - 7n^3$.

29. $x^{m+n} + x^n y^n + x^m y^m + y^{m+n}$.

30. $x^4 + 2x^3 y - a^2 x^2 + x^2 y^2 - 2axy^2 - y^4$.

APPLICATION OF THE THEORY OF DIVISORS.

§ 24. By Theorem I. we prove that

$x^n - a^n$ is divisible by $x - a$ *always*,

$x^n - a^n$ is divisible by $x + a$ when n is *even*,

$x^n + a^n$ is divisible by $x + a$ when n is *odd*.

By actual division we find in the above cases:

$$\frac{x^n - a^n}{x - a} = x^{n-1} + x^{n-2} a + \cdots + xa^{n-2} + a^{n-1} \qquad (1)$$

$$\frac{x^n - a^n}{x + a} = x^{n-1} - x^{n-2} a + \cdots + xa^{n-2} - a^{n-1} \qquad (2)$$

$$\frac{x^n + a^n}{x + a} = x^{n-1} - x^{n-2} a + \cdots - xa^{n-2} + a^{n-1} \qquad (3)$$

EXAMPLES.

1. Resolve into factors $x^3 - y^3$.

Here $x - y$ is one factor, and by (1) the other is

$x^2 + xy + y^2$.

2. Resolve $a^3 + (b - c)^3$.

Here $a + (b - c)$ is one factor, and by (3) the other is
$a^2 - a(b - c) + (b - c)^2$.

3. Resolve $x^{15} + 1024\,y^{10}$.

This equals $(x^3)^5 + [(2y)^2]^5$, one factor of which is
$x^3 + (2y)^2$, and by (3) the other factor is
$(x^3)^4 - (x^3)^3(4y^2) + (x^3)^2(4y^2)^2 - x^3(4y^2)^3 + (4y^2)^4$
$= x^{12} - 4x^9 y^2 + 16x^6 y^4 - 64x^3 y^6 + 256 y^8$.

4. Resolve $(x - 2y)^3 + (2x - y)^3$ into factors.

Here, by (3), we have
$$\frac{(x - 2y)^3 + (2x - y)^3}{x - 2y + 2x - y}$$
$$= (x - 2y)^2 - (x - 2y)(2x - y) + (2x - y)^2.$$
$\therefore$ the factors are $3(x - y)(7x^2 - 13xy + 7y^2)$.

5. Resolve $x^5 + x^4 y + x^3 y^2 + x^2 y^3 + xy^4 + y^5$.

By (1) we see that this equals
$$\frac{x^6 - y^6}{x - y} = \frac{(x^3 + y^3)(x^3 - y^3)}{x - y}$$
$$= (x + y)(x^2 - xy + y^2)(x^2 + xy + y^2).$$

6. Resolve $x^{11} - x^{10}a + x^9 a^2 - x^8 a^3 + x^7 a^4 - x^6 a^5 + x^5 a^6$
$\quad - x^4 a^7 + x^3 a^8 - x^2 a^9 + xa^{10} - a^{11}$.

This equals
$$\frac{x^{12} - a^{12}}{x + a} = \frac{(x^6 + a^6)(x^6 - a^6)}{x + a}$$
$$= \frac{(x^6 + a^6)(x^3 - a^3)(x^3 + a^3)}{x + a}$$
$$= (x^2 + a^2)(x^4 - x^2 a^2 + a^4)(x - a)(x^2 + xa + a^2)(x^2 - xa + a^2).$$

Ex. 37.

Factor the following:

1. $x^6 - y^6$; $x^3 - 1$; $x^3 + 8$; $8a^3 - 27x^3$; $8 + a^3x^3$.

2. $x^5 - a^{10}$; $27a^3 - 64$; $a^{12} - b^3$; $x^{10} - 32y^5$.

3. Find a factor which, multiplied by
   $a^4 + a^3b + a^2b^2 + ab^3 + b^4$, will give $a^5 - b^5$.

4. By what factor must $x^3 - 4x^2y + 16xy^2 - 64y^3$ be multiplied to give $x^4 - 256y^4$?

5. Factor $x^7 + x^6y + x^5y^2 + x^4y^3 + x^3y^4 + x^2y^5 + xy^6 + y^7$.

Find the factors of the following:

6. $(3y^2 - 2x^2)^3 - (3x^2 - 2y^2)^3$; $a^8 - 16b^4$.

7. $x^3 - y^3 - x(x^2 - y^2) + y(x - y)^2$.

8. $b(x^3 - a^3) + ax(x^2 - a^2) + a^3(x - a)$.

9. $b(m^3 + a^3) + am(m^2 - a^2) + a^3(m + a)$.

10. $x^6 - y^6 + 2xy(x^4 + x^2y^2 + y^4)$.

11. $(a^2 - bc)^3 + 8b^3c^3$; $x^{4m} - a^{4n}$.

12. $x^3 - 3ax^2 + 3a^2x - a^3 + b^3$.

13. $(x^3 + 8y^3)(x + y) - 6xy(x^2 - 2xy + 4y^2)$.

14. $8x^3 - 6xy(2x + 3y) + 27y^3$.

15. $1 - 2x + 4x^2 - 8x^3$.

16. $a^5 + a^4bc + a^3b^2c^2 + a^2b^3c^3 + ab^4c^4 + b^5c^5$.

§ 25. The principles illustrated in Chap. II. may be applied to factor various algebraic expressions, as in the following cases:

Examples.

1. Find the factors of

$$(a+b+c)(ab+bc+ca)-(a+b)(b+c)(c+a).$$

1. Observe that the expression is *symmetrical* with respect to a, b, c.

2. If there be any *monomial* factor, a must be one. Putting $a=0$, the expression vanishes; hence, a is a factor, and, by symmetry, b and c are also factors.

Therefore, abc is a factor.

3. There can be no other *literal* factor, because the given expression is of only *three* dimensions, and abc is of three dimensions.

4. But there may be a *numerical* factor, m suppose, so that we have

$$(a+b+c)(ab+bc+ca)-(a+b)(b+c)(c+a)=mabc.$$

To find m, put $a=b=c=1$ in this equation, and $m=1$.

Therefore, the expression $= abc$.

2. Resolve $a^2(b-c)+b^2(c-a)+c^2(a-b)$.

1. For $a=0$ this does not vanish; hence, a is not a factor, and, by symmetry, neither is b nor c.

2. Try a *binomial* factor; this will likely be of the form $b-c$; put $b-c=0$; that is, $b=c$ in the given expression, and there results

$$a^2(c-c)+c^2(c-a)+c^2(a-c), \text{ which } =0.$$

Therefore, $b-c$ is a factor, and, by symmetry, $c-a$ and $a-b$ are factors. Since the given expression is only of *three* dimensions, there can be no other *literal* factor; but there may be a *numerical* factor, m suppose, so that

$$a^2(b-c)+b^2(c-a)+c^2(a-b)=m(a-b)(b-c)(c-a).$$

To find the value of m, give a, b, c in this equation, any values which will not reduce either side to zero; let $a=1$, $b=2$, $c=0$, and we have $2=m(-2)$, or $m=-1$; so that the given expression

$$=-(a-b)(b-c)(c-a), \text{ or } (a-b)(b-c)(a-c).$$

3. Resolve

$$a^3(b+c^2)+b^3(c+a^2)+c^3(a+b^2)+abc(abc+1).$$

Here we see at once that there is no *monomial* factor. Put $b+c^2=0$, that is, $b=-c^2$, and the expression becomes

$$a^3(-c^2+c^2)-c^6(c+a^2)+c^3(a+c^4)-c^3a(-c^3a+1),$$

which $=0$.

$\therefore b+c^2$ is a factor; and, by symmetry, $c+a^2$ and $a+b^2$ are also factors; and proceeding as in former examples, we find $m=1$.

$\therefore$ the expression $=(b+c^2)(c+a^2)(a+b^2)$.

4. Resolve into factors the expression

$$(a-b)^3+(b-c)^3+(c-a)^3.$$

As before, we find that there are no monomial factors. Let $a-b=0$, or $a=b$; and, substituting b for a, the expression becomes zero. Hence, $a-b$ is a factor; by symmetry, $b-c$ and $c-a$ are factors. Hence, the factors are $m(a-b)(b-c)(c-a)$.

To find m, let $a=0$, $b=1$, $c=2$, and we have $6=2m$, or $m=3$.

The factors are, therefore, $3(a-b)(b-c)(c-a)$.

5. Resolve into factors

$$a^3(b-c)+b^3(c-a)+c^3(a-b).$$

As before, we find that there are no monomial factors.

Let $a - b = 0$, or $a = b$; substituting b for a, the expression becomes zero. Therefore, $a - b$ is a factor; by symmetry, $b - c$ and $c - a$ are factors.

Now the product of these three factors is of *three* dimensions, while the expression itself is of *four* dimensions. There must, therefore, be another factor of *one* dimension. It cannot be a monomial factor, for the expression has no such factors. It cannot be a binomial factor, such as $a + b$, for then, by symmetry, $b + c$ and $c + a$ would also be factors, which would give an expression of *six* dimensions. It cannot be a trinomial factor, unless a, b, and c are similarly involved. For instance, if $a - b + c$ were a factor, then, by symmetry, $b - c + a$ and $c - a + b$ would also be factors, and the dimensions would be *six* instead of *four*. The other factor must therefore be $a + b + c$. Hence,

$$a^3(b - c) + b^3(c - a) + c^3(a - b)$$
$$= m(a - b)(b - c)(c - a)(a + b + c).$$

To find m, put $a = 0$, $b = 1$, and $c = 2$, and we have $-6 = 6m$; therefore, $m = -1$.

Hence, the factors are $-(a - b)(b - c)(c - a)(a + b + c)$, or $(a - b)(a - c)(b - c)(a + b + c)$.

6. Prove that

$$a^3 + b^3 + c^3 + 3(a + b)(b + c)(c + a)$$

is exactly divisible by $a + b + c$, and find all the factors.

Let $a + b + c = 0$, or $a = -(b + c)$; substituting this value for a, we have

$$-(b + c)^3 + b^3 + c^3 + 3bc(b + c),$$
$$\text{or} -(b + c)^3 + (b + c)^3,$$

which $= 0$; and therefore $a + b + c$ is a factor.

As before, we find that there are no monomial factors. Since $a+b+c$, the factor already obtained, is of *one* dimension, the other factor must be of *two* dimensions, and as it must be symmetrical with respect to x, y, and z, it must be of the form

$$m(a^2+b^2+c^2)+n(ab+bc+ca),$$

in which m and n are independent of each other, and of a, b, and c.

To determine their values, put $c=0$, so that
$$a^3+b^3+c^3+3(a+b)(b+c)(c+a)$$
$$=(a+b+c)[m(a^2+b^2+c^2)+n(ab+bc+ca)]$$
becomes
$$a^3+b^3+3ab(a+b)=(a+b)[m(a^2+b^2)+nab].$$

But
$$a^3+b^3+3ab(a+b)=(a+b)^3.$$
$$\therefore (a+b)^3=(a+b)[m(a^2+b^2)+nab].$$
$$\therefore (a+b)^2=m(a^2+b^2)+nab.$$
That is,
$$a^2+b^2+2ab=m(a^2+b^2)+nab.$$

Now this is true for *all* values of a and b.
$$\therefore m=1 \text{ and } n=2.$$
$$\therefore a^3+b^3+c^3+3(a+b)(b+c)(c+a)$$
$$=(a+b+c)[a^2+b^2+c^2+2(ab+bc+ca)]$$
$$=(a+b+c)(a+b+c)^2$$
$$=(a+b+c)^3.$$

7. Simplify
$$a(b+c)^2+b(a+c)^2+c(a+b)^2-(a+b)(a-c)(b-c)$$
$$-(a-b)(a-c)(b+c)+(a-b)(b-c)(a+c).$$

Let $a=0$, and the expression becomes
$$bc^2+cb^2+bc(b-c)-bc(b+c)-bc(b-c),$$
which equals zero; therefore a is a factor; by symmetry, b and c are also factors.

The expression is of three dimensions, and abc is of three dimensions, there cannot therefore be any other literal factor.

Hence the expression equals $mabc$.

To find m, let $a = b = c = 1$, and we have

$$4 + 4 + 4 = m; \ m = 12.$$

$\therefore$ the expression $= 12abc$.

In the preceding examples the factors have been *linear*, but the principle applies equally well to those of higher dimensions.   (See Th. II. *Cor.*)

8. Examine whether $x^n + 1$ is a factor of

$$x^{3n} + 2x^{2n} + 3x^n + 2.$$

Let $x^n + 1 = 0$, or $x^n = -1$, and substituting, the expression vanishes; therefore, $x^n + 1$ is a factor.

9. Examine whether $a^2 + b^2$ is a factor of

$$2a^4 + a^3b + 2a^2b^2 + ab^3.$$

Let $a^2 + b^2 = 0$, or $a^2 = -b^2$.   Substituting, we have

$$2b^4 - ab^3 - 2b^4 + ab^3,$$

which $= 0$, and therefore $a^2 + b^2$ is a factor.

10. Prove that $a^3 + b^3$ is a factor of

$$a^5 + a^4b + a^3b^2 + a^2b^3 + ab^4 + b^5.$$

Let $a^3 + b^3 = 0$, or $a^3 = -b^3$.   Substituting, we have

$$- a^2b^3 - ab^4 - b^5 + a^2b^3 + ab^4 + b^5,$$

which $= 0$, and therefore $a^3 + b^3$ is a factor.

Ex. 38.

Resolve into factors:

1. $(x + y + z)^3 - (x^3 + y^3 + z^3).$

2. $bc(b - c) - ca(a - c) - ab(b - a).$

3. $(a^2 - b^2)^3 + (b^2 - c^2)^3 + (c^2 - a^2)^3$.

4. $x(y+z)^2 + y(z+x)^2 + z(x+y)^2 - 4xyz$.

5. $(a+b)^3 - (b+c)^3 + (c-a)^3$.

6. $a(b-c)^3 + b(c-a)^3 + c(a-b)^3$.

7. $(a+b+c)(ab+bc+ca) - abc$.

8. $a^3(c-b^2) + b^3(a-c^2) + c^3(b-a^2) + abc(abc-1)$.

9. $a^2(b+c) + b^2(c+a) + c^2(a+b) + 2abc$.

10. $(a-b)(c-h)(c-k) + (b-c)(a-h)(a-k)$
$+ (c-a)(b-h)(b-k)$.

11. $x^4 y^2 + x^2 y^4 + x^4 z^2 + x^2 z^4 + y^4 z^2 + y^2 z^4 + 2 x^2 y^2 z^2$.

12. $(a-b)^5 + (b-c)^5 + (c-a)^5$.

13. $ab(a+b) + bc(b+c) + ca(c+a) + (a^3 + b^3 + c^3)$.

14. $a^4(c-b^3) + b^4(a-c^3) + c^4(b-a^3) + abc(a^2 b^2 c^2 - 1)$.

15. $x^4(y^2 - z^2) + y^4(z^2 - x^2) + z^4(x^2 - y^2)$.

16. $x^4 + y^4 + z^4 - 2x^2 y^2 - 2y^2 z^2 - 2z^2 x^2$.

17. $(b-c)(x-b)(x-c) + (c-a)(x-c)(x-a)$
$+ (a-b)(x-a)(x-b)$.

18. $(a+b)^3 + (b+c)^3 + (c+a)^3$
$+ 3(a+2b+c)(b+2c+a)(c+2a+b)$.

19. Show that $a^5 + a^2 b^2 - ab^2 - b^3$ has $a^2 - b$ for a factor.

20. Show that $(x+y)^7 - x^7 - y^7$
$= 7xy(x+y)(x^2 + xy + y^2)^2$.

21. Examine whether $x^2 - 5x + 6$ is a factor of
$x^3 - 9x^2 + 26x - 24$.

22. Show that $a - b + c$ is a factor of
$a^2(b+c) - b^2(c+a) + c^2(a+b) + abc$.

23. Show that $a^2 + 3b$ is a factor of
$$a^4 - 4a^3b^3 + 3a^2b^4 + 3a^2b - 12ab^4 + 9b^5,$$
and find the other factor.

24. Find the factors of $a^4(b-c) + b^4(c-a) + c^4(a-b)$.

FACTORING A POLYNOME BY TRIAL DIVISORS.

§ 26. To find, if possible, a rational linear factor of the polynome
$$x^n + bx^{n-1} + cx^{n-2} + \cdots + hx + k$$

in which b, c, ..., h, k, are all integral, substitute successively for x every measure (both positive and negative) of the term k, till one is found, say r, that makes the polynome vanish, then $x - r$ will be a factor of the polynome.

EXAMPLES.

1. Factor $x^3 + 9x^2 + 16x + 4$.

The measures of 4 are ± 1, ± 2, and ± 4. Since every coefficient of the given polynome is positive, the positive measures of 4 need not be tried. Using the others, it will be found that -2 makes the polynome vanish. Thus,

$$
\begin{array}{r|rrrr}
 & 1 & 9 & 16 & 4 \\
-2 & & -2 & -14 & -4 \\
\hline
 & 1 & 7 & 2; & 0
\end{array}
$$

Hence, the factors are $(x+2)(x^2 + 7x + 2)$.

The labor of substitution may often be lessened by arranging the polynome in ascending powers of x, and using the reciprocals of the measures of k instead of the measures themselves. Should a fraction occur during the course of the work, further trial of that measure of k will be needless.

2. Factor $x^3 - 10x^2 - 63x + 60$.

The measures of 60 are ± 1, ± 2, ± 3, ± 4, ± 5, etc. Neither $+1$ nor -1 will make the polynome vanish. Try 2; thus,

$$
\begin{array}{r|rrrr}
 & 60 & -63 & -10 & 1 \\
1 & & 30 & & \\
\hline
2 & 30 & -16\frac{1}{2} & &
\end{array}
$$

A fraction occurring, we need go no further. -2 will also give a fraction, as may easily be seen. Next try 3; thus,

$$
\begin{array}{r|rrrr}
 & 60 & -63 & -10 & 1 \\
1 & & 20 & & \\
\hline
3 & 20 & -14\frac{1}{3} & &
\end{array}
$$

A fraction again occurring, we may stop. -3 will also give a fraction. Next try 4; thus,

$$
\begin{array}{r|rrrr}
 & 60 & -63 & -10 & 1 \\
1 & & 15 & -12 & \\
\hline
4 & 15 & -12 & -5\frac{1}{2} &
\end{array}
$$

-4 will also give a fraction. Next, trying 5, we find it fails, and we then try -5; thus,

$$
\begin{array}{r|rrrr}
 & 60 & -63 & -10 & 1 \\
-1 & & -12 & 15 & -1 \\
\hline
5 & 12 & -15 & 1\,; & 0
\end{array}
$$

The remainder vanishes. The factors are, therefore,
$$(x + 5)(x^2 - 15x + 12).$$

§ 27. When k has a large number of factors, the number that need actually be tried can often be considerably lessened by the following means:

For x substitute successively three or more consecutive terms of the progression ....., 3, 2, 1, 0, -1, -2, -3, ... Let ....., k_3, k_2, k_1, k, k_{-1}, k_{-2}, k_{-3}, ..., denote the corresponding values of the polynome; and let r denote a measure of k positive or negative.

The substitution of r for x need not be tried unless $r-1$ measure k_1, $r-2$ measure k_2, ..., and also $r+1$ measure k_{-1}, $r+2$ measure k_{-2}, ... If no measure of k fulfil these conditions, the polynome will have no linear factor.

If p denote a positive or arithmetical measure of k, the preceding criterion may be conveniently expressed as follows:

1. The substitution of $+p$ for x need not be tried unless $p-1$ measure k_1, $p-2$ measure k_2, ..., and also $p+1$ measure k_{-1}, $p+2$ measure p_{-2}, ...

2. The substitution of $-p$ for x need not be tried unless $p+1$ measure k_1, $p+2$ measure k_2, ..., and also $p-1$ measure k_{-1}, $p-2$ measure k_{-2}, ...

In trying for measures, the signs of k_2, k_1, k, ..., may evidently be neglected.

If k_t vanish, t positive or negative, then $x-t$ will be a factor of the polynome, and should be divided out before proceeding to test for other factors.

EXAMPLES.

1. Find the factors of $x^3 - 10x^2 - 63x + 60$.

Here $k=60$, $k_1=-12$, $k_2=-98$, $k_{-1}=112$, $k_{-2}=138$.

Tabulating the trial measures, we get

	98	1,	2,				
	12	2,	3,	4,			
k;	60	3,	4,	5,	6,	10,	12,
	112	4,			7,		
	138						

98		7,				
12	4,	6,				
k ; 60	3,	4,	5,	6,	10,	
112	2,	4,				
138	1,	3,				

In the upper or positive table, no measure of 60 gives a full column; hence, no positive integer substituted for x will make the given polynome vanish.

In the lower or negative table, 5 is the only measure of 60 that gives a full column; hence, -5 is the only negative integer that need be tried for x. Substituting -5 for x, the polynome vanishes; hence, $x + 5$ is a factor of $x^3 - 10x^2 - 63x + 60$.

In constructing the above tables it is evident that 12 is the highest measure of 60 that need be tried in the upper table, for the next measure, 15, would give 14 as a trial-measure of 12 (the absolute value of k_{-1}), and higher measures would give higher trial-measures. Similarly, 10 is the highest measure that need be tried in the lower table.

Since it can make no difference in the *full* columns which of the lines of measures is made the basis from which to construct these columns, it will be found best to construct the tables by the measures of that one of the k's which has the fewest number of them.

2. Find the factors of $x^4 + 12x^3 - 40x^2 + 67x - 120$.

$k = -120$, $k_1 = -80$, $k_2 = -34$, $k_{-1} = -238$.

Selecting the measures of 34 for trial-measures, and tabulating, we get

34	1,	2,	17,	34,	17,	34,
80	2,				16,	
k ; 120	3,				15,	
238					14,	

Here, in the only column that is full, 15 stands in the
line of 120, the absolute value of k, and as the column is decreasing the sign of the 15 must be minus;
hence, the only measure of k that need be tried is
-15. On substituting -15 for x, we get

$$
\begin{array}{r|rrrrr}
     & -120 & 67 & -40 & 12 & 1 \\
-1   &   & 8 & -5 & 3 & -1 \\
\hline
15   & -8   & 5 & -3 & 1\,; & 0
\end{array}
$$

Hence, the only linear factor of the given polynome is
$x + 15$, and, as is seen from the substitution, the
other factor is $x^3 - 3x^2 + 5x - 8$.

3. Factor $x^4 - 27x^2 + 14x + 120$.

$k = 120,\ k_1 = 108,\ k_2 = 56,\ k_{-1} = 80$.

$$
\begin{array}{r|l|l}
56   & 1, 2, 4, 7, 8, 14, 28, 56, & 4, 7, 8, 14, 28, 56, \\
108 & 2, 3,\quad\quad 9,   & 3, 6,\quad\quad 27, \\
k\,;\ 120 & 3, 4,\quad\quad 10,   & 2, 5, \\
80   & 4, 5,   & 1, 4,
\end{array}
$$

The positive or increasing columns give 3 and 4 to try;
the negative or decreasing columns give -2 and -5.
Using these in order, we get

$$
\begin{array}{r|rrrrr}
    & 120 & 14 & -27 & 0 & 1 \\
1   &   & 40 & 18 & -3 & -1 \\
\hline
3   & 40 & 18 & -3 & -1\,; & 0 \quad\therefore\ x-3 \text{ is a factor.} \\
1   &   & 10 & 7 & 1 \\
\hline
4   & 10 & 7 & 1\,; & 0 & \quad\therefore\ x-4 \text{ is a factor.} \\
-1 &   & -5 & -1 \\
\hline
2   & 5   & 1\,; & 0 & & \quad\therefore\ x+2 \text{ is a factor,}
\end{array}
$$

and there remains $x + 5$, a factor.
Hence, the factors are $(x - 3)(x - 4)(x + 2)(x + 5)$.

4. Factor $x^4 - px^3 + (q-1)x^2 + px - q$.

$k = -q$, $k_1 = 1 - p + (q-1) + p - q = 0$,
$\qquad k_{-1} = 1 + p + (q-1) - p - q = 0$.

Since both k_1 and k_{-1} vanish, the polynome is divisible by both $x - 1$ and $x + 1$.

	1	$-p$	$q-1$	p	$-q$
1		1	$-p+1$	$q-p$	q
	1	$-p+1$	$q-p$	q ;	0
-1		-1	$+p$	$-q$	
	1	$-p$	q ;	0	

Hence, the other factor is $x^2 - px + q$.

5. Factor $x^4 + 2ax^3 + (a^2+a)x^2 + 2a^2x + a^3$.

$k = a^3$, $k_1 = 1 + 2a + (a^2+a) + 2a^2 + a^3 = (a+1)^3$,
$\qquad k_{-1} = 1 - 2a + (a^2+a) - 2a^2 + a^3 = a^3 - a^2 - a - 1$.

The positive measures of k are 1, a, a^2, a^3. Of these 1 may be rejected at once, since neither k_1 nor k_2 vanish; and a^2 and a^3 may also be rejected, since k_1 or $(a+1)^3$ is not divisible by either $a^2 \pm 1$ or $a^3 \pm 1$. But k_1 is divisible by $a+1$, and k_{-1} is divisible by $a - 1$; thus, we need try the substitution of only $-a$ for x.

	1	$2a$	a^2+a	$2a^2$	a^3
$-a$		$-a$	$-a^2$	$-a^2$	$-a^3$
	1	a	a	a^2 ;	0
$-a$		$-a$	0	$-a^2$	
	1	0	a ;	0	

Hence, the factors are $(x+a)^2(x^2+a)$.

6. Factor $x^3 - (a+c)x^2 + (b+ac)x - bc$.

$k = -bc$,
$k_1 = 1 - (a+c) + (b+ac) - bc = 1 - a + b - c + ac - bc$,
$k_{-1} = -1 - (a+c) - (b+ac) - bc = -(1 + a + b + c + ac + bc)$.

The factors of k_1, other than 1, are b and c. k_1 is not divisible by either $b \pm 1$ nor by $c + 1$. However, k_1 is divisible by $c - 1$, and k_{-1} is at the same time divisible by $c + 1$; hence, we need try the substitution of only c for x.

$$
\begin{array}{c|cccc}
 & 1 & -(a+c) & (b+ac) & -bc \\
c & & c & -ac & bc \\
\hline
 & 1 & -a & b\,; &
\end{array}
$$

Hence, the factors are $(x - c)(x^2 - ax + b)$.

Ex. 39.

1. $a^3 - 9a^2 + 16a - 4$.

2. $x^3 - 9x^2 + 26x - 24$.

3. $x^3 - 7x^2 + 16x - 12$.

4. $x^3 - 12x + 16$.

5. $x^3 + 3x^2 + 5x + 3$.

6. $x^4 + 4x^3 + 10x^2 + 12x + 9$.

7. $x^3 - 3x + 2$.

8. $x^4 + 2x^2 + 9$.

9. $m^3 - 3m^2n + 4mn^2 - 2n^3$.

10. $x^3 + 2x^2 + 2$.

11. $m^3 - 5m^2n + 8mn^2 - 4n^3$.

12. $b^3 + b^2c + 7bc^2 + 39c^3$.

13. $m^4 - 4mn^3 + 3n^4$.

14. $a^4 - 7a^3b + 28ab^3 - 16b^4$.

15. $x^3 - 11x^2 + 39x - 45$.

16. $x^3 + 5x^2 + 7x + 2$.

17. $a^3 - 3a^2 - 193a + 195$.

18. $p^3 - 3p^2 - 6p + 8$.

19. $a^4 + 3a^3 - 3a^2 - 7a + 6$.

20. $a^{6n} - 6a^{4n} + 11a^{2n} - 6$.

21. $a^4 - 41a^2b^2 + 16b^4$.

22. $a^4 - a^2b^2 - 2ab^3 + 2b^4$.

23. $p^3 - 4p^2 + 6p - 4$.

24. $x^{3n} + 4x^{2n} - 5$.

25. $y^4 - 5y^3 + 8y^2 - 8$.

26. $a^4 - 2a^3 + 3a^2 - 2a + 1$.

27. $a^3 + a^2b^2 + ab^2 - 3b^3$.

28. $2a^{3n} - a^{2n} - a^n + 2$.

29. $x^4 - 18x^3 + 113x^2 - 288x + 252$.

30. $x^4 - 9x^3y + 29x^2y^2 - 39xy^3 + 18y^4$.

§ **28.** To find, if possible, a rational linear factor of the
polynome $ax^n + bx^{n-1} + cx^{n-2} + \cdots + hx + k,$
in which $a, b, c \ldots h, k$ are all integral.

First Method. Multiply the polynome by a^{n-1}.

$$(ax)^n + b(ax)^{n-1} + ac(ax)^{n-2} + \cdots + a^{n-2}h(ax) + a^{n-1}k;$$

or, writing y for ax,

$$y^n + by^{n-1} + acy^{n-2} + \cdots + a^{n-2}hy + a^{n-1}k.$$

Factor this polynome by the method of the last article,
replace y by ax, and divide the result by a^{n-1}.

EXAMPLE.

1. Factor $3x^4 + 5x^3 - 33x^2 + 43x - 20$.

Multiply by 3^3, and express in terms of $3x$.

$$(3x)^4 + 5(3x)^3 - 99(3x)^2 + 387(3x) - 540; \text{ or,}$$
$$y^4 + 5y^3 - 99y^2 + 387y - 540.$$

Here $k = -540$, $k_1 = 1 + 5 - 99 + 387 - 540 = -246$,
$k_{-1} = 1 - 5 - 99 - 387 - 540 = -1030$.

246	1, 2, **3**, 6, 41, 82, 123, 246.
540	2, 3, **4**,
1030	**3**,

246	3, 6, 41, ...	(Trying by factors of 246,
540	**2**, 5,	instead of by factors of
1030	**1**,	540, for convenience.)

The only factors of 540 in full columns are **4** in the
upper table and **2** in the lower one; hence, we need
try only the substitutions 4 and -2.

	-540	387	-99	5	1
1		-135	63	-9	-1
4	-135	63	-9	-1;	0

Hence, $y-4$ is a factor. The substitution of -2 need not now be tried, since we see that 135 is not a multiple of 2. The other factor is, therefore,

$$y^3 + 9y^2 - 63y + 135.$$

Replacing y by $3x$, and dividing by 27,

$$\tfrac{1}{27}(3x-4)(27x^3 + 81x^2 - 189x + 135)$$
$$= (3x-4)(x^3 + 3x^2 - 7x + 5),$$

which are the factors.

§ 29. *Second Method.* Write m for "a measure of a," and r for "a measure of k, positive or negative":

For x substitute every value of $r \div m$ till one, say $r' \div m'$, be found which makes the polynome vanish; then $m'x - r'$ will be a factor. Should a fraction be met with in the course of substitution, further trial of that value of $r \div m$ will be useless.

Should k have more factors than a, it will generally be better to arrange the polynome in ascending powers of x and use values of $m \div r$ instead of $r \div m$, making r positive and m positive or negative.

To reduce the number of trial measures, for x substitute successively three or more consecutive terms of the progression ....., 3, 2, 1, 0, -1, -2, -3, ..., then denoting the corresponding values of the polynome by ....., k_3, k_2, k_1, k, k_{-1}, k_{-2}, k_{-3}, ...

The substitution of r for x need not be tried unless $r-m$ measure k_1, $r-2m$ measure k_2, ..., and also $r+m$ measure k_{-1}, $r+2m$ measure k_{-2}, ...

If p denote a positive or arithmetical measure of k, this criterion may be expressed as follows:

1. The substitution of $+p$ for x need not be tried unless $p-m$ measure k_1, $p-2m$ measure k_2, ..., and also $p+m$ measure k_{-1}, $p+2m$ measure k_{-2}, ...

2. The substitution of $-p$ for x need not be tried unless $p + m$ measure k_1, $p + 2m$ measure k_2, ..., and also $p - m$ measure k_{-1}, $p - 2m$ measure k_{-2}, ...

It must be remembered that here m may be either positive or negative, as may also be one or more of the quantities, $p + m$, $p - m$, $p + 2m$, $p - 2m$, etc.

EXAMPLES.

1. Factor $36x^3 + 171x^2 - 22x + 480$.

$$k = 480, \qquad k_1 = 665, \qquad k_2 = 1408,$$
$$k_{-1} = 637, \qquad k_{-2} = 920,$$

and m may have any of the values,

$$\pm 1, \pm 2, \pm 3, \pm 4, \pm 6, \pm 9, \pm 12, \pm 18, \pm 36.$$

In forming the table of trial-measures, write out the measures of 1408, that is, k_2; they are

1, 2, 4, 8, 11, 16, 22, 32, 44, 64, 88, 128, 176, 352, 1408.

Taking each of these in succession, add to it each value of m separately. Should the sum appear among the measures of 665, that is, k_1, which are

1, 5, 7, 19, 35, 95, 133, 665,

enter these measures of k_2 and k_1 in a column in the table, writing above them the value m used. However, should the sum not be a measure of 665, another value of m must be tried. When all the values of m have been tried with one measure of 1408, another measure must be taken till all have been used. This having been done, proceed to test which of the columns can be filled up with measures of 480, 637, and 920, respectively, these being the values in this case of k, k_{-1}, k_{-2}.

The table will then appear thus:

m;	+4,	+6,	+3,	+1,	+3,	+3,	+3,
1408	1,	1,	2,	4,	4,	16,	32,
665	5,	7,	5,	5,	7,	19,	35,
k; 480			8,	6,	10,		
637				7,	13		
920				8,			
1113	. . . .						

m;	−2,	−6,	−3,	−9,	−3,	−9,	−1,	−3,	−9,	−4,	−6,	−9,	−3,
1408	1,	1,	2,	2,	4,	4,	8,	8,	8,	11,	11,	16,	22,
665	−1,	−5,	−1,	−7,	1,	−5,	7,	5,	−1,	7,	5,	7,	19,
k; 480	−3,		−4,	−16,	−2,		6,	2,	−10,	3,	−1,	−2,	16,
637			−7,					−1,		−1,	−7,		13,
920			−10,					−4,		−5,			10,
1113	. . . .						−7,						7,

There still remain five full columns, while the given polynome, being of the third degree, cannot have more than three linear factors. To reduce the number of these columns, and, as a consequence, the number of trial-measures, extend the table by calculating k_{-3} and the corresponding column-numbers for the full columns. $k_{-3} = 1113$, and the column-numbers are 9, −13, −7, −9, and 7. Of these, 9, −13, and −9 must be rejected, not being measures of 1113. This leaves only −7 and 7, to which correspond $\dfrac{-3}{2}$ and $\dfrac{-3}{16}$ as the values of $\dfrac{m}{r}$ to be tried in substitution for $\dfrac{1}{x}$. (See table above.) Making trial of these two, the polynome is found to vanish for $\dfrac{-3}{16}$ but not for $\dfrac{-3}{2}$.

The actual work of substitution will be as follows.

Arrangement in ascending powers of x:

	480	-22	171	36
-3		-720	1113	-1926
2	240	-371	642 ;	-945
	480	-22	171	36
-3		-90	21	-36
16	30	-7	12 ;	0

Hence, the factors are $3x+16$ and $12x^2 - 7x + 30$. The latter factor cannot be resolved, for it does not contain $3x+16$, and the only other factor, viz., $3x+2$, left for trial by the tables above, has been tried and has failed.

2. Factor $10x^4 - x^3(15y+4z) - x^2(40y^2 - 6yz) + x(60y^3 + 16y^2z) - 24y^3z$.

Here $m = \pm 1, \pm 2, \pm 5,$ or ± 10.

$$k = -24y^3z.$$
$$k_1 = 10 - (15y+4z) - (40y^2 - 6yz)$$
$$+ (60y^3 + 16y^2z) - 24y^3z$$
$$= 10 - 15y - 40y^2 + 60y^3$$
$$- 2z(2 - 3y - 8y^2 + 12y^3)$$
$$= (5 - 2z)(2 - 3y - 8y^2 + 12y^3).$$
$$k_{-1} = (5 + 2z)(2 + 3y - 8y^2 - 12y^3),$$

as may easily be found by making the calculation.

We get at a glance $2z$ a factor of k, $2z-5$ a factor of k_1, and $2z+5$ a factor of k_{-1}; hence, taking $m = 5$, we are directed to try the substitution $\dfrac{2z}{5}$ for x.

	$10, -(15y+4z),$	$-(40y^2-6yz),$	$(60y^3+16y^2z),$	$-24y^3z$
$2z$		$4z$	$-6yz$	$-16y^2z$ $24y^3z$
5	2 $-3y$	$-8y^2$	$12y^3$;	0

Hence, $5x - 2z$ is one factor, the other being

$$2x^3 - 3x^2y - 8xy^2 + 12y^3.$$

Seeking to determine the factors of this, we obtain

$$m = \pm 1 \text{ or } \pm 2, \ k = 12, \ k_1 = 3, \ k_2 = 0,$$
$$k_{-1} = 15, \ k_{-2} = 0.$$

The vanishing of k_2 shows that $x - 2y$ is a factor; and the vanishing of k_{-2} shows that $x + 2y$ is also a factor. Dividing these out, the remaining factor is found to be $2x - 3y$; so that the proposed polynome resolves into

$$(5x - 2z)(x - 2y)(x + 2y)(2x - 3y).$$

The factor $5x - 2z$ might easily have been got by the method of § 23, but the present solution shows we are independent of that section. It may also be obtained by rearranging the polynome in terms of y.

Ex. 40.

Factor :

1. $2x^3 - 20x^2 + 38x - 20$; $x^3 - 7x^2y + 16xy^2 - 12y^3$.

2. $12x^3 + 5x^2y + xy^2 + 3y^3$; $8x^3 - 14x + 6$.

3. $3x^3 - 15ax + a^2x - 5a^3$; $2x^3 + 9x^2y + 7xy^2 - 3y^3$.

4. $2b^4 - 7b^3c - 4b^2c^2 + bc^3 - 4c^4$;
    $15a^3 + 47a^2b + 13ab^2 - 12b^3$.

5. $4p^4 + 8p^3q + 7p^2q^2 + 8pq^3 + 3q^4$.

6. $150x^4 - 725x^3y + 931x^2y^2 + 920xy^3 - 1152y^4$.

7. $36x^4 - 6(9-7y)x^3 - 7(9+14y)x^2y + 3(49-40y)xy^2 + 180y^3$.

8. $10x^4 - x^3(15y+4z) + x^2(40y^2+6yz) + x(60y^3-16y^2z) - 24y^3z$.

§ 30. If the polynome $ax^n + bx^{n-1} + \cdots + hx + k$, in which $a, b, \ldots, h, k$ are all integral and n greater than 3, have no rational linear factors, it may have rational quadratic factors. Let m denote a positive measure of a, and r denote a measure, positive or negative, of k. The rational quadratic factors of the polynome, if there be any, must be of the form $mx^2 + qx - r$. To determine such factors we may proceed as follows :

For x substitute successively three or more consecutive terms of the progression ....., 3, 2, 1, 0, -1, -2, -3, ..., and denote the corresponding values of the polynome by ....., k_3, k_2, k_1, k, k_{-1}, k_{-2}, k_{-3}, ... Let r_3 denote "a measure of k_3, positive or negative"; r_2 denote "a measure of k_2, positive or negative"; etc. Then $mx^2 + qx - r$ need not be tried as a factor of the polynome, unless an arithmetical progression with q as common difference can be formed from among the values of ....., $9m + r_3$, $4m + r_2$, $m + r_1$, r, $m + r_{-1}$, $4m + r_{-2}$, $9m + r_{-3}$, ..., in which the coefficients of m are the *squares* of the terms of the series ....., 3, 2, 1, 0, -1, -2, -3, ...

EXAMPLES.

1. Factor $x^4 - 3x^3 - 13x^2 + 36x - 18$.

$$m = 1, \qquad k = -18,$$
$$k_1 = 3, \qquad k_2 = -6, \qquad k_3 = -9,$$
$$k_{-1} = -63, \quad k_{-2} = -102, \quad k_{-3} = -81, \quad k_{-4} = 78.$$

Trying for rational linear factors as by § 28, it will be found there are none. We therefore proceed to seek for rational quadratic factors. To do this, we first tabulate the arithmetical values of r_3, r_2, r_1, ...

9	1,	3,	9,						$9m$
6	1,	2,	3,	6,					$4m$
3	1,	3,							m
$k:$ 18	1,	2,	3,	6,	9,	18,			0 $\quad m = 1$
63	1,	3,	7,	9,	21,	63,			m
102	1,	2,	3,	6,	17,	34,	51,	102,	$4m$
81	1,	3,	9,	27,	81,				$9m$
78	1,	2,	3,	6,	13,	26,	39,	78,	$16m$

Taking these *both positive and negative*, we next tabulate the values of $9m+r_3$, $4m+r_2$, $m+r_1$, ... This
done, we then proceed to select and arrange in columns any arithmetical progressions that are found to
run completely through the table, one term of the
progression in each line of the table in regular order,
thus:

	0,	6,	8,	10,	12,	18,								0,	12
	−2,	1,	2,	3,	5,	6,	7,	10,						2,	7
	−2,	0,	2,	4,										4,	2
r;	−18,	−9,	−6,	−3,	−2,	−1,	1,	2,	3,	6,	9,	18,		6,	−3
	−62,	−20,	−8,	−6,	−2,	0,	2,	4,	8,	10,	22,	64,		8,	−8
	−98,	−47,	−30,	−13,	−2,	1,	2,	3,	5,	6,	7,	10,	21, 38, 55, 106,	10,	−13
	−72,	−18,	0,	6,	8,	10,	12,	18,	36,	90,				12,	−18
	−62,	−23,	−10,	3,	10,	13,	14,	15,	17,	18,	19,	22,	29, 42, 55, 94,	14,	−23

There are two columns of progressions: in the first, $r=6$
and q or the common difference is 2, giving the trial
factor x^2+2x-6; in the second column, $r=-3$ and
$q=-5$, thus giving the trial factor x^2-5x+3. On
actual trial, it will be found these are the factors of
$x^4-3x^3-13x^2+36x-18$.

2. Factor $6x^5-53x^4+179x^3-299x^2+260x-96$.

Here m may be 1, or 2, or 3, or 6.

$$k=-96, \quad k_1=-3, \quad k_2=4, \quad k_3=-9, \quad k_4=-32,$$
$$k_{-1}=-893.$$

The factors of k, k_1, ..., are:

32	1, 2, 4, 8, 16, 32,		16m
9	1, 3, 9,		9m
4	1, 2, 4,		4m
3	1, 3,		m
k; 96	1, 2, 3, 4, 6, 8, 12, 16, 24, 32, 48, 96,		0
893	1, 19, 47, 893,		m

As this table yields no complete column in arithmetical progression, the given polynome has no rational linear factor. (§ 29.)

Forming the table for $m=1$, it will be found that it also does not yield any trial divisor.

The table for $m=2$ is:

	0,	16,	24,	28,	30,	31, 33, 34, 36, 40, 48, 61,								28
	9,	15,	17,	19,	22,	27,									19
	4,	6,	7,	9,	10,	12,									10
	−1,	1,	3,	5,											1
$r;$	−96,	−48,	−32,	−24,	−16,	−12,	−8,	−6,	−4,	−3,	−2,	−1, 1, 2, 3,			−8
	−891,	−45,	−17,	1,	3,	21,	49,	895,							−17

This gives the complete column set out at the right. In it, $r=-8$, and the common difference is -9; hence, we have $2x^2 - 9x + 8$ as a trial divisor. On actual trial it will be found to be a factor of the given polynome, the co-factor being $3x^3 - 13x^2 + 19x - 12$.

Ex. 41.

Factor:

1. $x^4 - 12x^3 + 47x^2 - 66x + 27$.

2. $x^4 - 6x^3 - 2x^2 + 36x - 24$.

3. $x^4 - 2x^3 - 25x^2 + 18x + 24$.

4. $x^5 - 31x^3 + 186x - 180$.

5. $1 - 45x^2 + 32x^3 + 281x^4 - 518x^5 + 252x^6$.

6. $a^6 - 38a^4y^2 + 28a^3y^3 + 345a^2y^4 - 564ay^5 + 180y^6$.

7. $2x^4 - 5x^3 - 17x^2 + 53x - 28$.

8. $6x^6 - 53x^5 + 83x^4 + 45x^3 - 257x^2 + 32x + 15$.

9. $6 - 47y + 108y^2 - 74y^3 + 12y^4$.

10. $6x^5 - 17x^4 + 5x^3 + 13x^2 - 2x - 2$.

CHAPTER IV.

MEASURES AND MULTIPLES.

§ 31. When one quantity is to be divided by another, the quotient can often be readily obtained by resolving the divisor or dividend, or *both*, into factors.

EXAMPLES.

1. Divide $a^2 - 2ab + b^2 - c^2 + 2cd - d^2$ by $a - b + c - d$.

Here we see at once that the dividend
$$= (a - b)^2 - (c - d)^2,$$
and hence quotient $= a - b - (c - d) = a - b - c + d$.

2. Divide the product of $a^2 + ax + x^2$ and $a^3 + x^3$
by $a^4 + a^2 x^2 + x^4$.

Here $a^3 + x^3 = (a + x)(a^2 - ax + x^2)$, and the divisor
$$= (a^2 + ax + x^2)(a^2 - ax + x^2).$$
Hence, the quotient is $a + x$.

3. Divide $a^3 + a^2 b + a^2 c - abc - b^2 c - bc^2$ by $a^2 - bc$.

The dividend is $a(a^2 - bc) + b(a^2 - bc) + c(a^2 - bc)$.
Hence the quotient $= a + b + c$.

4. $(a^3 + b^3 - c^3 + 3abc) \div (a + b - c)$.

Dividend $= a^3 + b^3 + 3ab(a + b) - c^3 - 3ab(a + b) + 3abc$
$$= (a + b)^3 - c^3 - 3ab(a + b - c),$$
which is exactly divisible by $a + b - c$.
Quotient $= a^2 + b^2 + c^2 - ab + bc + ca$.

5. Divide $x^5 - x^4y + x^3y^2 - x^2y^3 + xy^4 - y^5$ by $x^3 - y^3$.

The dividend is (§ 24) evidently $(x^6 - y^6) \div (x + y)$, and this divided by $x^3 - y^3$

$$= (x^3 + y^3) \div (x + y) = x^2 - xy + y^2.$$

6. Divide $b(x^3+a^3)+ax(x^2-a^2)+a^3(x+a)$ by $(a+b)(x+a)$.

Striking the factor $x + a$ out of dividend and divisor, we have $b(x^2 - ax + a^2) + ax(x - a) + a^3$
$$= b(x^2 - ax + a^2) + a(x^2 - ax + a^2)$$
$$= (a + b)(x^2 - ax + a^2).$$

Hence, quotient $= x^2 - ax + a^2$.

7. Divide $apx^4 + x^3(aq + bp) + x^2(ar + bq + pc)$
$\qquad + x(qc + br) + cr$ by $ax^2 + bx + c$.

Factoring the dividend (§ 22), we have
$$(ax^2 + bx + c)(px^2 + qx + r).$$

Hence the quotient equals the latter factor.

8. Divide $6x^4 - 13ax^3 + 13a^2x^2 - 13a^3x - 5a^4$
$\qquad$ by $2x^2 - 3ax - a^2$.

This can be done by § 15. The divisor is $2x^2 - a^2 - 3ax$, and we see at once that $3x^2 + 5a^2$ must be two terms of the quotient.

Multiplying diagonally into the first two terms of the divisor, and adding the products, we get $+ 7a^2x^2$; but $+ 13a^2x^2$ is required. Hence, $+ 6a^2x^2$ is still required, and as this must come from the third term multiplied into $- 3ax$, that third term must be $- 2ax$. Therefore, the quotient is $3x^2 + 5a^2 - 2ax$.

NOTE. By multiplying the terms $- 2ax$, $- 3ax$, diagonally into the x^2's and a^2's, respectively, we get the remaining terms of the dividend. It is, of course, necessary to test whether the division is exact.

9. Divide $2a^4 - a^3b - 12a^2b^2 - 5ab^3 + 4b^4$ by $a^2 - b^2 - 2ab$.

Here, as before, one factor is $a^2 - b^2 - 2ab$; hence *two* terms of the other factor are $2a^2 - 4b^2$. Multiplying, as in the last example, we get $-6a^2b^2$; but $-12a^2b^2$ is required. Hence, $-6a^2b^2$ is still needed, and $+3ab$ is the third term of the required quotient, which is therefore $2a^2 - 4b^2 + 3ab$.

10. Prove that
$$(1 + x + x^2 + \cdots + x^{n-1})(1 - x + x^2 - \cdots + x^{n-1})$$
$$= 1 + x^2 + x^4 + \cdots + x^{2n-2}.$$

Product
$$= \frac{1 - x^n}{1 - x} \times \frac{1 + x^n}{1 + x}$$
$$= \frac{1 - x^{2n}}{1 - x^2} = 1 + x^2 + x^4 + \cdots + x^{2n-2}.$$

11. Divide $(a^2 - bc)^3 + 8b^3c^3$ by $a^2 + bc$.
$$= (a^2 - bc)^3 + (2bc)^3 \div (a^2 - bc) + 2bc$$
$$= (a^2 - bc)^2 - (a^2 - bc) \times 2bc + (2bc)^2$$
$$= a^4 - 4a^2bc + 7b^2c^2.$$

12. Divide $1 + 2,357,947,691 x^9$ by $1 - 11x + 121x^2$.

Dividend $= 1 + (11x)^9$
$$= [1 - (11x)^3 + (11x)^6][1 + (11x)^3].$$
Divisor $= [1 + (11x)^3] \div (1 + 11x)$.
$\therefore$ quotient $= [1 - (11x)^3 + (11x)^6](1 + 11x)$.

Ex. 42.

Find the quotients in the following cases :

1. $1 - x + x^2 - x^3 \div 1 - x$.

2. $1 - 2x^4 + x^8 \div x^4 + 2x^2 + 1$.

3. $x^{16} + a^8 x^8 + a^{16} \div x^4 - a^2 x^2 + a^4.$

4. $x^4 + 4 x^2 y^2 - 32 y^4 \div x - 2 y.$

5. $1 - 4 x^2 + 12 x^3 - 9 x^4 \div 1 + 2 x - 3 x^2.$

6. $(a^2 - 2 ax + x^2)(a^3 + 3 a^2 x + 3 ax^2 + x^3) \div a^2 - x^2.$

7. $x^3 - y^3 + z^3 + 3 xyz \div x - y + z.$

8. $6 a^4 - a^3 b + 2 a^2 b^2 + 13 ab^3 + 4 b^4 \div 2 a^2 - 3 ab + 4 b^2.$

9. $4 x^4 - x^2 y^2 + 6 xy^3 - 9 y^4 \div 2 x^2 + 3 y^2 - xy.$

10. $a^4 + b^4 - c^4 - 2 a^2 b^2 \div a^2 - b^2 - c^2.$

11. $21 a^4 - 16 a^3 b + 16 a^2 b^2 - 5 ab^3 + 2 b^4 \div 3 a^2 - ab + b^2.$

12. $2 a^3 - 7 a^2 - 46 a - 21 \div 2 a^2 + 7 a + 3.$

13. $[a^3(b - c) + b^3(c - a) + c^3(a - b)] \div a + b + c.$

14. $x^3 - 3 ax^2 + 3 a^2 x - a^3 + b^3 \div x - a + b.$

15. $x^4 - y^4 + z^4 + 2 x^2 z^2 - 2 y^2 - 1 \div x^2 - y^2 + z^2 - 1.$

16. $x^4 - (a + c) x^3 + (b + ac) x^2 - bcx \div x - c.$

17. $x^3 + x^2 y + xy^2 + y^3 \div x + y.$

18. $x^7 - x^6 y + x^5 y^2 - x^4 y^3 + x^3 y^4 - x^2 y^5 + xy^6 - y^7 \div x^4 + y^4.$

19. $a^4 + b^4 - c^4 - 2 a^2 b^2 - 2 c^2 - 1 \div a^2 - b^2 - c^2 - 1.$

20. $a^4 - ab^3 - ac^3 - 2 a^3 b + 2 b^4 + 2 bc^3 + 3 a^3 c - 3 b^3 c - 3 c^4$
$\div a + 3 c - 2 b.$

21. $a^2 b - bx^2 + a^2 x - x^3 \div (x + b)(a - x).$

22. $a(b - c)^3 + b(c - a)^3 + c(a - b)^3 \div a^2 - ab - ac + bc.$

23. $a^2 b^2 + 2 abc^2 - a^2 c^2 - b^2 c^2 \div ab + ac - bc.$

24. $x^3 + y^3 + 3 xy - 1 \div x + y - 1.$

25. $x^6 - x^3 - 2 \div x^2 - x + 1.$

26. $a^4 - 29\,a^2 - 50\,a - 21 \div a^2 - 5\,a - 7.$

27. $(2x - y)^2 a^4 - (x + y)^2 a^2 x^2 + 2(x + y)\,ax^4 - x^6$
$\div (2x - y)\,a^2 + (x + y)\,ax - x^3.$

28. $(x^3 - 1)\,a^3 - (x^3 + x^2 - 2)\,a^2 + (4x^2 + 3x + 2)\,a - 3\,(x + 1)$
$\div (x - 1)\,a^2 - (x - 1)\,a + 3.$

§ 32. The **Highest Common Factor** of two algebraic quantities may, in general, be readily found by factoring. The H.C.F. is often discovered by taking the sum or the difference (or sum *and* difference) of the given expressions, or of some multiples of them.

EXAMPLES.

1. Find the H.C.F. of $(b - c)\,x^2 + (2ab - 2ac)\,x + a^2 b - a^2 c$
 and $(ab - ac + b^2 - bc)\,x + a^2 c + ab^2 - a^2 b - abc.$

Taking out the common factor $b - c$, we get

$(b - c)(x^2 + 2ax + ab)$ and $(b - c)[(a - b)x - a^2 + ab].$

Therefore, $b - c$ is the H.C.F. of the given expressions.

2. Find the H.C.F. of $1 - x + y + z - xy + yz - zx - xyz$
 and $1 - x - y - z + xy + yz + zx - xyz.$

Their difference is

$2y + 2z - 2xy - 2zx = 2(1 - x)\,(y + z).$

Their sum is

$2 - 2x + 2yz - 2xyz = 2(1 - x)(1 + yz).$

Therefore, the H.C.F. is $(1 - x).$

3. Find the H.C.F. of $x^5 + 3\,x^4 - 8x^2 - 9x - 3$
 and $x^5 - 2x^4 - 6x^3 + 4x^2 + 13x + 6.$

The annexed method of finding the H.C.F. depends on the principle that, if a quantity measures two other quantities, it will measure any multiple of their sum or difference.

1	$+3$	0	-8	-9	-3	(a)
1	-2	-6	$+4$	$+13$	$+6$	(b)
	5	$+6$	-12	-22	-9	$(c)\ [=(a)-(b)]$

2	$+6$	0	-16	-18	-6	$(a)\times 2$
1	-2	-6	$+4$	$+13$	$+6$	(b)
3	$+4$	-6	-12	-5		(d)

15	$+18$	-36	-66	-27	$(c)\times 3$
15	$+20$	-30	-60	-25	$(d)\times 5$
	-2	-6	-6	-2	
	1	$+3$	$+3$	$+1$	(f)

25	$+30$	-60	-110	-45	$(c)\times 5$
27	$+36$	-54	-108	-45	$(d)\times 9$
	-2	-6	-6	-2	
	1	$+3$	$+3$	$+1$	(g)

H.C.F. $=(x+1)^3$.

The coefficients are written in two lines, (a) and (b). They are then subtracted so as to cancel the first terms. (a) is next multiplied by 2, and added to cancel the last terms. If (c) and (d) had been the same, their terms would have been the coefficients of the H.C.F. Since they are not, we proceed with them as with (a) and (b) till they become the same. When (a) and (b) do not contain the same number of terms, it is more convenient to find only (c), and then use this with the quantity containing the same number of terms. The general rule is to operate on lines containing the same, or nearly the same, number of terms.

4. Find the H.C.F. of

$$3x^3 + 2x^2 - 14x + 8 \text{ and } 6x^3 - 11x^2 + 13x - 12.$$

3	$+2$	-14	$+8$	(a)
6	-11	$+13$	-12	(b)

6	$+4$	-28	$+16$	$(a) \times 2$

15	-41	$+28$	(c)	$(b) - (a)$

$$(5 - 7)(3 - 4)$$

$$\text{H.C.F.} = 3x - 4. \qquad (d)$$

If (a) and (b) have a common factor, its first term must measure 3 and 6, and its last term must measure 8 and 12. (c) is not, therefore, the H.C.F. Resolve (c) into factors. $5x - 7$ is not a factor of (a) and (b). If, therefore, (a) and (b) have a common factor, it is $3x - 4$. On trial, $3x - 4$ is found to be a factor of (a), and, therefore, it is the H.C.F. of (a) and (b).

5. If $x^2 + px + q$ and $x^2 + rx + s$ have a common factor, prove that this factor is

$$x + \frac{q - s}{p - r}.$$

If $x - a$ be the common factor, then the remainders, on dividing the given expressions by $x - a$, must be zero; that is, $a^2 + pa + q = 0$,

and $a^2 + ra + s = 0$, or $(p - r)a = s - q$.

Hence, $a = \dfrac{s - q}{p - r}$, and $x - a = x - \dfrac{s - q}{p - r} = x + \dfrac{q - s}{p - r}$.

6. What value of a will make $a^2x^2 + (a + 2)x + 1$, and $a^2x^2 + a^2 - 5$ have a common measure.

They cannot have a monomial factor. Neither can they have one of two dimensions unless $(a + 2)$ vanishes; that is, unless $a = -2$, in which case the expressions become $4x^2 + 1$ and $4x^2 - 1$, which have no common factor. Hence, if the given quantities have a common factor, it must be of the form $x + m$; dividing $a^2x^2 + a^2 - 5$ by $x + m$, we have for remainder

$$a^2m^2 + a^2 - 5 = 0 \text{ or } m^2 = \frac{5 - a^2}{a^2}; \therefore m = \frac{1}{a}\sqrt{(5 - a^2)},$$

in which $\sqrt{(5 - a^2)}$ must be possible and integral; hence, $a^2 = 4$ ($a^2 = 1$ gives values to m which on *trial* fail) and $a = \pm 2$, of which the positive value must be taken; and, therefore, $2x + 1$ is the common factor.

7. If the H.C.F. of a and b be c, the L.C.M. of
$$(a + b)(a^3 - b^3) \text{ and } (a - b)(a^3 + b^3) \text{ is } \frac{a^6 - b^6}{c^2}.$$

Let $a = mc$, $b = nc$, and $\therefore a^3 = m^3c^3$, $b^3 = n^3c^3$.

Thus $(a + b) = c(m + n)$; $(a - b) = c(m - n)$,
  and $(a^3 + b^3) = c^3(m^3 + n^3)$; $(a^3 - b^3) = c^3(m^3 - n^3)$.

Hence, $(a + b)(a^3 - b^3) = c^4(m + n)(m^3 - n^3)$,
  and $(a - b)(a^3 + b^3) = c^4(m - n)(m^3 + n^3)$.

The H.C.F. of the last expressions is $c^4(m^2 - n^2)$; hence,

the L.C.M. $= c^4(m^6 - n^6) = \dfrac{c^6(m^6 - n^6)}{c^2} = \dfrac{a^6 - b^6}{c^2}.$

8. If $(x - a)^2$ measures $x^3 + qx + r$, find the relation between q and r.

Let $x + m$ be the other factor; then
$$x^3 + qx + r = (x - a)^2(x + m)$$
$$= x^3 + (m - 2a)x^2 + (a^2 - 2am)x + ma^2.$$

Equating coefficients, $m - 2a = 0$, $a^2 - 2am = q$, $ma^2 = r$.

Hence, $m = 2a$, and $\therefore a^2 - 4a^2 = q$, $2a^3 = r$,

and $a^2 = -\dfrac{q}{3}$, or $a^6 = -\dfrac{q^3}{27}$; and $a^3 = \dfrac{r}{2}$ or $a^6 = \dfrac{r^2}{4}$.

Therefore, $\dfrac{r^2}{4} = -\dfrac{q^3}{27}$, or $\dfrac{r^2}{4} + \dfrac{q^3}{27} = 0$.

Or thus: Dividing $x^3 + qx + r$ by $(x-a)^2$ we find the remainder $(q + 3a^2)x + r - 2a^3$, and as this will be the same for *all* values of x, we have, by equating coefficients,

$q + 3a^2 = 0$, and $r - 2a^3 = 0$,

or $q^3 - 27a^6$ and $r^2 = 4a^6$;

therefore $\dfrac{r^2}{4} + \dfrac{q^3}{27} = 0$, as before.

Ex. 43.

Find the H.C.F. of the following:

1. $2x^4 + 3x^3 + 5x^2 + 9x - 3$; $3x^4 - 2x^3 + 10x^2 - 6x + 3$.

2. $x^3 + (a+1)x^2 + (a+1)x + a$; $x^3 + (a-1)x^2 - (a-1)x + a$.

3. $px^3 - (p-q)x^2 + (p-q)x + q$; $px^3 - (p+q)x^2 + (p+q)x - q$.

4. $ax^3 - (a-b)x^2 - (b-c)x - c$; $2ax^3 + (a+2b)x^2 + (b+2c)x + c$.

5. $1 - 3\frac{2}{5}x - 3\frac{1}{3}x^2 + \frac{1}{3}x^3 - x^4$; $1 - 1\frac{1}{15}x - 3x^2 + 1\frac{1}{15}x^3 + x^4$.

6. $ac^{2a} + bc^{2b} + (a+b)c^{a+b}$; $a^c c^a + a^c c^b + c^a b^c + b^c c^b$.

7. $a^2x^3 + a^5 - 2abx^3 + b^2x^3 + a^3b^2 - 2a^4b$
   and $2a^2x^4 - 5a^4x^2 + 3a^6 - 2b^2x^4 + 5a^2b^2x^2 - 3a^4b^2$.

8. $(ax + by)^2 - (a-b)(x+z)(ax+by) + (a-b)^2 xz$
   and $(ax - by)^2 - (a+b)(x+z)(ax-by) + (a+b)^2 xz$.

9. $a(b^2 - c^2) + b(c^2 - a^2) + c(a^2 - b^2)$
   and $a(b^3 - c^3) + b(c^3 - a^3) + c(a^3 - b^3)$.

10. $a + a^{2m} + a^m + 1$ and $a^{3m} - a^{2m} + a^m - 1$.

11. If $x^3 + ax^2 + bx + c$ and $x^2 + a'x + b'$ have a common factor of one dimension in x, it must be one of the factors of $(a - a')x^2 + (b - b')x + c$.

12. Determine the H.C.F. of $(a-b)^5 + (b-c)^5 + (c-a)^5$ and $(a^2 - b^2)^5 + (b^2 - c^2)^5 + (c^2 - a^2)^5$.

13. Find the H.C. F. of
$$2(y^3-2y^2-y+2)x^3+3(y^2-1)x^2-(2y^3-y^2-2y+1)$$
and $3(y^3-4y^2+5y-2)x^2$
$$+7(y^2-2y+1)x-(3y^3-5y^2+y+1).$$

14. If x^2+px+q and x^2+mx+n have a common linear factor, show that
$$(n-q)^2+n(m-p)^2=m(m-p)(n-q).$$

15. Find the L.C.M. of x^3-3x^2+3x-1, x^3-x^2-x+1, x^4-2x^3+2x-1, and $x^4-2x^3+2x^2-2x+1$.

16. Find the L.C.M. of
$$x^3+6x^2+11x+6, \quad x^3+7x^2+14x+8,$$
$$x^3+8x^2+19x+12, \text{ and } x^3+9x^2+26x+24.$$

17. Find the value of y which will make
$$2(y^2+y)x^2+(11y-2)x+4 \text{ and}$$
$$2(y^3+y^2)x^3+(11y^2-2y)x^2+(y^2+5y)x+5y-1$$
have a common measure.

18. The product of the H.C.F. and L.C.M of two quantities is equal to half the sum of their squares; one of them is $2x^3-11x^2+17x-6$; find the other.

19. If $x+a$ and $x-a$ are both measures of x^3+px^2+qx+r, show that $pq=r$.

20. If x^3+qx+r and x^3+mx+n have a common measure $(x-a)^2$, show that $q^2n^3=m^2r^3$.

21. If the H.C.F. of x^3+px+q and x^2+mx+n be $x+a$, their L.C.M. is
$$x^4+(m-a)x^3+px^2+(a^3+mp)x+a(m-a)(a^2+p).$$

22. If x^2+qx+1 and x^3+px^2+qx+1 have a common factor of the form $x+a$, show that
$$(p-1)^2-q(p-1)+1=0.$$

23. If $x^3 + px^2 + q$ and $x^2 + mx + n$ have $x + a$ for their H.C.F., show that their L.C.M. is
$$x^4 + (m - a + p)x^3 + p(m - a)x^2 + a^2(a - p)x$$
$$+ a^2(a - p)(m - a).$$

24. If $x^2 + px + 1$ and $x^3 + px^2 + qx + 1$ have $x - a$ for a common factor, show that $a = \dfrac{1}{1 - q}$.

25. Find the H.C.F. of $(a^2 - b^2)^3 + (b^2 - c^2)^3 + (c^2 - a^2)^3$ and $a^5(b - c) + b^5(c - a) + c^5(a - b)$.

26. If a be the H.C.F. of b and c, β the H.C.F. of c and a, γ the H.C.F. of a and b, and δ the H.C.F. of a, b, and c, then the L.C.M. of a, b, and c is $\dfrac{abc\delta}{a\beta\gamma}$.

27. If $x + c$ be the H.C.F. of the $x^2 + ax + b$ and $x^2 + a'x + b'$, their L.C.M. will be
$$x^3 + (a + a' - c)x^2 + (aa' - c^2)x + (a - c)(a' - c)c.$$

28. Show that the L.C.M. of the quantities in Exam. 2 (solved above) will be a complete square if
$$x = y^2 + z^2 - y^2 z^2.$$

29. Find the H.C.F. of $x^8 + 2x^6 + 3x^4 - 2x^2 + 1$ and $6x^8 + x^7 + 17x^5 - 7x^3 - 2$.

FRACTIONS.

§ 33. When required to reduce a fraction to its lowest terms, we can often apply some of the preceding methods of factoring to discover the H.C.F. of the numerator and denominator.

EXAMPLES.

1. $\dfrac{ac + by + ay + bc}{af + 2bx + 2ax + bf} = \dfrac{c(a + b) + y(a + b)}{f(a + b) + 2x(a + b)} = \dfrac{c + y}{f + 2x}$.

2. $\dfrac{a^4 - ba^3 - a^2b^2 + ab^3}{a^5 - ba^4 - ab^4 + b^5} = \dfrac{a[a^3 + b^3 - ab(a+b)]}{a(a^4 - b^4) - b(a^4 - b^4)}$

$= \dfrac{a(a+b)(a-b)^2}{(a-b)(a^4 - b^4)} = \dfrac{a}{a^2 + b^2}.$

3. $\dfrac{x^5 + x^4 y + x^3 y^2 + x^2 y^3 + xy^4 + y^5}{x^5 - x^4 y + x^3 y^2 - x^2 y^3 + xy^4 - y^5}.$

Here the numerator is evidently $(x^6 - y^6) \div (x - y)$, and the denominator is

$\dfrac{x^6 - y^6}{x + y}.$ The result is therefore $\dfrac{x+y}{x-y}.$

4. $\dfrac{(x+y)^5 - x^5 - y^5}{(x+y)^4 + x^4 + y^4} = \dfrac{5x^4 y + 10x^3 y^2 + 10x^2 y^3 + 5xy^4}{(x+y)^4 - x^2 y^2 + (x^2 + y^2)^2 - x^2 y^2}$

$= \dfrac{5xy[x^3 + y^3 + 2xy(x+y)]}{(x^2 + y^2 + xy)[(x+y)^2 + xy + x^2 + y^2 - xy]}$

$= \dfrac{5xy(x+y)(x^2 + xy + y^2)}{2(x^2 + xy + y^2)^2} = \dfrac{5xy(x+y)}{2(x^2 + xy + y^2)}.$

5. $\dfrac{x^2 - 12x + 35}{x^3 - 10x^2 + 31x - 30}.$

Here we see at once that the numerator $=(x-5)(x-7)$; and it is plain that $x - 7$ is not a factor of the denominator; we therefore try $x - 5$ (Horner's division), and find the quotient to be $x^2 - 5x + 6$.

Hence, the result $= \dfrac{x-7}{x^2 - 5x + 6}.$

6. $\dfrac{x^4 + 2x^2 + 9}{x^4 - 4x^3 + 8x - 21}.$

The factors of the numerator are at once seen to be $x^2 + 2x + 3$ and $x^2 - 2x + 3$, of which the latter is one factor of the denominator, the other being (Horner's division) $x^2 - 2x - 7$.

Hence, the result is $\dfrac{x^2 + 2x + 3}{x^2 - 2x - 7}.$

Ex. 44.

Reduce the following to their lowest terms:

1. $\dfrac{x^2 - 7x + 6}{x^3 - 2x^2 - 8x - 96}$; $\dfrac{3xy^2 - 13xy + 14x}{7y^3 - 17y^2 + 6y}$.

2. $\dfrac{x^4 + a^2x^2 + a^4}{x^4 + ax^3 - a^3x - a^4}$; $\dfrac{x^2 + x - 12}{x^3 - 5x^2 + 7x - 3}$.

3. $\dfrac{x^3 - 3x + 2}{x^3 + 4x^2 - 5}$; $\dfrac{x^4 + 2x^2 + 9}{x^4 - 4x^3 + 4x^2 - 9}$.

4. $\dfrac{2 + bx}{2b + (b^2 - 4)x - 2bx^2}$; $\dfrac{x^3 + 2x^2 + 2x}{x^5 + 4x}$.

5. $\dfrac{5a^5 + 10a^4x + 5a^3x^2}{a^3x + 2a^2x^2 + 2ax^3 + x^4}$; $\dfrac{20x^4 + x^2 - 1}{25x^4 + 5x^3 - x - 1}$.

6. $\dfrac{x^7 - x^6y + x^5y^2 - x^4y^3 + x^3y^4 - x^2y^5 + xy^6 - y^7}{x^7 + x^6y + x^5y^2 + x^4y^3 + x^3y^4 + x^2y^5 + xy^6 + y^7}$

7. $\dfrac{3a^2x^4 - 2ax^2 - 1}{4a^3x^6 - 2a^2x^4 - 3ax^2 + 1}$; $\dfrac{x^2 + \left(\dfrac{a}{b} + \dfrac{b}{a}\right)xy + y^2}{x^2 + \left(\dfrac{a}{b} - \dfrac{b}{a}\right)xy - y^2}$.

8. $\dfrac{a^2(b - c) + b^2(c - a) + c^2(a - b)}{abc(a - b)(b - c)(c - a)}$.

9. $\dfrac{(a + b + c)^2}{a^3(b - c) + b^3(c - a) + c^3(a - b)}$.

10. From Exam. 4 (solved above) show that

$$\dfrac{(a-b)^4 + (b-c)^4 + (c-a)^4}{(a-b)^5 + (b-c)^5 + (c-a)^5} = \dfrac{(a-b)^2 + (b-c)^2 + (c-a)^2}{5(a-b)(b-c)(c-a)}.$$

11. $\dfrac{(x + y)^5 - x^5 - y^5}{(x + y)^7 - x^7 - y^7}$.

12. Show that

$$\frac{(a-b)^7+(b-c)^7+(c-a)^7}{(a-b)^5+(b-c)^5+(c-a)^5}$$
$$=\tfrac{7}{10}[(a-b)^2+(b-c)^2+(c-a)^2].$$

§ 34. In reducing complex fractions it is often convenient to multiply both terms of the complex fraction by the L.C.M. of all the *denominators* involved.

EXAMPLES.

1. Simplify $\dfrac{\frac{1}{2}(x+1\frac{1}{3})-\frac{2}{3}(1-\frac{3}{4}x)}{1\frac{3}{4}-\frac{1}{3}(x+4\frac{1}{4})}$.

Here the L.C.M. of all the denominators involved is 12; hence, multiplying both terms of the complex fraction by 12, and removing brackets, we have

$$\frac{6x+8-8+6x}{21-4x-17}=\frac{12x}{4-4x}=\frac{3x}{1-x}.$$

2. $\dfrac{a-\dfrac{a-b}{1+ab}}{1+\dfrac{a(a-b)}{1+ab}}$.

Here, multiplying both terms by $1+ab$, we get

$$\frac{a(1+ab)-a+b}{1+ab+a(a-b)}=\frac{b(a^2+1)}{a^2+1}=b.$$

3. $\dfrac{1}{x-1+\dfrac{1}{1+\dfrac{x}{4-x}}}$.

Here, multiplying both terms of the fraction which follows $x-1$ by $4-x$, the given fraction becomes at once

$$\frac{1}{x-1+\dfrac{4-x}{4}};$$

and now, multiplying both terms by 4, we have

$$\frac{4}{4x-4+4-x}=\frac{4}{3x}.$$

It may be observed that when the fraction is reduced to the form $\frac{a}{b} \div \frac{c}{d}$, we may strike out any factor common to the two *denominators*, and also any factor common to the two *numerators;* it is sometimes more convenient to do this than to multiply directly by the L.C.M. of all the denominators.

4. Simplify $\left(\dfrac{a+b}{a-b}+\dfrac{a-b}{a+b}\right) \div \left(\dfrac{a^2+b^2}{a^2-b^2}-\dfrac{a^2-b^2}{a^2+b^2}\right).$

Here the numerator of the first fraction is $(a+b)^2 + (a-b)^2$, and the denominator is a^2-b^2; the numerator of second fraction is $(a^2+b^2)^2-(a^2-b^2)^2$, and the denominator is a^4-b^4; the former denominator cancels this to a^2+b^2, which, of course, becomes a multiplier of the first numerator.

Hence, we have $\dfrac{(a^2+b^2)[(a+b)^2+(a-b)^2]}{(a^2+b^2)^2-(a^2-b^2)^2} = \dfrac{(a^2+b^2)^2}{2a^2b^2}.$

Occasionally, we at once discover a common complex factor; strike this out, and simplify the result.

5. $\dfrac{\dfrac{1}{a}+\dfrac{1}{b}+\dfrac{1}{c}}{\dfrac{1}{a^2}+\dfrac{1}{b^2}-\dfrac{1}{c^2}+\dfrac{2}{ab}}.$ Here the denominator

$$=\left(\frac{1}{a}+\frac{1}{b}\right)^2-\frac{1}{c^2}=\left(\frac{1}{a}+\frac{1}{b}+\frac{1}{c}\right)\left(\frac{1}{a}+\frac{1}{b}-\frac{1}{c}\right),$$

and cancelling the common factor, we have

$\dfrac{1}{\dfrac{1}{a}+\dfrac{1}{b}-\dfrac{1}{c}}$, and, multiplying by abc, this

$$=\frac{abc}{bc+ca-ab}.$$

Ex. 45.

Simplify the following:

1. $\dfrac{1 - \frac{1}{2}[1 - \frac{1}{3}(1-x)]}{1 - \frac{1}{3}[1 - \frac{1}{2}(1-x)]}$; $\dfrac{\dfrac{a+b}{a-b} + \dfrac{a-b}{a+b}}{\dfrac{a+b}{a-b} - \dfrac{a-b}{a+b}}$.

2. $\dfrac{\dfrac{x}{x+y} + \dfrac{x}{x-y}}{\dfrac{2x}{x^2 - y^2}}$; $\dfrac{\dfrac{1}{1-a} - \dfrac{1}{1+a}}{\dfrac{a}{1-a} + \dfrac{1}{1+a}}$.

3. $\dfrac{1}{1 + \dfrac{a}{1 + a + \dfrac{2a^2}{1+a}}}$.

4. $\dfrac{\dfrac{a^2 + b^2}{2a^2} - \dfrac{2b^2}{a^2 + b^2}}{\dfrac{a^2 + b^2}{2b^2} - \dfrac{2a^2}{a^2 + b^2}}$; $\dfrac{\dfrac{1}{a} + \dfrac{1}{ab^3}}{b - 1 + \dfrac{1}{b}}$.

5. $\dfrac{\dfrac{a+b}{c+d} + \dfrac{a-b}{c-d}}{\dfrac{a+b}{c-d} + \dfrac{a-b}{c+d}}$; $\dfrac{a + b + \dfrac{b^2}{a}}{a + b + \dfrac{a^2}{b}}$.

6. $\dfrac{3xyz}{yz - zx - xy} - \dfrac{\dfrac{x-1}{x} + \dfrac{y-1}{y} + \dfrac{z-1}{z}}{\dfrac{1}{x} + \dfrac{1}{y} + \dfrac{1}{z}}$.

7. $\dfrac{\dfrac{2}{a^2} + \dfrac{2}{b^2} + \dfrac{2}{c^2} + \dfrac{a^4 + b^4 + c^4}{a^2 b^2 c^2}}{\dfrac{a}{bc} + \dfrac{b}{ac} + \dfrac{c}{ab}}$.

8. $\dfrac{a^3 + a^2 b + ab^2 + b^3}{a^3 - a^2 b + ab^2 - b^3} \div \dfrac{a^2 + 2ab + b^2}{a^2 - b^2}.$

9. $\left(\dfrac{a+b}{a-b} + \dfrac{a^2 + b^2}{a^2 - b^2}\right) \div \left(\dfrac{a-b}{a+b} - \dfrac{a^3 - b^3}{a^3 + b^3}\right).$

10. $\dfrac{\dfrac{1}{a} + \dfrac{1}{b+c}}{\dfrac{1}{a} - \dfrac{1}{b+c}}\left(1 + \dfrac{b^2 + c^2 - a^2}{2bc}\right).$

11. $\dfrac{\dfrac{2(1-x)}{1+x} + \dfrac{(1-x)^2}{(1+x)^2} + 1}{\dfrac{2(1+x)}{1-x} + \left(\dfrac{1+x}{1-x}\right)^2 + 1}$; $\dfrac{\left(\dfrac{x-a}{x+a}\right)^2 + \left(\dfrac{x+a}{x-a}\right)^2 - 2}{\left(\dfrac{x-a}{x+a}\right)^2 + 2 + \left(\dfrac{x+a}{x-a}\right)^2}.$

12. $\dfrac{\dfrac{x}{y} + 1 + \dfrac{y}{x}}{\dfrac{x}{y} - 1 + \dfrac{y}{x}} \div \dfrac{1 - \dfrac{y^3}{x^3}}{1 + \dfrac{y^3}{x^3}}.$

13. $\dfrac{3\left(\dfrac{a-b}{a+b}\right)^2 - \left(\dfrac{a-b}{a+b}\right)^3 - 3\left(\dfrac{a-b}{a+b}\right) + 1}{3\left(\dfrac{a+b}{a-b}\right) - 3\left(\dfrac{a+b}{a-b}\right)^2 + \left(\dfrac{a+b}{a-b}\right)^3 - 1}.$

14. $\dfrac{x^5 - x^4 y + x^3 y^2 - x^2 y^3 + xy^4 - y^5}{x^5 + x^4 x + x^3 y^2 + x^2 y^3 + xy^4 + y^5} \div \left(\dfrac{x-y}{x+y}\right)^2.$

15. $\left(\dfrac{1-x^2}{1-x^3} + \dfrac{1-x}{1-x+x^2}\right) \div \left(\dfrac{1+x}{1+x+x^2} - \dfrac{1-x^2}{1+x^3}\right).$

16. Find the value of

$$\dfrac{a}{2na - 2nx} + \dfrac{b}{2nb - 2nx} \text{ when } x = \tfrac{1}{2}(a+b).$$

17. Find the value of $\sqrt{[1 - \sqrt{(1-x)}]}$

when $x = 2\left(\dfrac{1-b}{1+b}\right)^2 - \left(\dfrac{1-b}{1+b}\right)^4.$

18. Find the value of

$$\frac{\sqrt{(a+bx)}+\sqrt{(a-bx)}}{\sqrt{(a+bx)}-\sqrt{(a-bx)}} \text{ when } x = \frac{2ac}{b(1+c^2)}.$$

§ 35. When the sum of several fractions is to be found, it is generally best, instead of reducing at once all the fractions to a common denominator, to take two (or more) of them together, and combine the results.

EXAMPLES.

1. Find the sum of

$$\frac{x+y}{2x-2y} - \frac{y-x}{2x+2y} - \frac{x^2-y^2}{x^2+y^2}.$$

Here, taking the first two together, we have

$$\frac{(x+y)^2+(x-y)^2}{2(x^2-y^2)} = \frac{x^2+y^2}{x^2-y^2};$$

now add this to $-\dfrac{x^2-y^2}{x^2+y^2}$

and we get $\dfrac{(x^2+y^2)^2-(x^2-y^2)^2}{x^4-y^4} = \dfrac{4x^2y^2}{x^4-y^4}.$

2. Find the sum of

$$\frac{1+x}{1-x} + \frac{4x}{1+x^2} + \frac{8x}{1+x^4} - \frac{1-x}{1+x}.$$

Here, taking the first and the last together, we have

$$\frac{(1+x)^2-(1-x)^2}{1-x^2} = \frac{4x}{1-x^2};$$

taking this result with the second fraction, we have

$$4x\left(\frac{1}{1+x^2}+\frac{1}{1-x^2}\right) = \frac{8x}{1-x^4};$$

now take this result with the remaining fraction, and we get

$$8x\left(\frac{1}{1-x^4}+\frac{1}{1+x^4}\right) = \frac{16x}{1-x^8}.$$

3. $\dfrac{x^{3n}}{x^n - 1} - \dfrac{x^{2n}}{x^n + 1} - \dfrac{1}{x^n - 1} + \dfrac{1}{x^n + 1}$.

Taking in pairs those whose denominators are alike, we have

$$\frac{x^{3n} - 1}{x^n - 1} - \frac{x^{2n} - 1}{x^n + 1} = x^{2n} + x^n + 1 - (x^n - 1) = x^{2n} + 2.$$

The work is often made easier by *completing the divisions* represented by the fractions.

4. Find the sum of $1 + \dfrac{2x + 1}{2(x - 1)} - \dfrac{4x + 5}{2x + 2}$.

By dividing numerators by denominators, this

$$= 1 + 1 + \frac{3}{2x - 2} - 2 - \frac{1}{2x + 2} = \frac{3}{2x - 2} - \frac{1}{2x + 2}$$

$$= \frac{3x + 3 - x + 1}{2x^2 - 2} = \frac{x + 2}{x^2 - 1}.$$

5. $\dfrac{x}{x - 2} + \dfrac{x - 9}{x - 7} - \dfrac{x + 1}{x - 1} - \dfrac{x - 8}{x - 6}$.

We have, by division,

$$1 + \frac{2}{x - 2} + 1 - \frac{2}{x - 7} - 1 - \frac{2}{x - 1} - 1 + \frac{2}{x - 6},$$

or $\dfrac{2}{x - 2} + \dfrac{2}{x - 6} - \dfrac{2}{x - 7} - \dfrac{2}{x - 1}$

$$= \frac{2(2x - 8)}{(x - 2)(x - 6)} - \frac{2(2x - 8)}{(x - 1)(x - 7)}$$

$$= (4x - 16)\left(\frac{1}{x^2 - 8x + 12} - \frac{1}{x^2 - 8x + 7}\right)$$

$$= (80 - 20x) \div (x^4 - 16x^3 + 83x^2 - 152x + 84).$$

[Denominator $= (x^2 - 8x)^2 + 19(x^2 - 8x) + 84$.]

6. Find the value of $\dfrac{x+2a}{x-2a}+\dfrac{x+2b}{x-2b}$ when $x=\dfrac{4ab}{a+b}$.

By division,

$$1+\dfrac{4a}{x-2a}+1+\dfrac{4b}{x-2b}=2+4\left(\dfrac{a}{x-2a}+\dfrac{b}{x-2b}\right);$$

but the quantity in the brackets

$$=\dfrac{(a+b)x-4ab}{(x-2a)(x-2b)}=0,\ \text{since}\ (a+b)x=4ab.$$

Hence, the value of the given expression is 2.

Ex. 46.

Simplify the following :

1. $\dfrac{x-a}{5}+\dfrac{x^2+ax+a^2}{x+a}-\dfrac{x^3-a^3}{x^2-a^2}.$

2. $\dfrac{a^3+b^3}{a^2-ab+b^2}+\dfrac{a^3-3a^2b+3ab^2-b^3}{a^3-b^3}-\dfrac{a(a-b)-b(a-b)}{a^2+ab+b^2}.$

3. $\left(\dfrac{1}{a+x}+\dfrac{1}{a-x}+\dfrac{2a}{a^2+x^2}\right)\left(\dfrac{1}{a+x}-\dfrac{1}{a-x}-\dfrac{2x}{a^2+x^2}\right).$

4. $\dfrac{a}{a+b}+\dfrac{b}{a-b}-\dfrac{ab}{ab-b^2}+\dfrac{ab}{a^2+ab}.$

5. $\dfrac{3+2x}{2-x}-\dfrac{2-3x}{2+x}+\dfrac{16x-x^2}{x^2-4}.$

6. $\dfrac{1}{4a^3(a+x)}+\dfrac{1}{4a^3(a-x)}+\dfrac{1}{2a^2(a^2+x^2)}.$

7. $\dfrac{1}{2}\left(\dfrac{3x+2y}{3x-2y}\right)-\dfrac{1}{2}\left(\dfrac{3x-2y}{3x+2y}\right).$

8. $\dfrac{x+1}{2x-1}-\dfrac{x-1}{2x+1}-\dfrac{1-3x}{x(1-2x)}+\dfrac{x}{x(4x^2-1)}+\dfrac{1}{x(16x^4-1)}.$

9. $\dfrac{1}{2x+2}-\dfrac{4}{x+2}+\dfrac{9}{2(x+3)}-\dfrac{x-1}{(x+2)(x+3)}.$

10. $\dfrac{2(x+y)}{x-y} - \dfrac{2(y-x)}{x+y} - \dfrac{4(x^2-y^2)}{x^2+y^2} + \dfrac{4(x^4+y^4)}{x^4-y^4}.$

11. $(a-b)\left[\dfrac{1}{(x+a)^2} + \dfrac{1}{(x+b)^2}\right] + 2\left(\dfrac{1}{x+a} - \dfrac{1}{x+b}\right).$

12. $\left(\dfrac{a+x}{a-x} + \dfrac{4ax}{a^2+x^2} + \dfrac{8a^3x}{a^4+x^4} - \dfrac{a-x}{a+x}\right)$

$\qquad \div \left(\dfrac{a^2+x^2}{a^2-x^2} + \dfrac{4a^2x^2}{a^4+x^4} - \dfrac{a^2-x^2}{a^2+x^2}\right).$

13. $\dfrac{5x-4}{9} + \dfrac{12x+2}{11x-8} - \dfrac{10x+17}{18}.$

14. $\dfrac{a}{a^2+b^2} + \dfrac{a}{a^2-b^2} + \dfrac{a^2}{(a-b)(a^2+b^2)} - \dfrac{2a^3-b^3-ab^2}{a^4-b^4}.$

15. $\dfrac{12x+10a}{3x+a} + \dfrac{117a+28x}{9a+2x} - 18.$

16. $\dfrac{4x-17}{x-4} - \dfrac{8x-30}{2x-7} + \dfrac{10x-3}{2x-5} - \dfrac{5x-4}{x-1}.$

17. Find the value of $\dfrac{a+b+2c}{a+b-2c} + \dfrac{a+b+2d}{a+b-2d}$

$\qquad$ when $a+b = \dfrac{4cd}{c+d}.$

18. $\dfrac{x^{3n}}{x^n-y^n} - \dfrac{y^n x^{2n}}{x^n+y^n} - \dfrac{y^{3n}}{x^n-y^n} + \dfrac{y^{3n}}{x^n+y^n}.$

19. $\dfrac{(a-b)^{3n}}{(a-b)^n-1} - \dfrac{(a-b)^{2n}}{(a-b)^n+1} - \dfrac{1}{(a-b)^n-1} + \dfrac{1}{(a-b)^n+1}.$

20. $\dfrac{1}{(a^2-b^2)(x^2+b^2)} + \dfrac{1}{(b^2-a^2)(x^2+a^2)} - \dfrac{1}{(x^2+a^2)(x^2+b^2)}.$

21. $\dfrac{1+x}{1-x^3} + \dfrac{1-x}{1+x^3} - \dfrac{2}{1-x^2} - \dfrac{2x^2}{x^6+1}.$

22. $\dfrac{a^3+a^2b+ab^2+b^3}{a^3-a^2b-ab^2+b^3} \times \dfrac{(a+b)^2-3ab}{(a-b)^2+3ab} \times \dfrac{(a-b)^3-a^3+b^3}{(a+b)^3-a^3-b^3}.$

§ **36.** The following are additional examples in which a knowledge of factoring and of the principle of symmetry is of advantage.

EXAMPLES.

1. $\dfrac{x^2-(y-z)^2}{(x+z)^2-y^2}+\dfrac{y^2-(z-x)^2}{(y+x)^2-z^2}+\dfrac{z^2-(x-y)^2}{(z+y)^2-x^2}.$

Cancelling the common factor $x-y+z$ in the two terms of the first fraction, there results $\dfrac{x+y-z}{x+y+z}$; hence, by symmetry, the denominators of the other two fractions will be $x+y+z$, and the numerators will be $y+z-x, z+x-y$. Hence, the sum of the three numerators $=x+y+z$, and the result $=1$.

2. Simplify $\dfrac{ab}{(c-a)(c-b)}+\dfrac{bc}{(a-b)(a-c)}+\dfrac{ca}{(b-c)(b-a)}.$

The L.C.M. of denominators is evidently
$(a-b)(b-c)(c-a).$

This gives for numerator of first fraction $-ab(a-b)$; and, by symmetry, the other numerators are
$-bc(b-c),\ -ca(c-a).$

Hence, we have $-\dfrac{ab(a-b)+bc(b-c)+ca(c-a)}{(a-b)(b-c)(c-a)}$

$=-\dfrac{(a-b)(b-c)(a-c)}{(a-b)(b-c)(c-a)}=1.$

3. Reduce the following to a single fraction:

$$\dfrac{a}{(a-b)(a-c)(x-a)}+\dfrac{b}{(b-a)(b-c)(x-b)}$$
$$+\dfrac{c}{(c-a)(c-b)(x-c)}.$$

Here the L.C.M. is $(a-b)(b-c)(c-a)(x-a)(x-b)(x-c)$;
the numerator of the first fraction is
$$-a(b-c)(x-b)(x-c),$$
and, therefore, by symmetry, that of second is
$$-b(c-a)(x-c)(x-a),$$
and that of third is
$$-c(a-b)(x-a)(x-b);$$
and their sum is
$$-[a(b-c)(x-b)(x-c)+b(c-a)(x-c)(x-a)$$
$$+c(a-b)(x-a)(x-b)].$$

This vanishes if $a=b$; hence, $a-b$ is a factor, and therefore, by symmetry, $b-c$ and $c-a$ are also factors. Now the product of these is of the third degree, while the whole expression rises only to the fourth; *hence, x^2 cannot be involved.* The other factor must therefore be of the form $mx+n$, *in which m is a number.*

To determine n, put $x=0$, and the expression becomes $abc(a-b+b-c+c-a)=0$; hence, $n=0$, or the other factor is mx.

To determine m, put $a=0$, $b=1$, $c=-1$, and m will be found to be 1. The numerator is, therefore,
$$x(a-b)(b-c)(c-a),$$
and the result is
$$\frac{x}{(x-a)(x-b)(x-c)}.$$

3. Simplify
$$\frac{a+b}{(b-c)(c-a)}+\frac{b+c}{(c-a)(a-b)}+\frac{c+a}{(a-b)(b-c)}.$$

L.C.M. of denominators is $(a-b)(b-c)(c-a)$; hence, first numerator is a^2-b^2, and, by symmetry, second numerator is b^2-c^2, and third numerator is c^2-a^2; the sum of these $=0$, which is the required result.

4. Reduce

$$\frac{2}{x-y}+\frac{2}{y-z}+\frac{2}{z-x}+\frac{(x-y)^2+(y-z)^2+(z-x)^2}{(x-y)(y-z)(z-x)}.$$

Here the numerator becomes

$$2(y-z)(z-x)+2(x-y)(z-x)$$
$$+2(x-y)(y-z)+(x-y)^2+(y-z)^2+(z-x)^2,$$

which is evidently

$$[(x-y)+(y-z)+(z-x)]^2=0.$$

5. $a^3\left(\dfrac{a^3+2b^3}{a^3-b^3}\right)^3+b^3\left(\dfrac{2a^3+b^3}{b^3-a^3}\right)^3.$

Observe that the denominators become the same by changing the sign between the fractions, and that the expression is symmetrical with respect to a and b. The numerator of the first fraction is

$$a^{12}+6a^9b^3+12a^6b^6+8a^3b^9,$$

and, by symmetry, that of the other is

$$-b^{12}-6b^9a^3-12b^6a^6-8b^3a^9.$$

Their sum is, therefore,

$$a^{12}-b^{12}+6a^3b^3(a^6-b^6)-8a^3b^3(a^6-b^6)$$
$$=(a^6-b^6)(a^6+b^6+6a^3b^3-8a^3b^3)$$
$$=(a^6-b^6)(a^3-b^3)^2=(a^3+b^3)(a^3-b^3)^3,$$

and since the *denominator* of the given expression is $(a^3-b^3)^3$, therefore the result is a^3+b^3.

Ex. 47.

Simplify the following:

1. $x\left(\dfrac{x-2y}{x+y}\right)^3+y\left(\dfrac{2x-y}{x+y}\right)^3.$

2. $a\left(\dfrac{a+2b}{a-b}\right)^3+b\left(\dfrac{2a+b}{b-a}\right).$

3. $\dfrac{a+b}{(b-c)(c-a)}+\dfrac{b+c}{(c-a)(a-b)}+\dfrac{c+a}{(a-b)(b-c)}.$

4. $\dfrac{1}{(a-b)(a-c)} + \dfrac{1}{(b-a)(b-c)} + \dfrac{1}{(c-a)(c-b)}$.

5. $\dfrac{a-b}{a+b} + \dfrac{b-c}{b+c} + \dfrac{c-a}{c+a} + \dfrac{(a-b)(b-c)(c-a)}{(a+b)(b+c)(c+a)}$.

6. $\dfrac{a^2}{(a+b)(a+c)(x+a)} + \dfrac{b^2}{(a+b)(b-c)(x+b)}$
$- \dfrac{c^2}{(a+c)(b-c)(x+c)}$.

7. $\dfrac{x^2}{(x-y)(x-z)} + \dfrac{y^2}{(y-x)(y-z)} + \dfrac{z^2}{(z-x)(z-y)}$.

8. $\dfrac{a^3}{(a-b)(a-c)} + \dfrac{b^3}{(b-a)(b-c)} + \dfrac{c^3}{(c-a)(c-b)}$.

9. $\dfrac{1}{\left(\dfrac{a}{b}-1\right)\left(\dfrac{a}{c}-1\right)} + \dfrac{1}{\left(\dfrac{b}{a}-1\right)\left(\dfrac{b}{c}-1\right)} + \dfrac{1}{\left(\dfrac{c}{a}-1\right)\left(\dfrac{c}{b}-1\right)}$.

10. $x^3\left(\dfrac{x^3-2y^3}{x^3+y^3}\right)^3 + y^3\left(\dfrac{2x^3-y^3}{x^3+y^3}\right)^3$.

11. $\dfrac{1}{(b+c-2a)(c+a-2b)} + \dfrac{1}{(c+a-2b)(a+b-2c)}$
$+ \dfrac{1}{(a+b-2c)(b+c-2a)}$.

12. $\dfrac{b^2-c^2}{(b+c)^2} + \dfrac{c^2-a^2}{(c+a)^2} + \dfrac{a^2-b^2}{(a+b)^2}$.

13. $\dfrac{a^2}{(a-b)(a-c)(x-a)} + \dfrac{b^2}{(b-a)(b-c)(x-b)}$
$+ \dfrac{c^2}{(c-a)(c-b)(x-c)}$.

14. $\dfrac{x(y+z)}{(x-y)(z-x)} + \dfrac{y(z+x)}{(y-z)(x-y)} + \dfrac{z(x+y)}{(z-x)(y-z)}$.

15. $\dfrac{(a+b)^2+(b-c)^2+(a+c)^2}{(a+b)(b-c)(a+c)} - \dfrac{2}{a+c} - \dfrac{2}{b-c} + \dfrac{2}{a+b}.$

16. $\dfrac{1}{x(x-a)(x-b)} + \dfrac{1}{a(b-a)(x-a)} + \dfrac{1}{b(b-a)(x-b)}.$

RATIOS.

§ 37. If $\dfrac{a}{b} = \dfrac{c}{d}$, therefore, $ad = bc$.

Now, dividing $ad = bc$ by ca, we have $\dfrac{b}{a} = \dfrac{d}{c}$ (1)

$ad = bc$ by cd, we have $\dfrac{a}{c} = \dfrac{b}{d}$ (2)

$ad = bc$ by ab, we have $\dfrac{d}{b} = \dfrac{c}{a}$ (3)

Also, $\dfrac{ma+nc}{mb+nd} =$ each of the given fractions.   (4)

For $\dfrac{ma+nc}{mb+nd} = \dfrac{mb\left(\frac{a}{b}\right) + nd\left(\frac{c}{d}\right)}{mb+nd} = \dfrac{(mb+nd)\frac{a}{b}}{mb+nd}$

$= \dfrac{a}{b}$ or $\dfrac{c}{d}$.

A very important case of this is $m = 1$, $n = \pm 1$; hence,

$\dfrac{a}{b} = \dfrac{c}{d} = \dfrac{a+c}{b+d} = \dfrac{a-c}{b-d}$ (5)

Also $\dfrac{a-b}{a+b} = \dfrac{c-d}{c+d}$ (6)

For, by (2) and (5),

$\dfrac{a}{c} = \dfrac{b}{d} = \dfrac{a-b}{c-d} = \dfrac{a+b}{c+d}$ $\therefore \dfrac{a-b}{a+b} = \dfrac{c-d}{c+d}.$

Or thus:

$\dfrac{a-b}{a+b} = \dfrac{\frac{a}{b}-1}{\frac{a}{b}+1} = \dfrac{\frac{c}{d}-1}{\frac{c}{d}+1} = \dfrac{c-d}{c+d}.$

Generally, to prove that, if $\dfrac{a}{b} = \dfrac{c}{d}$, any fraction whose numerator and denominator are homogeneous functions of a and b, and are of the same degree, will be equal to a similar fraction formed with c instead of a, and d instead of b.

Express the first fraction in terms of $\dfrac{a}{b}$, and for $\dfrac{a}{b}$ substitute its equivalent $\dfrac{c}{d}$, and reduce the result.

By (2), the fractions may be formed of a and c, and b and d.

$$\text{If } \frac{a}{b} = \frac{c}{d} = \frac{e}{f}, \ \frac{ma + nc + pe}{mb + nd + pf} = \frac{a}{b} \text{ or } \frac{c}{d} \text{ or } \frac{e}{f} \tag{7}$$

$$\frac{ma + nc + pe}{mb + nd + pf} = \frac{mb\left(\dfrac{a}{b}\right) + nd\left(\dfrac{c}{d}\right) + pf\left(\dfrac{e}{f}\right)}{mb + nd + pf}$$

$$= \frac{(mb + nd + pf)\dfrac{a}{b}}{mb + nd + pf} = \frac{a}{b}.$$

$$\text{If } \frac{a}{b} = \frac{c}{d} \text{ and } \frac{m}{n} = \frac{p}{q},$$

$$\frac{ma \pm pc}{nb \pm qd} = \frac{pa \pm mc}{qb \pm nd} = \frac{ma}{nb} \text{ or } \frac{pa}{qb} \text{ or } \ldots\ldots \tag{8}$$

$$\text{For, } \frac{ma}{nb} = \frac{pc}{qd} = \frac{ma \pm pc}{nb \pm qd} \text{ by (5),}$$

$$\frac{pa}{qb} = \frac{mc}{nd} = \frac{pa \pm mc}{qb \pm nd}.$$

But $\dfrac{ma}{nb} = \dfrac{pa}{qb}$, hence the equality stated in (8).

$$\text{If } \frac{a}{b} = \frac{c}{d} = \frac{e}{f} \text{ and } \frac{m}{n} = \frac{p}{q} = \frac{r}{s},$$

$$\frac{ma \pm pc \pm re}{nb \pm qd \pm sf} = \frac{pa \pm rc \pm me}{qb \pm sd \pm nf} = \ldots\ldots = \frac{ma}{nb} = \ldots\ldots \tag{9}$$

If an upper sign be taken in a numerator, the corresponding upper sign must be taken in the denominator; if a lower sign, the corresponding lower sign; otherwise, all the signs are independent of each other.

EXAMPLES.

1. If $\dfrac{a}{b} = \dfrac{c}{d}$, show that $\dfrac{5a - 4b}{7a + 5b} = \dfrac{5c - 4d}{7c + 5d}$.

The given fraction $= \dfrac{5\dfrac{a}{b} - 4}{7\dfrac{a}{b} + 5} = \dfrac{5\dfrac{c}{d} - 4}{7\dfrac{c}{d} + 5} = \dfrac{5c - 4d}{7c + 5d}$.

2. If $\dfrac{a}{b} = \dfrac{c}{d}$, show that $\dfrac{2a^3 + 3a^2b}{3a^2b - 4b^3} = \dfrac{2c^3 + 3c^2d}{3c^2d - 4d^3}$.

Dividing the given fraction by b^3, we have

$$\dfrac{2\dfrac{a^3}{b^3} + 3\dfrac{a^2}{b^2}}{3\dfrac{a^2}{b^2} - 4},$$

and this becomes, on substituting for $\dfrac{a}{b}$ its equal $\dfrac{c}{d}$,

$$\dfrac{2\dfrac{c^3}{d^3} + 3\dfrac{c^2}{d^2}}{3\dfrac{c^2}{d^2} - 4} = \dfrac{2c^3 + 3c^2d}{3c^2d - 4d^3}.$$

3. If $3a = 2b$, find the value of $\dfrac{a^3 + b^3}{a^2b - ab^2}$.

This $= \left(\dfrac{a^3}{b^3} + 1\right) \div \left(\dfrac{a^2}{b^2} - \dfrac{a}{b}\right)$ [by dividing both numerator and denominator by b^3]. But, from the given relation

$\dfrac{a}{b} = \dfrac{2}{3}$, we have, by substituting for $\dfrac{a}{b}$,

$(\tfrac{8}{27} + 1) \div (\tfrac{4}{9} - \tfrac{2}{3}) = 35 \div (-6) = -5\tfrac{5}{6}$.

4. If $\dfrac{a}{b} = \dfrac{c}{d}$, prove that $\dfrac{a^3 + b^3}{c^3 + d^3} \times \dfrac{b}{d} = \left(\dfrac{a+b}{c+d}\right)^4$.

We have $\dfrac{a}{c} = \dfrac{b}{d} = \dfrac{a+b}{c+d}$.

Also $\dfrac{a^3 + b^3}{c^3 + d^3} = \dfrac{b^3}{d^3}\left(\dfrac{a^3}{b^3} + 1\right) \div \left(\dfrac{c^3}{d^3} + 1\right) = \dfrac{b^3}{d^3}$,

and this multiplied by $\dfrac{b}{d}$ gives $\dfrac{b^4}{d^4} = \left(\dfrac{a+b}{c+d}\right)^4$.

5. If $\dfrac{x^3 + ax^2 - bx + c}{x^3 - ax^2 + bx + c} = \dfrac{x^2 + ax - b}{x^2 - ax + b}$, show that $x = \dfrac{b}{a}$.

Multiplying both terms of second fraction by x, it becomes $\dfrac{x^3 + ax^2 - bx}{x^3 - ax^2 + bx}$;

now each of the given fractions

$= \dfrac{\text{difference of numerators}}{\text{difference of denominators}} = \dfrac{c}{c} = 1$.

Hence, $x^2 + ax - b = x^2 - ax + b$ or $2ax = 2b$.

Therefore, $x = \dfrac{b}{a}$.

6. If $\dfrac{a}{b} = \dfrac{c}{d} = \dfrac{e}{f}$, show that $\dfrac{ac + ce + ea}{bd + df + fb} = \dfrac{a^2 + c^2 + e^2}{b^2 + d^2 + f^2}$.

For $\dfrac{ac}{bd} = \dfrac{ce}{df} = \dfrac{ea}{fb} = \dfrac{ac + ce + ea}{bd + df + fb}$.

By (7) making $m = n = p = 1$.

Also $\dfrac{a^2}{b^2} = \dfrac{c^2}{d^2} = \dfrac{e^2}{f^2} = \dfrac{a^2 + c^2 + e^2}{b^2 + d^2 + f^2}$. By (7).

But $\dfrac{ac}{bd} = \dfrac{a^2}{b^2}$, hence the required equality.

The problem is a particular case of (9), with all the signs $+$, and a for m, b for n, c for p, etc.

If the fractions *given* equal to one another have not monomial terms, instead of seeking to express the proposed quantity in terms of one fraction, and then substituting an equivalent fraction, it is often better to assume a single letter to represent the common value of the fractions given equal, and to work in terms of this assumed letter.

7. If $\dfrac{a+b}{3(a-b)} = \dfrac{b+c}{4(b-c)} = \dfrac{c+a}{5(c-a)}$,

prove that $32a + 35b + 27c = 0$.

Assume each of the given fractions $= x$, so that

$$a+b = 3(a-b)x,\ b+c = 4(b-c)x,\ c+a = 5(c-a)x,$$

or $\dfrac{a+b}{3} + \dfrac{b+c}{4} + \dfrac{c+a}{5} = x(a-b+b-c+c-a) = 0$.

Hence, adding these fractions, we have

$$32a + 35b + 27c = 0.$$

This example might also be worked as a particular case of (7); thus,

$$\frac{a+b}{3(a-b)} = \frac{b+c}{4(b-c)} = \frac{c+a}{5(c-a)}$$

$$= \frac{20(a+b)+15(b+c)+12(c+a)}{60(a-b)+60(b-c)+60(c-a)} = \frac{32a+35b+27c}{0}.$$

Hence, $32a + 35b + 27c = 0 \times \dfrac{a+b}{3(a-b)} = 0$.

8. If $\dfrac{a^2}{b^2} + \dfrac{c^2}{f^2} = \dfrac{2c}{d}\left(\dfrac{a}{b} - \dfrac{c}{d} + \dfrac{e}{f}\right)$,

prove that $\left(\dfrac{a+c+e}{b+d+f}\right)^2 = \dfrac{a^2+c^2+e^2}{b^2+d^2+f^2}$.

Transposing terms, etc., we have

$$\frac{a^2}{b^2} - \frac{2ac}{bd} + \frac{c^2}{d^2} + \frac{e^2}{f^2} - \frac{2ce}{df} + \frac{c^2}{d^2} = 0,$$

or $\left(\dfrac{a}{b} - \dfrac{c}{d}\right)^2 + \left(\dfrac{e}{f} - \dfrac{c}{d}\right)^2 = 0$.

That is, the sum of two essentially positive quantities $= 0$; therefore each of them must $= 0$. Hence, we have

$$\frac{a}{b} - \frac{c}{d} = 0, \text{ and } \frac{e}{f} - \frac{c}{d} = 0;$$

hence, $\frac{a}{b} = \frac{c}{d} = \frac{e}{f}$; therefore, $\frac{a^2}{b^2} = \frac{a^2 + c^2 + e^2}{b^2 + d^2 + f^2}$.

Also $\frac{a}{b} = \frac{a + c + e}{b + d + f}$; hence, $\frac{a^2}{b^2} = \left(\frac{a + c + e}{b + d + f}\right)^2$;

therefore, $\left(\frac{a + c + e}{b + d + f}\right)^2 = \frac{a^2 + c^2 + e^2}{b^2 + d^2 + f^2}$.

Ex. 48.

1. If $\frac{a}{b} = \frac{c}{d}$, prove that $\frac{a^2 - ab + b^2}{ab - 4b^2} = \frac{c^2 - cd + d^2}{cd - 4d^2}$.

2. If $\frac{a}{b} = \frac{c}{d}$, prove that $\frac{a^2 - c^2}{b^2 - d^2} = \left(\frac{a - c}{b - d}\right)^2 = \left(\frac{a + c}{b + d}\right)^2$.

3. Given the same, show that each of these fractions
$$= \sqrt{\left(\frac{a^2 + c^2}{b^2 + d^2}\right)}.$$

4. If $2x = 3y$, write down the value of
$$\frac{2x^3 - x^2y + y^3}{x^2y + xy^2 + 2y^3} \text{ and of } \frac{x^4 - 3x^3y + 2y^4}{(x^2 - y^2)^2}.$$

5. If $\frac{a}{b} = \frac{c}{d} = \frac{e}{f}$, show that $\frac{a}{b} = \frac{ma - nc - pe}{mb - nd - pf}$.

6. From the same relations prove that
$$\frac{a^3}{b^3} = \left(\frac{a - mc - ne}{b - md - nf}\right)^3.$$

7. If $\frac{1 + x}{1 - x} = \frac{b}{a}\left(\frac{1 + x + x^2}{1 - x + x^2}\right)$, then $x^3 = (b - a) \div (b + a)$.

8. If $\dfrac{\sqrt{(a+x)}+\sqrt{(a-x)}}{\sqrt{(a+x)}-\sqrt{(a-x)}} = a$, prove that $x = \dfrac{2a^2}{1+a^2}$.

9. If $\dfrac{mx+a+b}{nx+a+c} = \dfrac{mx-c-d}{nx-b-d}$, prove that $x = \dfrac{b-c}{n-m}$.

10. If $\dfrac{a-b}{ay+bx} = \dfrac{b-c}{bz+cx} = \dfrac{c-a}{cy+az} = \dfrac{a+b+c}{ax+by+cz}$,

then each of these fractions $= \dfrac{1}{x+y+z}$,

$a+b+c$ not being zero.

11. If $\dfrac{a+b}{a-b} = \dfrac{b+c}{2(b-c)} = \dfrac{c+a}{3(c-a)}$, then $8a+9b+5c=0$.

12. If $\dfrac{\sqrt{a}+\sqrt{(a-x)}}{\sqrt{a}-\sqrt{(a-x)}} = \dfrac{1}{a}$, show that $\dfrac{a-x}{a} = \left(\dfrac{1-a}{1+a}\right)^2$.

13. If $\dfrac{x^2-yz}{x(1-yz)} = \dfrac{y^2-xz}{y(1-xz)}$, and x, y, z be unequal, show that each of these fractions is equal to $x+y+z$.

14. If $\dfrac{x^2+2x+1}{x^2-2x+3} = \dfrac{y^2+2y+1}{y^2-2y+3}$, show that each of these fractions $= (xy-1) \div (xy-3)$.

15. If $\dfrac{25x^2-16}{10x+8} = \dfrac{3(x^2-4)}{2x-4}$, show that $\dfrac{x+1}{x+5} = \dfrac{3}{5}$.

16. If $y = \dfrac{4bc}{b+c}$, show that $\dfrac{y+2b}{y-2b} + \dfrac{y+2c}{y-2c} = 2$.

17. If $\dfrac{1}{4}\left(\dfrac{a^2+b^2}{a^2-b^2}\right) = \dfrac{1}{5}\left(\dfrac{b^2+c^2}{b^2-c^2}\right) = \dfrac{1}{6}\left(\dfrac{c^2+a^2}{c^2-a^2}\right)$, prove that $25a^2 + 27b^2 + 22c^2 = 0$.

18. If $\dfrac{a^2}{x^2-yz} = \dfrac{b^2}{y^2-zx} = \dfrac{c^2}{z^2-xy}$,

show that $a^2x + b^2y + c^2z = (a^2+b^2+c^2)(x+y+z)$.

19. If $\dfrac{x}{a+b-c} = \dfrac{y}{b+c-a} = \dfrac{z}{c+a-b}$,

then will $(a-b)x + (b-c)y + (c-a)z = 0$.

20. If $\dfrac{a}{b} = \dfrac{c}{d} = \dfrac{e}{f}$, then $\left(\dfrac{a^2+c^2+e^2}{b^2+d^2+f^2}\right)^2 = \dfrac{a^4+c^4+e^4}{b^4+d^4+f^4}$.

21. If $\dfrac{bx+ay}{a-b} = \dfrac{cy+bz}{b-c} = \dfrac{az+cx}{c-a}$,

show that $(a+b+c)(x+y+z) = ax+by+cz$.

22. If $\dfrac{x^3-5x^2a-a^3+5xa^2}{x^3+x^2a+xa^2+a^3} = \dfrac{x-a}{x+a}$, show that each of

these expressions $= 1$.

23. If $\dfrac{1}{6}\left(\dfrac{a-b}{a+b}\right) = \dfrac{1}{5}\left(\dfrac{b-c}{b+c}\right) = \dfrac{1}{10}\left(\dfrac{c-a}{c+a}\right)$, and a, b, c be

unequal, show that $16a + 11b + 15c = 0$.

24. If $\left(\dfrac{x+yz}{y+zx}\right)^2 = \dfrac{1-y^2}{1-x^2}$, prove that $x^2+y^2+z^2+2xyz = 1$.

25. If $\dfrac{a}{x-y} = \dfrac{b}{y-z} = \dfrac{c}{z-x}$, show that $a+b+c = 0$.

26. If $\dfrac{a}{b} = \dfrac{c}{d}$, prove that $\dfrac{a+b}{a-b} = \dfrac{\sqrt{(ac)}+\sqrt{(bd)}}{\sqrt{(ac)}-\sqrt{(bd)}}$.

27. If $\dfrac{a}{b} = \dfrac{c}{d} = \dfrac{e}{f}$, then each is equivalent to $\dfrac{la+mc+ne}{lb+md+nf}$.

Hence, show that

$$\dfrac{a}{2z+2x-y} = \dfrac{b}{2x+2y-z} = \dfrac{c}{2y+2z-x},$$

when $\dfrac{x}{2a+2b-c} = \dfrac{y}{2b+2c-a} = \dfrac{z}{2c+2a-b}$.

28. If $\dfrac{a}{b} = \dfrac{c}{d}$, prove that $\left(\dfrac{a-b}{c-d}\right)^n = \sqrt{\left(\dfrac{a^{2n}+b^{2n}}{c^{2n}+d^{2n}}\right)}$.

29. If $\dfrac{x}{a(y+z)} = \dfrac{y}{b(x+z)} = \dfrac{z}{c(x+y)}$,

prove that $\dfrac{x}{a}(y-z) + \dfrac{y}{b}(z-x) + \dfrac{z}{c}(x-y) = 0.$

30. If $\dfrac{a}{lx(ny-mz)} = \dfrac{b}{my(lz-nx)} = \dfrac{c}{nz(mx-ly)}$,

then will $\dfrac{a}{lx}(l-x) + \dfrac{b}{my}(m-y) + \dfrac{c}{nz}(n-z) = 0.$

31. If $z = \dfrac{\sqrt{(ay^2-a^2)}}{y}$, and $y = \dfrac{\sqrt{(ax^2-a^2)}}{x}$,

show that $x = \dfrac{\sqrt{(az^2-a^2)}}{z}.$

32. If $\dfrac{x^2-yz}{a^2} = \dfrac{y^2-xz}{b^2} = \dfrac{z^2-xy}{c^2} = 1$,

show that $x+y+z = \dfrac{a^2x + b^2y + c^2z}{a^2 + b^2 + c^2}.$

33. If $\dfrac{m}{x} = \dfrac{n}{y} = \dfrac{r}{z}$, and $\dfrac{x^2}{a^2} = \dfrac{y^2}{b^2} = \dfrac{z^2}{c^2} = 1$,

prove that $\dfrac{m^2}{a^2} + \dfrac{n^2}{b^2} + \dfrac{r^2}{c^2} = 3\dfrac{m^2+n^2+r^2}{x^2+y^2+z^2}.$

34. If $\dfrac{a}{b} = \dfrac{c}{d} = \dfrac{e}{f}$, then $\dfrac{a^{3n}-c^{3n}}{b^{3n}-d^{3n}} = \dfrac{a^n c^n e^n - (a^n - c^n + e^n)^3}{b^n d^n f^n - (b^n - d^n + f^n)^3}.$

35. If $\dfrac{a_1}{b_1} = \dfrac{a_2}{b_2} = \dfrac{a_3}{b_3} = \cdots\cdots = \dfrac{a_n}{b_n}$,

then $\dfrac{a_1 a_2 - a_2 a_3 + \cdots\cdots (-1)^{n-1} a_{n-1} a_n}{b_1 b_2 - b_2 b_3 + \cdots\cdots (-1)^{n-1} b_{n-1} b_n}$

$= \dfrac{a_1 \sqrt{a_2 a_3} + a_2 \sqrt{a_3 a_4} + \cdots\cdots}{b_1 \sqrt{b_2 b_3} + b_2 \sqrt{b_3 b_4} + \cdots\cdots}.$

36. If $\dfrac{A+B+C}{abc} = \dfrac{A}{a} + \dfrac{B}{b} + \dfrac{C}{c}$,

and $(A+B+C)(a+b+c) = Aa + Bb + Cc$,

then will $\dfrac{A}{1+a^2} + \dfrac{B}{1+b^2} + \dfrac{C}{1+c^2} = 0$,

and also $\dfrac{A}{a+\dfrac{1}{a}} + \dfrac{B}{b+\dfrac{1}{b}} + \dfrac{C}{c+\dfrac{1}{c}} = 0$.

37. If $\dfrac{xh}{a^2} = \dfrac{yk}{b^2} = \dfrac{zl}{c^2}$, and $\dfrac{x^2}{a^2} = \dfrac{y^2}{b^2} = \dfrac{z^2}{c^2} = 1$,

then will $\dfrac{1}{3}\left(\dfrac{x}{h} + \dfrac{y}{k} + \dfrac{z}{c}\right)^2 = \dfrac{a^2}{h^2} + \dfrac{b^2}{k^2} + \dfrac{c^2}{l^2}$.

COMPLETE SQUARES, ETC.

1. What quantity must be added to $x^2 + px$ to make it a complete square?

Let r be the quantity.

Then $x^2 + px + r = $ complete square $= (x + \sqrt{r})^2$
$= x^2 + 2x\sqrt{r} + r$.

Equating coefficients, we have

$$2\sqrt{r} = p; \text{ hence, } r = \frac{p^2}{4} = \left(\frac{p}{2}\right)^2.$$

Or thus: Since $(a + x)^2 = a^2 + 2ax + x^2$, we observe (see § 12) that *four times the product of the extremes is equal to the square of the mean;* hence,

$$4x^2r = p^2x^2; \text{ therefore, } r = \left(\frac{p}{2}\right)^2, \text{ as before.}$$

Or, we may extract the square root and equate the remainder to zero ; thus :

$$x^2 + px + r \ (x + \frac{p}{2}$$

$$x^2$$
$$\overline{}$$

$$2x + \frac{p}{2} \qquad px + r$$

$$px + \frac{p^2}{4}$$
$$\overline{}$$

$$r - \frac{p^2}{4}$$

Now, if the expression be a complete square, this remainder must vanish ; hence, we have

$$r = \frac{p^2}{4} = \left(\frac{p}{2}\right)^2.$$

2. Find the relation connecting a, b, c, if $ax^2 + bx + c$ is a complete square.

Assume $ax^2 + bx + c = (\sqrt{a} \cdot x + \sqrt{c})^2$
$= ax^2 + 2\sqrt{(ac)}x + c$.

Now, since this holds for all values of x, we have
$2\sqrt{(ac)} = b$, or $b^2 = 4ac$, the relation required.

3. Determine the relation amongst a, b, c, in order that $a^2x^2 + bx + bc + b^2$ may be a perfect square.

As in Exam. 1, we have $4a^2x^2(bc + b^2) = b^2x^2$;

hence, $\dfrac{1}{4a^2} - \dfrac{c}{b} = 1$.

Or thus :

Assume $a^2x^2 + bx + bc + b^2 = (ax + \sqrt{bc + b^2})^2$
$= a^2x^2 + 2ax\sqrt{bc + b^2} + bc + b^2$.

Equating coefficients, we have $b = 2a\sqrt{bc + b^2}$;

hence, $\dfrac{1}{4a^2} - \dfrac{c}{b} = 1$, as before.

.The same result may also be obtained by extracting the square root and equating the remainder to zero.

4. Show that if $x^4 + ax^3 + bx^2 + cx + d$ be a complete square, the coefficients satisfy the equation
$c^2 - a^2d = 0$.

Is it necessary that the coefficients satisfy any other equation?

Extracting the square root of $x^4 + ax^3 + bx^2 + cx + d$ in the usual manner, we have for the final remainder

$$\left[c - \frac{a}{2}\left(b - \frac{a^2}{4} \right) \right] x + d - \frac{1}{4}\left(b - \frac{a^2}{4} \right)^2.$$

Now, if the expression be a complete square, this remainder must vanish; and, that it may vanish for general values of x, we must have

$$c - \frac{a}{2}\left(b - \frac{a^2}{4} \right) = 0 \tag{1}$$

$$d - \frac{1}{4}\left(b - \frac{a^2}{4} \right)^2 = 0 \tag{2}$$

Eliminating $b - \frac{a^2}{4}$, we have $c^2 - a^2d = 0$ (3)

The coefficients must satisfy the equations (1) and (2), and therefore either of these equations, together with the equation (3), which results from them.

The same result may be obtained by assuming

$$x^4 + ax^3 + bx^2 + cx + d = (x^2 + \tfrac{1}{2}ax + \sqrt{d})^2$$
$$= x^4 + ax^3 + 2x^2\sqrt{d} + \tfrac{1}{4}a^2x^2 + ax\sqrt{d} + d.$$

Equating coefficients, we have $2\sqrt{d} + \tfrac{1}{4}a^2 = b$ (1)
and $a\sqrt{d} = c$ (2)
From (2) we have $c^2 - a^2d = 0$, as before.

5. What must be the value of m and of n
if $4x^4 - 12x^3 + 25x^2 - 4mx + 8n$ is a perfect square?
Assume the expression $= [(2x^2 - 3x + \sqrt{(8n)}]^2$
$= 4x^4 - 12x^3 + 4x^2\sqrt{(8n)} + 9x^2 - 6x\sqrt{(8n)} + 8n.$
Equating coefficients, we have $6\sqrt{(8n)} = 4m$ $\quad\quad$ (1)
and $\quad\quad\quad\quad\quad\quad\quad 4\sqrt{(8n)} + 9 = 25$ $\quad\quad$ (2)
Therefore, $n = 2$, $m = 6$.
Or thus: Extracting the square root in the ordinary
way, the remainder is found to be
$(-4m + 24)x + 8n - 16$; hence, we must have
$4m + 24 = 0$, or $m = 6$; and $8n - 16 = 0$, or $n = 2$.

6. If $ax^3 + bx^2 + cx + d$ be a complete cube, show that
$ac^3 = db^3$ and $b^2 = 3ac$.
Assume $ax^3 + bx^2 + cx + d = (a^{\frac{1}{3}}x + d^{\frac{1}{3}})^3$
$= ax^3 + 3a^{\frac{2}{3}}d^{\frac{1}{3}}x^2 + 3a^{\frac{1}{3}}d^{\frac{2}{3}}x + d.$
Equating coefficients, $b = 3a^{\frac{2}{3}}d^{\frac{1}{3}}$ $\quad\quad$ (1)
$\quad\quad\quad\quad\quad\quad\quad\quad c = 3a^{\frac{1}{3}}d^{\frac{2}{3}}$ $\quad\quad$ (2)
Dividing (1) by (2), $\dfrac{b}{c} = \dfrac{a^{\frac{1}{3}}}{d^{\frac{1}{3}}}$; hence, $ac^3 = db^3$.
Also, $\quad\quad\quad\quad\quad\quad b^2 = 9a^{\frac{4}{3}}d^{\frac{2}{3}}$ $\quad\quad$ (3)
Dividing (3) by (2), $\dfrac{b^2}{c} = 3a$; hence, $b^2 = 3ac$.

7. Find the relations subsisting between a, b, c, d, e, when
$ax^4 + bx^3 + cx^2 + dx + e$ is a complete *fourth* power.
Assume $ax^4 + bx^3 + cx^2 + dx + e = (a^{\frac{1}{4}}x + e^{\frac{1}{4}})^4$
$= ax^4 + 4a^{\frac{3}{4}}e^{\frac{1}{4}}x^3 + 6a^{\frac{1}{2}}e^{\frac{1}{2}}x^2 + 4a^{\frac{1}{4}}e^{\frac{3}{4}}x + e.$
Equating coefficients, we have
$\quad\quad\quad\quad b = 4a^{\frac{3}{4}}e^{\frac{1}{4}},$
$\quad\quad\quad\quad c = 6a^{\frac{1}{2}}e^{\frac{1}{2}},$
$\quad\quad\quad\quad d = 4a^{\frac{1}{4}}e^{\frac{3}{4}};$
whence, $\quad\quad bd = 16ae$ $\quad\quad$ (1)
$\quad\quad\quad bc = 24a^{\frac{5}{4}}e^{\frac{3}{4}} = 6a \times 4a^{\frac{1}{4}}e^{\frac{3}{4}} = 6ad$ $\quad\quad$ (2)
$\quad\quad\quad cd = 24a^{\frac{3}{4}}e^{\frac{5}{4}} = 6e \times 4a^{\frac{3}{4}}e^{\frac{1}{4}} = 6be$ $\quad\quad$ (3)

8. Show that $x^4 + px^3 + qx^2 + rx + s$ can be resolved into two rational quadratic factors if s be a perfect square, negative, and equal to

$$\frac{r^2}{p^2 - 4q}.$$

Since $-s$ is a perfect square, let it be n^2.

Assume $x^4 + px^3 + qx^2 + rx - n^2$

$= (x^2 + mx + n)(x^2 + m'x - n)$

$= x^4 + (m + m')x^3 + mm'x^2 - n(m - m')x - n^2.$

Equating coefficients, we have

$$m + m' = p,$$
$$mm' = q,$$
$$m - m' = \frac{r}{n},$$
$$m^2 + 2mm' + m'^2 = p^2,$$
$$4mm' = 4q.$$

Hence, $(m - m')^2 = p^2 - 4q = \dfrac{r^2}{n^2}.$

Therefore, $\dfrac{r^2}{p^2 - 4p^2} = n^2 = -s.$

Ex. 49.

1. What is the condition that $(a - x)(b - x) - c^2$ may be a perfect square ?

2. Find the value of n which will make $2x^2 + 8x + n$ a perfect square.

3. Find a value of x which will make
$x^4 + 6x^3 + 11x^2 + 3x + 31$ a perfect square.

4. Extract the square root of
$(a - b)^4 - 2(a^2 + b^2)(a - b)^2 + 2(a^4 + b^4).$

5. Find the values of m and n which will make
$4x^4 - 4x^3 + 5x^2 - mx + n$ a perfect square.

6. What must be added to $x^4 - \sqrt{(4x^4 - 16x^2 + 16)} - 4x^2$ in order to make it a complete square?

7. The expression $x^4 + x^3 - 16x^2 - 4x + 48$ is resolvable into two factors of the form $x^2 + mx + 6$ and $x^2 + nx + 8$. Determine the factors.

8. Find the value of c which will make
$$4x^4 - cx^3 + 5x^2 + \frac{cx}{2} + 1 \text{ a complete square.}$$

9. Obtain the square root of
$$4[(a^2 - b^2)cd + ab(c^2 - d^2)]^2 + [(a^2 - b^2)(c^2 - d^2) - 4abcd]^2.$$

10. If $(a - b)x^2 + (a + b)^2 x + (a^2 - b^2)(a + b)$ be a complete square, then $a = 3b$, or $b = 3a$.

11. Find the simplest quantity which, subtracted from $a^2x^2 + 4abx + 4acx + 5bc + b^2c^2$, will give for remainder an exact square.

12. $x^4 - 4x^3 - x^2 + 16x - 12$ is resolvable into quadratic factors of the form $x^2 + mx + p$ and $x^2 + nx + q$; find them.

13. Find the values of m which will make $x^2 + max + a^2$ a factor of $x^4 - ax^3 + a^2x^2 - a^3x + a^4$.

14. Show that if $x^4 + ax^3 + bx^2 + cx + d$ be a perfect square, the coefficients satisfy the relations $8c = a(4b - a^2)$ and $64d = (4b - a^2)^2$.

15. Investigate the relations between the coefficients in order that $ax^2 + by^2 + cz^2 + dxy + eyz + fxz$ may be a complete square.

16. If $x^3 + ax^2 + bx + c$ be exactly divisible by $(x + d)^2$, show that $\frac{1}{2}(b^2 - d^2) = \frac{c}{d} = d(a - 2d)$.

17. Determine the relations among a, b, c, d, when $ax^3 - bx^2 + cx - d$ is a complete cube.

18. The polynome $ax^3 + 3bx^2 + 3cx + d$ is exactly divisible by $(a - x)^2$; show that
$$(ad - bc)^2 = 4(ac - b^2)(bd - c^2).$$

19. Find the relation between p and q when $x^3 + px^2 + q$ is exactly divisible by $(x - a)^2$.

20. If $x^2 + nax + a^2$ be a factor of $x^4 + ax^3 + a^2x^2 + a^3x + a^4$, show that $n^2 - n - 1 = 0$.

21. If $x^4 + ax^3 + bx^2 + cx + d$ be the product of two complete squares, show that
$$(4b - a^2)^2 = 64d, \quad (4b - a^2)a = 8c, \quad a\sqrt{(3a^2 - 2b)} = 3b.$$

22. Prove that $x^4 + px^3 + qx^2 + rx + s$ is a perfect square if $p^2s = r^2$ and $q = \dfrac{p^2}{4} + 2\sqrt{s}$.

23. If $ax^3 + 3bx^2 + 3cx + d$ contain $ax^2 + 2bx + c$ as a factor, the former will be a complete cube, and the latter a complete square.

24. If $m^2x^2 + px + pq + q^2$ be a perfect square, find p in terms of m and q.

25. Find the relation between p and q in order that $x^3 + px^2 + qx + r$ may contain $(x + 2)^2$ as a factor.

26. If $x^3 + px^2 + qx + r$ be algebraically divisible by $3x^2 + 2px + q$, show that the quotient is $x + \dfrac{p}{3}$.

CHAPTER V.

§ 38. PRELIMINARY EQUATIONS. Although the following exercise belongs in theory to this chapter, in practice the *numerical* examples should immediately follow Exercise I., and the literal examples Exercise III. Like those exercises, this one is merely a specimen of what the teacher should give till his pupils have thoroughly mastered this preliminary work. But few numerical examples are given, it being left to the teacher to supply these.

Ex. 50.

What values must x have that the following equations may be true?

1. $x - 5 = 0$; $x - 3\frac{1}{4} = 0$; $x - a = 0$; $x + 3 = 0$.

2. $x + 4\frac{1}{2} = 0$; $x + a = 0$; $x + 3 = 5$; $x - 4 = 6$.

3. $x - a = b$; $x + a = c$; $x - b = -c$; $6 - x = 3$.

4. $8 - x = 10$; $5 + x = 11$; $9 + x = 4$; $7 - x = -5$.

5. $8 + x = -6$; $a - x = 3b$; $2a = x + 3b$; $3a = 5b - x$.

6. $2x - 6 = 8$; $3x + 8 = 20$; $ax = a^2$; $mx = bm$.

7. $3x = c$; $ax = 5$; $ax = 0$; $(a + b)x = b + a$.

8. $(a - b)x = b - a$; $(a + b)x = (a + b)^2$; $(a - b)x = a^2 - b^2$.

9. $(a + b)x = b^2 - a^2$; $(a^2 - ab + b^2)x = a^3 + b^3$.

10. $(a^2 - b^2)x = a - b$; $(a^2 - b^2)x = a + b$; $(a^2 + b^2)x = 1$.

11. $a + x - b = a + b$; $\; x - a + b = b - x + a$.

12. $2a - x = x - 2b$; $\; ax + bx = c$; $\; ax - b = cx$.

13. $ax - b = bx - c$; $\; ax - ab = ac$.

14. $ax - a^2 = bx - b^2$; $\; ax - a^3 = bx - b^3$.

15. $ax - a^3 = b^3 - bx$; $\; ax + b + c = a + bx + cx$.

16. $a - bx - c = b - ax + cx$; $\; a + bx + cx^2 = ax - b + cx^2$.

17. $bx - cx^2 + e = ex - b - cx^2$; $\; 3x = \frac{2}{5}$; $\; 4x = \frac{6}{7}$.

18. $10x = \dfrac{1}{6} - 1$; $\; ax = \dfrac{b}{c}$; $\; ax = \dfrac{a^2}{b}$.

19. $abx = \dfrac{a}{b} + \dfrac{b}{a}$; $\; bcx = \dfrac{ac^2}{b} + \dfrac{ab^2}{c}$.

20. $\frac{1}{2}x = 5$; $\; \frac{2}{3}x = 8$; $\; 0.5x = 2$; $\; 0.3x = 0.06$.

21. $0.02x = 20$; $\; 0.\dot{3}x = 0.2$; $\; 0.\dot{4}x = 0.\dot{6}$.

22. $0.\dot{1}\dot{8}x = 1.8$; $\; \dfrac{x}{a} = b$; $\; \dfrac{ax}{b} = c$.

23. $\dfrac{ax}{b} = \dfrac{b}{c}$; $\; \dfrac{x}{a+b} = c$; $\; \dfrac{ax}{a+b} = b$.

24. $\dfrac{a+b}{a-b}x = \dfrac{a}{b}$; $\; \dfrac{a-b}{a+b}x = \dfrac{a+b}{b-a}$.

25. $\dfrac{a}{b-a}x = \dfrac{a}{a-b}$; $\; \dfrac{b-a}{a+b}x = \dfrac{a-b}{b+a}$.

26. $\dfrac{a+b}{a+c}x = \dfrac{a-c}{a+b}$; $\; \dfrac{1}{x} = \dfrac{1}{2}$; $\; \dfrac{2}{x} = \dfrac{3}{5}$.

27. $\dfrac{1}{x} = \dfrac{1}{ab}$; $\; \dfrac{1}{x} = \dfrac{a}{b}$; $\; \dfrac{a}{x} = \dfrac{b}{c}$; $\; \dfrac{7}{x} = \dfrac{1}{3} + \dfrac{1}{4}$.

28. $\dfrac{3}{20} + \dfrac{4}{5x} = \dfrac{33}{5x} - \dfrac{1}{3}$; $\; \dfrac{a}{x} + \dfrac{b}{c} = 0$.

29. $\dfrac{5}{x-7} = 6 - \dfrac{7}{x-7}$; $\; \dfrac{5}{3x-4} = 7 + \dfrac{9}{4-3x}$.

30. $(x-4)-(x+5)+x=3;\ 2x-(x-5)-(4-3x)=5.$

31. $2(3-x)+3(x-3)=0;\ 2(3x-4)-3(3-4x)+9(2-x)=10.$

32. $a(1-2x)-(2x-a)=1;\ x-5(a-x)=bx-5a.$

33. $mx(3a-4)+3mx-3a+1=0.$

34. $a(bx-c)+b(cx-a)+c(ax-b)=0.$

35. $a(ax-b)+b(cx-c)+c(cx-a)=0.$

36. $a(bx-a)+b(cx-b)+c(ax-c)=0.$

37. $a(x-2b)+b(x-2c)+c(x-2a)=a^2+b^2+c^2.$

38. $3\{3[3(3x-2)-2]-2\}-2=1.$

39. $9\{7[5(3x-2)-4]-6\}-8=1.$

40. $\frac{1}{3}\{\frac{1}{3}[\frac{1}{3}(\frac{1}{3}[x+2]+2)+2]+2\}=1.$

41. $\frac{1}{9}\{\frac{1}{7}[\frac{1}{5}(\frac{1}{3}[x+2]+4)+6]+8\}=1.$

42. $\frac{1}{2}\{\frac{1}{2}[\frac{1}{2}(\frac{1}{2}x-\frac{1}{2})-\frac{1}{2}]-\frac{1}{2}\}-\frac{1}{2}=0.$

43. $\frac{2}{3}\{\frac{2}{3}[\frac{2}{3}(\frac{2}{3}x-1\frac{1}{3})-1\frac{1}{3}]-1\frac{1}{3}\}-1\frac{1}{3}=0.$

44. $\frac{13}{18}\{\frac{8}{11}[\frac{5}{7}(\frac{3}{4}[\frac{2}{3}x+4]+8)+12]+20\}+32=58.$

45. $\frac{4}{5}\{\frac{3}{4}[\frac{2}{3}(\frac{1}{2}[x+7]-3)+6]-1\}=4.$

46. $r\{q[p(n[mx-a]-b)-c]-d\}-e=0.$

47. $(1+6x)^2+(2+8x)^2=(1+10x)^2.$

48. $9(2x-7)^2+(4x-27)^2=13(4x+15)(x+6).$

49. $(3-4x)^2+(4-4x)^2=2(5+4x)^2.$

50. $(9-4x)(9-5x)+4(5-x)(5-4x)=36(2-x)^2.$

§ 39. In solving fractional equations, the principles illustrated in the sections on fractions may frequently be applied with advantage, as in the following cases.

When an equation involves several fractions, we may *take two or more of them together.*

EXAMPLES.

1. Solve $\dfrac{8x+5}{14} + \dfrac{7x-3}{6x+2} = \dfrac{4x+6}{7}$.

Here, instead of multiplying through by the L.C.M. of the denominators, we combine the first fraction with the last, getting at once

$$\frac{7x-3}{6x+2} = \frac{7}{14} = \frac{1}{2}. \quad \therefore 7x-3 = 3x+1, \text{ and } x = 1.$$

2. Solve $\dfrac{2x+8\frac{1}{2}}{9} - \dfrac{13x-2}{17x-32} + \dfrac{x}{3} = \dfrac{7x}{12} - \dfrac{x+16}{36}$.

In this case, taking together all the fractions having only numerical denominators, we get

$$\frac{8x+34+12x-21x+x+16}{36} = \frac{13x-2}{17x-32};$$

or $\dfrac{25}{18} = \dfrac{13x-2}{17x-32}. \quad \therefore 425x - 800 = 234x - 36 ;$

hence, $x = 4$.

It is often advantageous to *complete* the *divisions* represented by the fractions.

3. Solve $\dfrac{4x-17}{9} - \dfrac{3\frac{2}{3} - 22x}{33} = x - \dfrac{6}{x}\left(1 - \dfrac{x^2}{54}\right)$.

Here, completing the divisions, we have

$$\frac{4x}{9} - \frac{17}{9} - \frac{1}{9} + \frac{2x}{3} = x - \frac{6}{x} + \frac{x}{9},$$

$$\frac{10x}{9} - 2 = x + \frac{x}{9} - \frac{6}{x}.$$

Therefore, $-2 = -\dfrac{6}{x}$, or $x = \mathbf{3}$.

4. $\dfrac{ax+b}{x-m} + \dfrac{cx+d}{x-n} = a + c.$

$\therefore a + \dfrac{am+b}{x-m} + c + \dfrac{cn+d}{x-n} = a + c.$

$\therefore (am+b)(x-n) + (cn+d)(x-m) = 0.$

$\therefore (am+b+cn+d)x = (a+c)mn + bn + dm.$

5. Similarly may be solved

$\dfrac{ax+b}{x-m} + \dfrac{cx+d}{x-n} + \dfrac{ex^2+fx-g}{(x-m)(x-n)} = a+c+e.$

$\therefore \dfrac{am+b}{x-m} + \dfrac{cn+d}{x-n} + \dfrac{[e(m+n)+f]x - emn - g}{(x-m)(x-n)} = 0.$

$\therefore (am+b)(x-n) + (cn+d)(x-m)$
$\quad + [e(m+n)+f]x - emn - g = 0.$

$\therefore [(a+c)m + b + (c+e)n + d + f]x$
$\quad = (a+b+e)mn + bn + dm + g.$

6. $\dfrac{132x+1}{3x+1} + \dfrac{8x+5}{x-1} = 52.$

$\therefore 44 - \dfrac{43}{3x+1} + 8 + \dfrac{13}{x-1} = 52,$ or $\dfrac{13}{x-1} = \dfrac{43}{3x+1}.$

$\therefore 39x + 13 = 43x - 43,$ and $x = 14.$

7. $\dfrac{25 - \frac{1}{3}x}{x+1} + \dfrac{16x + 4\frac{1}{5}}{3x+2} = 5 + \dfrac{23}{x+1}.$

Taking the last fraction with the first, and multiplying
the resulting equation by 15, we have

$\dfrac{240x+63}{3x+2} = 75 + \dfrac{5x-30}{x+1}.$

$\therefore 80 - \dfrac{97}{3x+2} = 75 + 5 - \dfrac{35}{x+1},$

or $\dfrac{97}{3x+2} = \dfrac{35}{x+1}.$ $\therefore 8x = 27,$ and $x = 3\frac{3}{8}.$

8. $\dfrac{x-a}{b+c}+\dfrac{x-b}{a+c}+\dfrac{x-c}{b+c}=3.$

$\therefore \dfrac{x-a}{b+c}-1+\dfrac{x-b}{a+c}-1+\dfrac{x-c}{b+a}-1=0.$

$\therefore \dfrac{x-(a+b+c)}{b+c}+\dfrac{x-(a+b+c)}{a+c}+\dfrac{x-(a+b+c)}{b+a}=0,$

which is satisfied by $x-(a+b+c)=0$;

$\therefore x=a+b+c.$

9. $\dfrac{m}{x-a}+\dfrac{n}{x-b}=\dfrac{m+n}{x-c}.$

$\therefore \dfrac{m(x-c)}{x-a}+\dfrac{n(x-c)}{x-b}=m+n.$

$\therefore \dfrac{m(a-c)}{x-a}+\dfrac{n(b-c)}{x-b}=0.$ (See Exam. 4.)

$\therefore [m(a-c)+n(b-c)]x=mb(a-c)+na(b-c).$

10. $\dfrac{1}{3}\left(\dfrac{y-2}{y-5}\right)+\dfrac{1}{6}\left(\dfrac{y-1}{y-7}\right)-\dfrac{2}{9}\left(\dfrac{y-1}{y-10}\right)=\dfrac{5}{18}.$

$\therefore \dfrac{1}{3}+\dfrac{1}{y-5}+\dfrac{1}{6}+\dfrac{1}{y-7}-\dfrac{2}{9}-\dfrac{2}{y-10}=\dfrac{5}{18}.$

$\therefore \dfrac{1}{y-5}+\dfrac{1}{y-7}-\dfrac{2}{y-10}=0.$

$\therefore \dfrac{y-10}{y-5}+\dfrac{y-10}{y-7}-2=0.$

$\therefore \dfrac{-5}{y-5}+\dfrac{-3}{y-7}=0.$

$\therefore -8y+50=0. \quad \therefore y=6\tfrac{1}{4}.$

Ex. 51.

1. $\dfrac{10x+17}{18}-\dfrac{12x+2}{13x-16}=\dfrac{5x-4}{9}.$

2. $\dfrac{6x+13}{15}-\dfrac{9x+15}{5x-25}+3=\dfrac{2x+15}{5}.$

3. $\dfrac{7\,x+1}{x-1}=\dfrac{35}{9}\left(\dfrac{x+4}{x+2}\right)+3\tfrac{1}{9}.$

4. $\dfrac{4\,x-7}{2\,x-9}+\dfrac{2-14\,x}{7}+\dfrac{3\frac{1}{3}+x}{14}=\dfrac{10-3\frac{6}{7}\,x}{2}-\dfrac{19}{21}.$

5. $\dfrac{2\,x+a}{3\,(x-a)}+\dfrac{3\,x-a}{2\,(x+a)}=2\tfrac{1}{6}.$

6. $\dfrac{x-4}{6\,x+5}+\dfrac{3\,x-13}{18\,x-6}=\dfrac{1}{3}.$

7. $\dfrac{3\,x+1}{2\,x-15}-\dfrac{x-11}{2\,x-10}=1.$

8. $\dfrac{x-9}{x-5}+\dfrac{x-5}{x-8}=2.$

9. $\dfrac{x-12}{x-7}+\dfrac{x-4}{x-12}=2+\dfrac{7}{x-7}.$

10. $\dfrac{3\,x-19}{x-13}+\dfrac{3\,x-11}{x+7}=6.$

11. $\dfrac{x-2}{2\,x+1}+\dfrac{x-1}{3\,(x-3)}=\dfrac{5}{6}$

12. $\dfrac{x+1}{4\,(x+2)}+\dfrac{x+4}{5\,x+13}=\dfrac{9}{20}.$

13. $\dfrac{5\,(2\,x^2+3)}{2\,x+1}+\dfrac{5-7\,x}{2\,x-5}=5\,x-6.$

14. $\dfrac{3}{x-7}+\dfrac{1}{x-9}=\dfrac{4}{x-8}.$

15. $\dfrac{7\,x+55}{2\,x+5}-\dfrac{3\,x}{2}=9-\dfrac{3\,x^2+8}{2\,x-4}.$

16. $\dfrac{17}{x-16}+\dfrac{15}{x-18}=\dfrac{32}{x-17}.$

17. $\dfrac{1-25x}{15} - \dfrac{3-2\frac{1}{2}x}{14(x-1)} = \dfrac{28-5x}{3} - \dfrac{10x-11}{30} + \dfrac{x}{3}.$

18. $\dfrac{1}{x-2} - \dfrac{2+2\frac{1}{2}x^2 - \frac{1}{2}x^3}{6-5x+x^2} = \frac{1}{2}x - \dfrac{2}{x-5}.$

19. $\dfrac{30+6x}{x+1} + \dfrac{60+8x}{x+3} = \dfrac{48}{x+1} + 14.$

20. $\dfrac{5x^2+x-3}{5x-4} = \dfrac{7x^2-3x-9}{7x-10}.$

21. $\dfrac{x}{x-2} + \dfrac{x-9}{x-7} = \dfrac{x+1}{x-1} + \dfrac{x-8}{x-6}.$

22. $\dfrac{x^2-3x-9}{x-5} + \dfrac{x^2-7x-17}{x-9} = \dfrac{2(x^2-6x-15)}{x-8}.$

23. $\dfrac{4x+7}{4x+5} + \dfrac{4x+9}{4x+7} = \dfrac{4x+6}{4x+4} + \dfrac{4x+10}{4x+8}.$

24. $\dfrac{2x-3}{2x-4} - \dfrac{2x-4}{2x-5} = \dfrac{2x-7}{2x-8} - \dfrac{2x-8}{2x-9}.$

25. $\dfrac{7x+6}{28} - \dfrac{2x+4\frac{2}{7}}{23x-6} + \dfrac{x}{4} = \dfrac{11x}{21} - \dfrac{x-3}{42}.$

26. $\dfrac{x-5}{x-6} + \dfrac{x-11}{x-12} = \dfrac{x-7}{x-8} + \dfrac{x-9}{x-10}$

27. $\dfrac{x-1\frac{25}{26}}{2} - \dfrac{2-6x}{13} = x - \dfrac{5x-\frac{1}{4}(10-3x)}{39}.$

28. $\dfrac{10-17x}{27(x^2-x+1)} + \dfrac{1+x}{2(x^2+1)} + \dfrac{7}{54(x+1)} = \dfrac{1}{9(x^2+1)}.$

29. $\dfrac{2x^2+x-30}{2x-7} + \dfrac{x^2+4x-4}{x-1} = \dfrac{x^2-17}{x-4} + \dfrac{2x^2+7x-13}{2x-3}.$

30. $\dfrac{x-a}{x-b} + \dfrac{x-b}{x-a} - \dfrac{(a-b)^2}{(x-a)(x-b)} = \dfrac{2(a-x)}{a+x}.$

31. $\dfrac{12x+10a}{3x+a}+\dfrac{28x+117a}{2x+9a}=18.$

32. $\dfrac{13\frac{1}{2}x-5}{13\frac{1}{2}x-6}+\dfrac{13\frac{1}{2}x-11}{13\frac{1}{2}x-12}=\dfrac{13\frac{1}{2}x-7}{13\frac{1}{2}x-8}+\dfrac{13\frac{1}{2}x-9}{13\frac{1}{2}x-10}.$

33. $\dfrac{1}{2(x-1)^2}+\dfrac{1}{2(x-1)}-\dfrac{x}{2(x^2+1)}=\dfrac{16x}{(x-1)(x^2+1)}.$

34. $\frac{1}{2}(\frac{2}{3}x+4)-\dfrac{7\frac{1}{2}-x}{3}=\dfrac{x}{2}\Big(\dfrac{6}{x}-1\Big).$

35. $\dfrac{3x}{2}-\dfrac{81x^2-9}{(3x-1)(x+3)}=3x-\dfrac{3}{2}\Big(\dfrac{2x^2-1}{x+3}\Big)-\dfrac{57-3x}{2}.$

36. $1+\dfrac{2x+1}{2(x-1)}-\dfrac{4x+5}{2(x+1)}=\dfrac{x^2-x+2}{x^2-2x+1}-1.$

37. $\dfrac{7x-30}{10\frac{1}{2}}-\dfrac{5x-7}{\frac{1}{2}x-3}-\dfrac{2-21x}{21}$

$$=\dfrac{42x-171}{63}-10+\dfrac{2x-9}{63-14x}-\tfrac{1}{7}(4-7x).$$

38. $\dfrac{18x-22}{13-2x}+6x+\dfrac{1+6x}{8}=13\frac{1}{4}-\dfrac{101-54x}{8}.$

39. $\dfrac{4-9x}{1-3x}-\dfrac{5-12x}{7-4x}=2-\dfrac{24x^2-5}{7-25x+12x^2}.$

40. $\dfrac{8x+25}{2x+5}+\dfrac{16x+93}{2x+11}=\dfrac{18x+86}{2x+9}+\dfrac{6x+26}{2x+7}.$

41. $\dfrac{1}{x+a+b}-\dfrac{1}{x-a+b}-\dfrac{1}{x+a-b}+\dfrac{1}{x-a-b}=0.$

§ 40. The results deduced in the sections on ratios may often be applied with advantage.

Examples.

1. $\dfrac{ax+b}{cx+d}=\dfrac{m}{n}.$

$\therefore \dfrac{(ax+b)\,d-(cx+d)\,b}{(cx+d)\,a-(ax+b)\,c}=\dfrac{md-nb}{na-mc}.$ (Page 156)

$\therefore x=\dfrac{md-nb}{na-mc}.$

2. $\dfrac{ax^{2}+bx+c}{mx^{2}+nx+p}=\dfrac{a}{m}.$

$\therefore \dfrac{(ax^{2}+bx+c)-ax^{2}}{(mx^{2}+nx+p)-mx^{2}}=\dfrac{a}{m}.$ (Page 155)

$\therefore \dfrac{bx+c}{nx+p}=\dfrac{a}{m},$ etc.

3. $\dfrac{3x+7}{x+4}=\dfrac{3x-13}{x-4}.$

By (5), page 155, each of these fractions

$=\dfrac{\text{difference of numerators}}{\text{difference of denominators}}.$

$\therefore \dfrac{20}{8}=\dfrac{3x+7}{x+4}=3-\dfrac{5}{x+4},$ or $\dfrac{1}{2}=\dfrac{5}{x+4};$ $\therefore x=6.$

4. $\dfrac{mx+a+b}{nx-c-d}=\dfrac{mx+a+c}{nx-b-d}.$

$\therefore \dfrac{mx+a+b}{mx+a+c}=\dfrac{nx-c-d}{nx-b-d};$ or by (2), page 155,

$\dfrac{mx+a+b}{b-c}=\dfrac{nx-c-d}{b-c};$

or $(n-m)x=a+b+c+d;$ $\therefore x=\cdots\cdots.$

5. $\dfrac{1}{x-6a}+\dfrac{2}{x+3a}+\dfrac{3}{x-2a}=\dfrac{6}{x-a}.$

Transposing, $\dfrac{1}{x-6a}+\dfrac{2}{x+3a}=\dfrac{6}{x-a}-\dfrac{3}{x-2a}.$

$\therefore \dfrac{3x-9a}{x^2-3ax-18a^2}=\dfrac{3x-9a}{x^2-3ax+2a^2}=\dfrac{0}{20a^2}$ by (5) of § 37.

$\therefore 3x-9a=0.$

$\therefore x=3a.$

Ex. 52.

1. $\dfrac{1+x}{1-x}=\dfrac{1}{a}.$

2. $\dfrac{x+a}{x-a}=m.$

3. $\dfrac{ax+b}{ax-b}=\dfrac{m}{n}.$

4. $\dfrac{a+x}{b+2x}=1.$

5. $\dfrac{a(b+x)}{a-x}=b.$

6. $\dfrac{a}{a-x}=\dfrac{b}{b-x}.$

7. $\dfrac{a+x}{a-x}=\dfrac{a+b}{a-b}.$

8. $\dfrac{x+m}{x-m}=\dfrac{a+b}{a-b}.$

9. $\dfrac{a+b}{1+cx}=\dfrac{a-b}{1-cx}.$

10. $\dfrac{a+bx}{a+b}=\dfrac{c+dx}{c+d}.$

11. $\dfrac{a+bx}{a-b}=\dfrac{c+dx}{c-d}.$

12. $\dfrac{a-x}{b-x}=\dfrac{a+x}{b+x}.$

13. $\dfrac{2x^2-5x+9}{2x^2-7x+3}=\dfrac{x^2-7x+5}{x^2-8x+2}.$

14. $\dfrac{ax+b-c}{ax-b+c}=\dfrac{(b-c)^2}{(b+c)^2}.$

15. $\dfrac{a^2}{a^2+ab+b^2}-\dfrac{a^2x}{a^3-b^3}=\dfrac{2c}{a-b}-2cx.$

16. $\dfrac{2x-7}{2x-3}=\dfrac{x+7}{x+11}.$

17. $\dfrac{4x-5}{2x+10}=\dfrac{10x-32}{5x-8\frac{1}{2}}.$

18. $\dfrac{57x-43}{19x+13}=\dfrac{39x-7}{13x+25}.$

19. $\dfrac{23x+5\frac{4}{5}}{115x-29}=\dfrac{36x-7}{180x+23}.$

20. $\dfrac{210x-73}{310x-66}=\dfrac{21x+7.3}{31x+8}.$

21. $\dfrac{mx-a-b}{mx-c-d}=\dfrac{mx-a-c}{nx-b-d}.$

22. $\dfrac{x^3 + ax^2 - bx + c}{x^3 - ax^2 + bx + c} = \dfrac{x^2 + ax - b}{x^2 - ax + b}.$

23. $\dfrac{a^3x^3 + a^2bx^2 - acx + d}{a^3x^3 - a^2bx^2 + acx + d} = \dfrac{a^2x^2 + abx - c}{a^2x^2 - abx + c}.$

24. $\dfrac{8x^3 + 12x^2 - 8x + 5}{8x^3 - 12x^2 + 8x + 5} = \dfrac{4x^2 + 6x - 4}{4x^2 - 6x + 4}.$

25. $\dfrac{1}{ax} + \dfrac{1}{bx} + \dfrac{1}{cx} = \tfrac{1}{2}(a + \overset{\cdot}{b} + c)^2 - \dfrac{1}{2}\left(\dfrac{a}{bcx} + \dfrac{b}{acx} + \dfrac{c}{abx}\right).$

26. $\dfrac{1 - ax}{bc} + \dfrac{1 - bx}{ac} + \dfrac{1 - cx}{ab} = \left(\dfrac{2}{a} + \dfrac{2}{b} + \dfrac{2}{c}\right)x.$

27. $\dfrac{(a - b)^2}{abc} - 1 + \dfrac{a}{b} = \dfrac{a^2 - b^2}{abc} + \left(1 + \dfrac{a}{b}\right)x.$

28. $\dfrac{x}{(a + b)^2} + \dfrac{ac}{(a - b)^3} - \dfrac{c}{(a^2 - b^2)(a + b)} = \dfrac{ax}{(a - b)^2}.$

29. $\dfrac{x + a}{(a - b)(c - a)} - \dfrac{x - b}{(a - b)(b - c)} + \dfrac{x + c}{(b - c)(c - a)}$

$= \dfrac{a + c}{(a - b)(b - c)(c - a)}.$

30. $x\left(\dfrac{x + 2a}{x - a}\right)^3 + a\left(\dfrac{a + 2x}{a - x}\right)^3 = 2a.$

31. $\dfrac{x + a}{x^2 + ax + a^2} - \dfrac{x - a}{x^2 - ax + a^2} = \dfrac{a^4}{x(x^4 + a^2x^2 + a^4)}.$

32. $\dfrac{x + a}{x + b} = \left(\dfrac{2x + a + c}{2x + b + c}\right)^2.$

33. $\dfrac{1}{(x + a)^2 - b^2} + \dfrac{1}{(x + b)^2 - a^2} = \dfrac{1}{x^2 - (a + b)^2} + \dfrac{1}{x^2 - (a - b)^2}.$

34. $41\left(\dfrac{6x + 45}{x + 1} + \dfrac{7x + 67}{x + 4}\right) + 130 = 39\left(\dfrac{8x + 57}{x + 2} + \dfrac{9x + 68}{x + 3}\right).$

§ **41.** Sometimes a factor independent of x can be discovered and rejected.

EXAMPLES.

1. $\dfrac{3\,abc}{a+b} - \dfrac{bx}{a} + \dfrac{a^2 b^2}{(a+b)^3} = 3\,cx - \dfrac{b^2 x}{a}\left(\dfrac{2a+b}{(a+b)^2}\right).$

Transpose $\dfrac{bx}{a}$ and factor ; then,

$$\dfrac{ab}{a+b}\left[3\,c + \dfrac{ab}{(a+b)^2}\right] = x\left[3\,c + \dfrac{b}{a}\left(1 - \dfrac{2\,ab+b^2}{(a+b)^2}\right)\right]$$

$$= x\left[3\,c + \dfrac{b}{a}\left(\dfrac{a^2}{(a+b)^2}\right)\right]$$

$$= x\left[3\,c + \dfrac{ab}{(a+b)^2}\right].$$

$$\therefore \dfrac{ab}{a+b} = x.$$

2. $\dfrac{x+a}{(a-b)(c-a)} - \dfrac{x-b}{(a-b)(b-c)} - \dfrac{x-c}{(b-c)(c-a)}$

$$= \dfrac{b+c}{(a-b)(b-c)(c-a)}.$$

Add, term by term, the identity (Th. III., page 67),

$$\dfrac{x-a}{(a-b)(c-a)} + \dfrac{x-b}{(a-b)(b-c)} + \dfrac{x-c}{(b-c)(c-a)} = 0.$$

$$\therefore \dfrac{2x}{(a-b)(c-a)} = \dfrac{b+c}{(a-b)(b-c)(c-a)}.$$

$$\therefore x = \dfrac{1}{2}\left(\dfrac{b+c}{b-c}\right).$$

3. $(x+a+b)^3 + (a+b)^3 - (x+b)^3 - (x+a)^3$
$\qquad + x^3 + a^3 + b^3 = abc.$

The left-hand member vanishes for $x=0$, and hence, by symmetry, for $a=0$ and $b=0$; therefore, it is of the form $mabx$, in which m is *numerical*.

Put $x=a=b$, and m is found to be 6.

Hence, the equation reduces to

$\qquad 6\,abx = abc$, and $\therefore x = \tfrac{1}{6}c.$

4. $\left(\dfrac{x-a}{x-b}\right)^3 = \dfrac{x-2a+b}{x-2b+a}.$

Let $x-b=m$, $x-a=n$, and hence, $m-n=a-b$; then we have

$$\frac{n^3}{m^3} = \frac{n-(m-n)}{m+(m-n)} = \frac{2n-m}{2m-n}.$$

$\qquad \therefore 2mn^3 - n^4 = 2m^3n - m^4.$

$\qquad \therefore m^4 - n^4 - 2mn(m^2 - n^2) = 0.$

$\qquad \therefore (m^2 - n^2)(m^2 + n^2 - 2mn) = 0.$

$\qquad \therefore (m+n)(m-n)^3 = 0.$

But $m-n = a-b$, and rejecting this factor, which does not contain x,

$\qquad m+n = 0.$

But $m+n = 2x - a - b.$

$\qquad \therefore 2x - a - b = 0.$

$\qquad \therefore x = \tfrac{1}{2}(a+b).$

Ex. 53.

1. $a(b-x) + b(c-x) = b(a-x) + cx.$

2. $(a+bx)(a-b) - (ax-b) = ab(x+1).$

3. $(a-b)(x-c) + (a+b)(x+c) = 2(bx+ad).$

4. $(a-b)(x-c) - (a+b)(x+c) + 2a(b+c) = 0.$

5. $(a-b)(a-c)(a+x)+(a+b)(a+c)(a-x)=0.$

6. $(a-b)(a-c+x)+(a+b)(a+c-x)=2a^2.$
   Solve in $(x-c)$.

7. $(m+a)(a+b-x)+(a-m)(b-x)=a(m+b).$

8. $m(a+b-x)=n(x-a-b).$

9. $(m+n)(m-n-x)+m(x-n)-n(x-m)=m^2-n^2.$

10. $\dfrac{m-x}{m}+\dfrac{n-x}{n}+\dfrac{p-x}{p}=3.$

11. $\dfrac{a^2b-x}{a}+\dfrac{b^2c-x}{b}+\dfrac{c^2a-x}{c}=0.$

12. $\dfrac{a-x}{bc}+\dfrac{b-x}{ca}+\dfrac{c-x}{ab}=0.$

13. $\dfrac{1-ax}{bc}+\dfrac{1-bx}{ca}+\dfrac{1-cx}{ab}=0.$
   Deduce the solution from that of No. 12.

14. $\dfrac{a-bx}{bc}+\dfrac{b-cx}{ca}+\dfrac{c-ax}{ab}=0.$

15. $(a+b+c)x-\dfrac{a^2+b^2}{a-b}=\dfrac{2abx}{a+b}+\left(\dfrac{a+b}{a-b}\right)c.$

16. $\dfrac{3abc}{a+b}+\dfrac{a^2b^2}{(a+b)^3}+\dfrac{(2a+b)b^2x}{a(a+b)^2}=\dfrac{(b+3ac)x}{a}.$

17. $\dfrac{10}{x}+\dfrac{4}{9}=\dfrac{9}{x}+\dfrac{2}{3}.$ Solve in $\dfrac{1}{x}.$

18. $\dfrac{7}{x}+\dfrac{1}{3}=\dfrac{23-x}{3x}+\dfrac{7}{12}-\dfrac{1}{4x}.$

19. $\dfrac{7}{3}+\dfrac{13}{5x}=\dfrac{2(5x-12)}{3x}-\dfrac{17}{20}+\dfrac{10}{x}.$

20. $\dfrac{10-x}{3}+\dfrac{13+x}{7}=\dfrac{7x+266}{x+21}-\dfrac{4x+17}{21}.$

21. $\dfrac{5}{x+3} + \dfrac{3}{2(x+3)} = \dfrac{1}{2} - \dfrac{7}{2(x+3)}.$

22. $\dfrac{6x+5}{8x-15} - \dfrac{1+8x}{15} = \dfrac{1-x}{3} - \dfrac{x-3}{5}.$

23. $\dfrac{a - \dfrac{1}{x}}{a + \dfrac{1}{x}} - \dfrac{1}{x} = \dfrac{x - \dfrac{1}{a}}{x + \dfrac{1}{a}} - \dfrac{1}{a}.$

24. $\dfrac{a^2}{b - \dfrac{c^2}{d - \dfrac{e^2}{x}}} = 1.$

25. $(x-1)(x-2) - (x-3)(x-4) = 3.$

26. $(x-3)(x-4) = (x-2)(x-6).$

27. $2(x-4)(3x+4) + (2x-3)(3x+2)$
$\qquad - 6(x-2)(2x-3) = 0.$

28. $(a-x)(b-x) = x^2.$ **30.** $(a-x)(b+x) = b^2 - x^2.$

29. $(a-x)(x-b) = x^2 - c^2.$ **31.** $(x-a)(x-b) = x^2 - a^2.$

32. $(a+x)(b+x) = (a-x)(b-x).$

33. $(ax+b)(bx+a) = (b-ax)(a-bx).$

34. $(a-x)(b-x) + (a-c-x)(x-b+c) = 0.$

35. $(a-x)(b-x) - (c-x)(d-x) = (c+d)x - cd.$

36. $(x-a)(x-b) - (x-c)(x-d) = (d-a)(d-b).$

37. $[(a^2 - b^2)x - ab][a - (a+b)x] + 2ab^2x$
$\qquad = [(a+b)^2 x + ab][b - (a-b)x].$

38. $(x+1)(x+2)(x+3) = (x-3)(x+4)(x+5).$

39. $(x+1)(x+2)(x+3)$
$\qquad = (x-1)(x-2)(x-3) + 3(x+1)(4x+1).$

40. $(x+1)(x+4)(x+7) = (x+2)(x+5)^2.$

41. $(x+2)(x+5)^2 = (x+3)^2(x+6).$

42. $(x-1)(x-4)(x-6) - x(x-2)(x-9) = 136.$

43. $(a+x)(b+x)(c+x)-(a-x)(b-x)(c-x)$
$\qquad = 2(x^3+abc).$

44. $\dfrac{(x-a)(x-b)(x-c)-(d-a)(d-b)(d-c)}{x-d}=(x-d)^2.$

45. $x(x-a)^2-(x-a+b)(x-a+c)(x-b-c)$
$\qquad = (a^2+bc)(b+c).$

46. $(x-a+b)(x-b+c)(x-c+d)-x^2(x-a+d)$
$\qquad = bc(d-a).$

47. $(x-a+b)(x-b+c)(x-c+d)$
$\qquad -x(x-a+c)(x-c+d)=bc(d-a).$

48. $(x-2a)(x-2b)(x-2c)-(x-a-b)(x-b-c)(x-c-a)$
$\qquad = (a+b+c)(a^2+b^2+c^2)-9abc.$

49. $x^3-(x-a+b)(x-b+c)(x-c+a)$
$\qquad = (a+b+c)(a^2+b^2+c^2)-2(a^2b+b^2c+c^2a)-3abc.$

50. $x\left(a-\dfrac{1}{x}\right)\left(b-\dfrac{1}{x}\right)\left(c-\dfrac{1}{x}\right)+\dfrac{1}{x^2}=\dfrac{a+b+c}{x}.$

51. $(x+a)(x+b)+(x+c)(x+a)=(x+b)(x+d)$
$\qquad +(x+d)(x+c).$

52. $(ax+b)(ax-c)-a(b-x)(ax+b)$
$\qquad = a^2(x-c)(x-b)-a(ax-c)(c-x).$

53. $\dfrac{2x-3}{x-4}+\dfrac{3x-2}{x-8}=\dfrac{5x^2-29x-4}{x^2-12x+32}.$

54. $\dfrac{5x-1}{3(x+1)}-\dfrac{3x+2}{2(x-1)}=\dfrac{x^2-30x+2}{6x^2-6}.$

55. $\dfrac{3x-7}{2x-9}-\dfrac{3(x+1)}{2(x+3)}=\dfrac{11x+3}{2x^2-3x-27}.$

56. $\dfrac{7x-5}{3x-2}+\dfrac{8x-7}{3x-1}+\dfrac{10x+7}{9x^2-9x+2}=5.$

57. $\dfrac{2x+7}{3x-7}+\dfrac{3x-6}{2x-5}+\dfrac{5(x-1)}{9x-25}=\dfrac{3x-2}{2x-5}+\dfrac{5x-8}{9x-25}+\dfrac{2x+2}{3x-7}.$

58. $\dfrac{4x-3}{1+x} - \dfrac{3}{1-x} = \dfrac{4x^2+2}{x^2-1}.$

59. $\dfrac{x-a}{x-m} - \dfrac{x-b}{x-n} = 0.$

60. $\dfrac{\frac{1}{4}-x}{\frac{1}{4}+x} + \dfrac{1}{4} = \dfrac{x}{\frac{1}{4}-2x} - \dfrac{1}{4}.$

61. $\dfrac{a}{c} + \dfrac{cx}{ax-b} = \dfrac{c}{a} + \dfrac{ax}{cx-b}.$

62. $\dfrac{\frac{3}{2}-\frac{1}{x}}{\frac{3}{2}+\frac{1}{x}} - \dfrac{\frac{2}{3}-\frac{1}{x}}{\frac{2}{3}+\frac{1}{x}} = \dfrac{\frac{3}{2}-\frac{2}{3}}{\frac{1}{2x}+1}.$

63. $\dfrac{2(x-1)}{x-7} + \dfrac{x+8}{x-4} = \dfrac{3(5x+16)}{5x-28}.$

64. $\dfrac{ax}{mx-p} + \dfrac{cx}{nx-q} = \dfrac{a}{m} + \dfrac{c}{n}.$

65. $\dfrac{ax+b}{mx-p} + \dfrac{cx+d}{nx-q} = \dfrac{a}{m} + \dfrac{c}{n}.$

66. $\dfrac{b-x}{a+x} + \dfrac{c-x}{a-x} = \dfrac{a(c-2x)}{a^2-x^2}.$

67. $\dfrac{a+b}{x-a} + \dfrac{b+c}{x-b} = \dfrac{a+c+2b}{x-c}.$

68. $\dfrac{ax+b}{ax-b} - \dfrac{bx}{ax+b} = \dfrac{ax}{ax-b} - \dfrac{(ax^2-2b)b}{a^2x^2-b^2}.$

69. $\dfrac{ax-b}{mx-p} + \dfrac{cx-d}{nx-q} + \dfrac{(bn+dm)x-(bq+dp)}{(mx-p)(nx-q)} = \dfrac{a}{m} + \dfrac{c}{n}.$

70. $\dfrac{m}{x-a}+\dfrac{n}{x-b}+\dfrac{p}{x-c}=\dfrac{m}{x-c}+\dfrac{n}{x-a}+\dfrac{p}{x-b}.$

71. $\dfrac{ax-2\,a}{ax-2\,b}=\dfrac{ax-2\,b}{ax+2\,a}.$ **73.** $\dfrac{2\,x^2-3\,x+5}{7\,x^2-4\,x+2}=\dfrac{2}{7}.$

72. $\dfrac{\dfrac{1}{a}-\dfrac{1}{x}}{\dfrac{1}{a}+\dfrac{1}{x}}=\dfrac{a-\dfrac{1}{x}}{a+\dfrac{1}{x}}.$ **74.** $\dfrac{ax^2-bx+c}{mx^2-nx+p}=\dfrac{a}{m}.$

75. $\dfrac{ax^3-bx^2+ax-d}{mx^3-nx^2+mx-q}=\dfrac{ax-b}{mx-n}.$

76. $\dfrac{\dfrac{1}{4}-x}{\dfrac{1}{4}+x}+\dfrac{1}{4}=\dfrac{x}{\dfrac{1}{4}+x}-\dfrac{1}{4}.$ **77.** $\dfrac{\frac{2}{3}x-\dfrac{2}{3}}{\dfrac{2}{3}-x}-\dfrac{2}{3}=\dfrac{2}{3}+\dfrac{\frac{2}{3}x+\dfrac{2}{3}}{x-\dfrac{2}{3}}.$

78. $\dfrac{21}{x-98}-\dfrac{71}{x-94}=\dfrac{21}{x+44}-\dfrac{71}{x-52}.$

79. $\dfrac{7}{x-6}+\dfrac{3}{x-11}=\dfrac{9}{x-7}+\dfrac{1}{x-12}.$

80. $\dfrac{9}{x-51}-\dfrac{9}{x-15}=\dfrac{2}{x-81}-\dfrac{2}{x+81}.$

81. $\dfrac{5}{x-6}+\dfrac{4}{x-9}=\dfrac{8}{x-7}+\dfrac{1}{x-10}.$

82. $\dfrac{1}{x-6}+\dfrac{8}{x-3}=\dfrac{5}{x-2}+\dfrac{4}{x-5}.$

83. $\dfrac{m-n}{x-a}-\dfrac{a-b}{x-m}=\dfrac{m-n}{x-b}-\dfrac{a-b}{x-n}.$

84. $\dfrac{a+b}{x-b}-\dfrac{a+c}{x-c}=\dfrac{b+d}{x-(a+b+2c+d)}-\dfrac{c+d}{x-(a+2b+c+d)}.$

85. $(x+a+b)^4 - (x+a)^4 - (x+b)^4 + x^4 - (a+b)^4 + a^4 + b^4$
$= 12ab[x^2+(a+b)^2].$

86. $\dfrac{a-x}{a^2-bc} + \dfrac{b-x}{b^2-ca} + \dfrac{c-x}{c^2-ab} = \dfrac{3x}{ab+bc+ca}.$

87. $\dfrac{(m-n)(x-a)}{b+c} + \dfrac{(n-p)(x-b)}{c+a} + \dfrac{(p-m)(x-c)}{a+b} = 0.$

88. $(x+a+b)^5 - (a+b)^5 - (x+b)^5 - (x+a)^5 + x^5 + a^5 + b^5$
$= 10abx(2x+a+b)(x+a+b).$

89. $\dfrac{ax-1}{a^2(c+b)} + \dfrac{bx-1}{b^2(c+a)} + \dfrac{cx-1}{c^2(a+b)} = \dfrac{3x}{ab+bc+ca}.$

90. $\dfrac{x-2a}{b+c-a} + \dfrac{x-2b}{c+a-b} + \dfrac{x-2c}{a+b-c} = 3.$

91. $\dfrac{x-2a}{b+c-a} + \dfrac{x-2b}{c+a-b} + \dfrac{x-2c}{a+b-c} = \dfrac{3x}{a+b+c}.$

92. $\dfrac{a-x}{a^2-bc} + \dfrac{b-x}{b^2-bc} + \dfrac{c-x}{c^2-ab} = \dfrac{3}{a+b+c}.$

93. $\dfrac{x+2ab}{a+b+c} + \dfrac{2ab-x}{b+c-a} = \dfrac{x-2ab}{a-b+c} + \dfrac{x+2ab}{a+b-c}.$

94. $\dfrac{a}{x+b-c} + \dfrac{b}{x+a-c} = \dfrac{a-c}{x+b} + \dfrac{b-c}{x+a}.$

95. $\dfrac{m^2(a-b)}{x-m} + \dfrac{n^2(b-c)}{x-n} + \dfrac{p^2(c-d)}{x-p}$

$+ \dfrac{q^2[pd+(n-p)c+(m-n)b-ma]}{x-q} = 0.$

96. $\dfrac{(x-2)(x-5)(x-6)(x-9)+(a+2)(a-4)(a-5)(a-11)}{x}$

$+ \dfrac{(b+1)(b+5)(b+8)(b+12)}{x} = (x-4)(x-7)(x-11)$

$+ \dfrac{(a^2-1)(a-8)(a-10)+(b+2)(b+3)(b+10)(b+11)}{x}.$

Equations Resolvable into Linear Equations.

§ **42.** In order that the product of two or more factors may vanish, it is necessary, *and it is sufficient,* that one of the factors should vanish. Thus, in order that $(x-a)(x-b)$ may vanish, either $x-a$ must vanish, or $x-b$ must vanish, and it is sufficient that one of them should do so.

Hence, the single equation $(x-a)(x-b)=0$ is really equivalent to the two disjunctive equations,

$$x-a=0 \text{ or } x-b=0,$$

for either of these will fulfil the conditions of the given equation, and that is all that is required.

Similarly, were it required to find what values of x would make the product $(x-a)(x-b)(x-c)$ vanish, they would be given by

$$x-a=0, \text{ or } x-b=0, \text{ or } x-c=0.$$

$$\therefore x=a \text{ or } b \text{ or } c.$$

Hence, the single equation $(x-a)(x-b)(x-c)=0$ is equivalent to the three disjunctive equations

$$x-a=0, \text{ or } x-b=0, \text{ or } x-c=0.$$

Examples.

1. Solve $x^2-x-20=0.$

The expression $=(x-5)(x+4)$, which will vanish if either of its factors does; that is, if $x-5=0$, or $x+4=0.$

$$\therefore x=5, \text{ or } x=-4.$$

2. Solve $x^4-x^3-x^2+x=0.$

This gives $x^3(x-1)-x(x-1)=x(x-1)(x^2-1)$
$=x(x-1)(x+1)(x-1),$
which vanishes for $x=0$, $x=1$, $x=-1.$

3. Solve $x^2(a-b)+a^2(b-x)+b^2(x-a)=0$.

$\therefore x^2(a-b)-x(a^2-b^2)+ab(a-b)=0$.

$\therefore (x-a)(x-b)(a-b)=0$.

If $a-b=0$, the given equation will hold irrespective of the values of $x-a$ and $x-b$, and therefore of the values of x; but if $a-b$ be not zero, then must either $x-a=0$, or $x-b=0$.

$\therefore x=a$, or $x=b$.

4. Solve $221\,x^2-5\,x-6=0$.

Here we have the factors $17x-3$ and $13x+2$; hence, the equation is satisfied by

$17x-3=0$, or $x=\frac{3}{17}$,

and also by $13x+2=0$, or $x=-\frac{2}{13}$.

5. Solve $(x-a)^3+(a-b)^3+(b-x)^3=0$.

The expression is equal to $3(x-a)(a-b)(b-x)$, and therefore vanishes for $x-a=0$, or $x=a$; and for $x-b=0$, or $x=b$.

Ex. 54.

1. If an equation in x have the factors $2x-4$ and $2x-6$, find the corresponding values of x.

2. If an equation give the factors $2x-1$ and $3x-1$, what are the corresponding values of x?

3. If an equation give the factors $3x^2-12$ and $4x-5$, find the corresponding values of x.

Find the values of x for which the following expressions will vanish :

4. x^2-2x+1 ; $4x^2-12x+9$.

5. $9x^2-4$; $x^2-(a+b)^2$; $x^2-2ax+a^2$.

6. $x^2 - 9x + 20$; $4x^2 - 18x + 20$.

7. $x^2 + x - 6$; $x^2 - x - 12$; $9x^2 - 9x - 28$.

8. $6x^2 - 12x + 6$; $6x^2 - 13x + 6$; $6x^2 - 20x + 6$.

9. $6x^2 - 5x - 6$; $6x^2 - 37x + 6$; $6x^2 + x - 12$.

10. A certain equation of the fourth degree gives the factors $x^2 - x - 2$ and $4x^2 - 2x - 2$. Find all the values of x.

Find the values of x in the following cases :

11. $x^3 - 2bx^2 - 3b^2x = 0$.

12. $x^3 - ax^2 - a^2x + a^3 = 0$.

13. $x^3 - 3x + 2 = 0$.

14. $x^4 - 2ax^3 + 2a^3x - a^4 = 0$.

15. $x^3 + (b + c)x^2 - bcx - b^2c - bc^2 = 0$.

16. $\dfrac{x-a}{x-b} + \dfrac{x-b}{x-a} - \dfrac{(a-b)^2}{(x-a)(x-b)} = \dfrac{x^2 - a^2}{(x-a)(x-b)}$.

17. $x^3 - bx^2 - a^2x + a^2b = 0$.

18. $3x^3 + 5abx^2 - 4a^2b^2x - 4a^3b^3 = 0$.

19. $x^3(a - b) + a^3(b - x) + b^3(x - a) = 0$.

20. $\dfrac{(x-b)(x-c)}{(a-b)(a-c)} + \dfrac{(x-c)(x-a)}{(b-c)(b-a)} = 1$.

21. $x\left(\dfrac{x-2a}{x+a}\right)^3 + a\left(\dfrac{2x-a}{x+a}\right)^3 = x^2 - a^2$.

22. $(x + a + b)^3 - x^3 - a^3 - b^3 = (x + a)(a^2 - b^2)$.

23. $\dfrac{ab}{(b-a)(x-a)} + \dfrac{bx}{(x-a)(a-b)} + \dfrac{ax}{(a-b)(b-x)} = \dfrac{1}{a-b}$.

24. Form the polynome which will vanish for $x = 5$ or -6 or 7.

25. Form the polynome which will vanish for $x = a$ or $4a$ or $3a$ or $-4a$.

26. Form the equation whose roots are 0, 1, -2, and 4.

§ 43. Employing the language of **Algebra**, the principle illustrated in the preceding section may be stated as follows:

DEFINITION. Any quantity which substituted for x makes the *expression* $f(x)$ vanish, is said to be a root of the *equation* $f(x) = 0$. Thus, if a be a root of the equation $f(x) = 0$, then $f(a) = 0$.

By Th. I., if $x - a$ is a factor of the *polynome* $f(x)^n$, then $f(a)^n = 0$, and a must be a root of the *equation* $f(x)^n = 0$; hence, in solving the equation, we are merely finding a value, or values, of x, which will make the corresponding polynome vanish. Suppose $f(x)^n = (x - a) \phi(x)^{n-1} = 0$, we are required to find a value, or values, of x which will make $(x - a) \phi(x)^{n-1}$ vanish. The polynome will certainly vanish if *one* of its factors vanishes, whether the other does or not, and will not vanish unless at least *one* of its factors vanishes. Hence, $(x - a) \phi(x)^{n-1}$ will vanish if $x - a = 0$, quite irrespective of the value of $\phi(x)^{n-1}$. Also, if $\phi(x)^{n-1} = 0$, the polynome will vanish, irrespective of the value of $x - a$. It follows, therefore, that if $f(x)^n$ can be resolved into two or more factors, each of these factors equated to zero will give one or more roots of the equation $f(x)^n = 0$.

When there can be found two or more values of x which satisfy the conditions of given equations, they are sometimes distinguished thus: x_1, x_2, x_3, etc., to read "one value of x," "a second value of x," "a third value of x," etc. Thus, if

$$(x - a)(x - b)(x - c) = 0,$$

$$\therefore x_1 = a, \quad x_2 = b, \quad x_3 = c.$$

EXAMPLES.

1. Solve $2x^3 - 13x^2 + 27x - 18 = 0$.

Factoring, $(x - 2)(x - 3)(2x - 3) = 0$.

$\therefore x_1 = 2,\ x_2 = 3,\ x_3 = 1\frac{1}{2}$.

2. $x^2 - (a + b)x + (a + c)b = (a + c)c$.

$\therefore x^2 - (a + b)x + (a + c)(b - c) = 0$.

$\therefore x^2 - [(a + c) + (b - c)]x + (a + c)(b - c) = 0$.

$\therefore [x - (a + c)][x - (b - c)] = 0$.

$\therefore x_1 = a + c,\ x_2 = b - c$.

3. $x = \dfrac{(a^2 + b^2)x - (a^2 - b^2)}{(a^2 - b^2)x - (a^2 + b^2)}$;

$\therefore \dfrac{x + 1}{x - 1} = \dfrac{a^2(x - 1)}{b^2(x + 1)}$. $\quad \therefore \left(\dfrac{x + 1}{x - 1}\right)^2 - \dfrac{a^2}{b^2} = 0$.

$\therefore \dfrac{x_1 + 1}{x_1 - 1} - \dfrac{a}{b} = 0$. $\qquad \therefore x_1 = \dfrac{a + b}{a - b}$.

$\therefore \dfrac{x_2 + 1}{x_2 - 1} + \dfrac{a}{b} = 0$. $\qquad \therefore x_2 = \dfrac{a - b}{a + b}$.

4. $\dfrac{(a - x)^2 + (b - x)^2}{(a - x)^2 + (a - x)(b - x) + (b - x)^2} = \dfrac{34}{49}$.

$\therefore \dfrac{(a-x)^2 + 2(a-x)(b-x) + (b-x)^2}{(a-x)^2 - 2(a-x)(b-x) + (b-x)^2} = \dfrac{2(49) - 34}{3(34) - 2(49)} = 16$.

$\therefore \left[\dfrac{(a - x) + (b - x)}{(a - x) - (b - x)}\right]^2 - 4^2 = 0$.

$\therefore \dfrac{(a - x_1) + (b - x_1)}{a - b} - 4 = 0$. $\quad \therefore x_1 = \tfrac{1}{2}(5b - 3a)$.

$\therefore \dfrac{(a - x_2) + (b - x_2)}{a - b} + 4 = 0$. $\quad \therefore x_2 = \tfrac{1}{2}(5a - 3b)$.

5. $\dfrac{(x - a)(x - b)}{(c - a)(c - b)} + \dfrac{(x - b)(x - c)}{(a - b)(a - c)} = 1$.

Subtract term by term from the identity (see page 67)

$$\frac{(x-a)(x-b)}{(c-a)(c-b)} + \frac{(x-b)(x-c)}{(a-b)(a-c)} + \frac{(x-c)(x-a)}{(b-c)(b-a)} = 1.$$

$$\therefore (x-c)(x-a) = 0. \quad \therefore x_1 = c, \quad x_2 = a.$$

6. Find the *rational* roots of $x^4 - 12x^3 + 51x^2 - 90x + 56 = 0.$

Factoring the left-hand member by the method of § 27,

$$(x-2)(x-4)(x^2 - 6x + 7) = 0.$$

$$\therefore x_1 = 2, \quad x_2 = 4, \quad \text{or } x^2 - 6x + 7 = 0.$$

Since $x^2 - 6x + 7$ cannot be resolved into rational factors, we know that it will not give rational roots; therefore, $x_1 = 2$, $x_2 \doteq 4$ are the only values that meet the condition of the problem.

In order that two expressions having a common factor may be equal, it is necessary either that the common factor should vanish, or else that the product of the remaining factors of one of the expressions should be equal to the product of the remaining factors of the other expression, and it is sufficient if one of these conditions be fulfilled. In symbols this is

If $(x-a)f(x) = (x-a)\phi(x), \quad \therefore x_1 = a \text{ or } f(x) = \phi(x).$

7. $x + \dfrac{1}{x} = a + \dfrac{1}{a}.$

$$\therefore x - a = \frac{1}{a} - \frac{1}{x}. \quad \therefore \frac{x-a}{1} = \frac{x-a}{ax}.$$

$$\therefore x - a = 0, \text{ or } ax = 1. \quad \therefore x_1 = a, \quad x_2 = \frac{1}{a}.$$

8. $(x+a+b)(x+b+c) = (x-3a+b)(2x-3a+2b-c).$

$$\therefore \frac{x+a+b}{x-3a+b} = \frac{2x-3a+2b-c}{x+b+c} = \frac{x-4a+b-c}{3a+c}.$$

Page 155, (5).

$$\therefore \frac{2(x-a+b)}{x-3a+b} = \frac{x-a+b}{3a+c}.$$

$$\therefore x_1 = a - b.$$

$$\tfrac{1}{2}(x_2 - 3a + b) = 3a + c. \quad \therefore x_2 = 9a - b + 2c.$$

9. $(x-2)(x-5)(x-6)(x-9)+(y+2)(y-4)(y-5)(y-11)$
$\quad + (z+1)(z+5)(z+8)(z+12)$
$\quad = x(x-4)(x-7)(x-11)+(y+1)(y-1)(y-8)(y-10)$
$\quad + (z+2)(z+3)(z+10)(z+11).$

Let $x' = x^2 - 11x$, $y' = y^2 - 9y$, and $z' = z^2 + 13z$.

$\therefore (x'+18)(x'+30)+(y'-22)(y'+20)+(z'+12)(z'+40)$
$\quad = x'(x'+28)+(y'-10)(y'+8)+(z'+22)(z'+30).$

$\therefore x'^2 + 48x' + 540 + y'^2 - 2y' - 440 + z'^2 + 52z' + 480$
$\quad = x'^2 + 28x' + y'^2 - 2y' - 80 + z'^2 + 52z' + 660.$

$\therefore 20x' = 0, \ x^2 - 11x = 0, \ x_1 = 0, \ x_2 = 11.$

Ex. 55.

What can you deduce from the following statements?

1. $A \cdot B = 0.$ 3. $(a-b)x = 0.$

2. $A \cdot B \cdot C = 0.$ 4. $12xy = 0.$

5. What is the difference between the equation
$$(x-5y)(x-4y+3) = 0$$
and the simultaneous equations
$$x-5y = 0 \text{ and } x-4y+3 = 0?$$

What values of x will satisfy the following equations?

6. $x(x-a) = 0.$ 11. $x(x^2 - a^2) = 0.$

7. $ax(x+b) = 0.$ 12. $a^2x^3 = b^2x.$

8. $(x-a)(bx-c) = 0.$ 13. $x^2 + (a-x)^2 = a^2.$

9. $ax^2 = 3ax.$ 14. $x^2 + (a-x)^2 = (a-2x)^2.$

10. $x^2 = (a+b)x.$ 15. $(a-x)^2 + (x-b)^2 = a^2 + b^2.$

16. $(a - x)(x - b) + ab = 0.$

17. $(a - x)^2 - (a - x)(x - b) + (x - b)^2 = a^2 + ab + b^2.$

18. $x^2 - (a - b)x - ab = 0.$

19. $x^3 - (a + b + c)x^2 + (ab + bc + ca)x - abc = 0.$

If x must be positive, what value, or values, of x will satisfy the following equations:

20. $(x - 5)(x + 4) = 0.$ **23.** $3x^2 - 10x + 3 = 0.$

21. $x^2 + 29x - 30 = 0.$ **24.** $x^4 - 13x^2 + 36 = 0.$

22. $x^2 - 17x - 84 = 0.$ **25.** $x^3 - 2x^2 - 5x + 6 = 0.$

Solve the following equations:

26. $(a - x)^2 + (x - b)^2 = (a - b)^2.$

27. $(a - x)^2 - (a - x)(x - b) + (x - b)^2 = (a - b)^2.$

28. $a^2(a - x)^2 = b^2(b - x)^2.$ **29.** $a^2(b - x)^2 = b^2(a - x)^2.$

30. $(x - a)^3 + (a - b)^3 + (b - x)^3 = 0.$

31. $(x - 1)^2 = a(x^2 - 1).$

32. $\dfrac{a - x}{x - b} = \dfrac{x - a}{c + x}.$ **33.** $\dfrac{a + b - x}{a - c - x} = \dfrac{a - c + x}{a + c - x}$

34. $(x - a + b)(x - a + c) = (a - b)^2 - x^2.$

35. $(x - a)^2 - b^2 + (a + b - x)(b + c - x) = 0.$

36. $(a + b + c)x^2 - (2a + b + c)x + a = 0.$

37. $\dfrac{a + b - x}{c} = \dfrac{a + b - c}{x}.$

38. $(a - x)^2 + (a - b)^2 = (a + b - 2x)^2.$

39. $x(a + b - x) + (a + b + c)c = 0.$

40. $(n - p)x^2 + (p - m)x + m - n = 0.$

41. $\dfrac{ax^2 - bx + c}{mx^2 - nx + p} = \dfrac{c}{p}.$ **42.** $\dfrac{ax^2 - bx + c}{mx^2 - nx + p} = \dfrac{a - b + c}{m - n + p}.$

43. $4x^2 + a^2 - b^2 - 2(a+b)x = (a-x)(b+x) - (a+x)(b-x).$

44. $(2a - b - x)^2 + 9(a-b)^2 = (a+b-2x)^2.$

45. $(2a + 2c - x)^2 = (2b + x)(3a - b + 3c - 2x).$

46. $(3a - 5b + x)(5a - 3b - x) = (7a - b - 3x)^2.$

47. $(3a - b + x)(3a + b - x) = (5a + 3b - 3x)^2.$

48. $a(a - b) - b(a - c)x + c(b - c)x^2 = 0.$

49. $(ab + bc + ca)(x^2 + x + 1) + (a-b)^2$
$= (2ac + b^2)(x^2 + x + 1) + (a-c)^2 x.$

50. $(x + 1)(x + 3)(x - 4)(x - 7)$
$+ (x - 1)(x - 3)(x + 4)(x + 7) = 96.$

51. $(x - 1)(x + 3)(x - 5)(x + 9)$
$+ (x + 1)(x - 3)(x + 5)(x - 9) + 18 = 0.$

52. $x + \dfrac{1}{x} = 3\tfrac{1}{3}.$

53. $x + \dfrac{1}{x} = \dfrac{a+b}{a-b} + \dfrac{a-b}{a+b}.$

54. $x - \dfrac{1}{x} = \dfrac{a}{b} - \dfrac{b}{a}.$

55. $\dfrac{a+x}{b+x} + \dfrac{b+x}{a+x} = 2\tfrac{1}{2}.$

56. $\dfrac{a-x}{x-b} + \dfrac{x-b}{a-x} = \dfrac{13}{6}.$

57. $\dfrac{a-x}{b+x} - \dfrac{b+x}{a-x} = \dfrac{m}{n} - \dfrac{n}{m}.$

58. $\dfrac{a}{x} + \dfrac{x}{a} = \dfrac{m}{n} + \dfrac{n}{m}.$

59. $\dfrac{(a-x)^2 + (x-b)^2}{(a-x)(x-b)} = \dfrac{5}{2}.$

60. $\dfrac{(x+a)^2 + (x-b)^2}{(x+a)^2 - (x-b)^2} = \dfrac{a^2 + b^2}{2ab}.$

61. $\dfrac{(a-x)^2 - (x-b)^2}{(a-x)(x-b)} = \dfrac{4ab}{a^2 - b^2}.$

62. $\dfrac{(a-x)^2 + (a-x)(x-b) + (x-b)^2}{(a-x)^2 - (a-x)(x-b) + (x-b)^2} = \dfrac{49}{19}.$

63. $\dfrac{2a^2 + a(a-x) + (a+x)^2}{2a^2 + a(a+x) + (a-x)^2} = \dfrac{3}{2}.$

64. $(5 - x)^4 + (2 - x)^4 = 17.$

65. $(x - a)^3 + (a - b)^3 + (b - x)^3 = x^2 - a^2.$

66. $(a - x)^4 + (x - b)^4 = (a - b)^4.$

67. $(x + a)^3 - (a + b)^3 + (b - x)^3 = (x + a)(x + b)(a + b).$

68. $x^3 - (x - b)^3 - (x - a + b)^3 - a^3 + (x - a)^3 + (a - b)^3$
$\quad + b^3 = (a - b)c^2.$

69. $(x + a)^3 - (x + b)^3 - (x - b)^3 - (2a)^3 + (x - a)^3$
$\quad + (a + b)^3 + (a - b)^3 = (a^2 - b^2)c.$

70. $(x - a + b)^3 - (x - a)^3 + (x - b)^3 - x^3 + a^3 - (a - b)^3 = b^3.$

71. $\dfrac{(a - x)^4 + (x - b)^4}{(a - x)^2 + (x - b)^2} = \frac{41}{20}(a - b)^2.$

72. $\dfrac{(a - x)^5 + (x - b)^5}{(a - x)^4 + (x - b)^4} = \frac{211}{97}(a - b).$

73. $2(a - x)^4 - 9(a - x)^3(x - b) + 14(a - x)^2(x - b)^2$
$\quad - 9(a - x)(x - b)^3 + 2(x - b)^4 = 0.$

74. $4(a - x)^4 - 17(a - x)^2(x - b)^2 + 4(x - b)^4 = 0.$

Find the rational roots of the following equations:

75. $x^4 - 12x^3 + 49x^2 - 78x + 40 = 0.$ Let $z = x^2 - 6x.$

76. $x^4 - 6x^3 + 7x^2 + 6x - 8 = 0.$

77. $x^4 - 10x^3 + 35x^2 - 50x + 24 = 0.$

78. $32x^4 - 48x^3 - 10x^2 + 21x + 5 = 0.$

79. $x^3 - 6x^2 + 5x + 12 = 0.$

80. $11x^4 + 10x^3 - 40x = 176.$

81. $\dfrac{5}{x} - \dfrac{4}{x - a} - \dfrac{9}{x - 2a} - \dfrac{4}{x - 3a} + \dfrac{5}{x - 4a} = 0.$

82. $\dfrac{14}{x+20} + \dfrac{5}{x+5} - \dfrac{4}{x-4} = \dfrac{14}{x-55} + \dfrac{5}{x-40} - \dfrac{4}{x-25}.$

83. $\dfrac{2x+5a}{x} - \dfrac{x+8a}{x-a} + \dfrac{x}{x-2a} = \dfrac{x-a}{x-3a} - \dfrac{x+5a}{x-4a} + \dfrac{2x-5a}{x-5a}.$

84. $\dfrac{x+4}{x+2} + \dfrac{x+2}{x} + \dfrac{x+4}{x-1} = \dfrac{x+3}{x-2} + \dfrac{x-1}{x-3} + \dfrac{x-3}{x-5}.$

85. $\dfrac{7}{x} - \dfrac{31}{x-1} + \dfrac{20}{x-2} + \dfrac{8}{x-3} + \dfrac{20}{x-4} - \dfrac{31}{x-5} + \dfrac{7}{x-6} = 0.$

86. $x^3 - \left(\dfrac{a-b}{a+b}\right)x^2 - \dfrac{2cx^2}{1+cx} = x + \left(\dfrac{a-b}{a+b}\right)\left(\dfrac{1-cx}{1+cx}\right).$

87. $\dfrac{4x^4 + 4a^4 - 33x^2a^2}{2x+a} = \frac{1}{2}(4x^3 - 8x^2a + 9xa^2 - 2a^3).$

88. $\dfrac{7}{x^2 - 11x + 28} + \dfrac{7}{x^2 - 17x + 70} = \dfrac{3\frac{1}{2}x^2}{x^2 - 14x + 40}.$

89. $\dfrac{8}{x^2 - 6x + 5} + \dfrac{8}{x^2 - 14x + 45} = \dfrac{x^4}{x^2 - 10x + 9}.$

90. $6(a-x)^4 - 25(a-x)^3(x-b) + 38(a-x)^2(x-b)^2$
    $- 25(a-x)(x-b)^3 + 6(x-b)^4 = 0.$

CHAPTER VI.

Simultaneous Linear Equations.

§ 44. There are three general methods of resolving simultaneous linear equations: first, by substitution; second, by comparison; third, by elimination. The last is often subdivided into the method by cross-multipliers, and the method by arbitrary multipliers.

In applying the elimination method, the work should be done with detached coefficients, each equation should be numbered, and a register of the operations performed should be kept.

Ex. Resolve
$$u + v + x + y + z = 15,$$
$$u + 2v + 4x + 8y + 16z = 57,$$
$$u + 3v + 9x + 27y + 81z = 179,$$
$$u + 4v + 16x + 64y + 256z = 453,$$
$$u + 5v + 25x + 125y + 625z = 975.$$

Register.	u	v	x	y	z			
	1	1	1	1	1	$=$	15	(1)
	1	2	4	8	16	$=$	57	(2)
	1	3	9	27	81	$=$	179	(3)
	1	4	16	64	256	$=$	453	(4)
	1	5	25	125	625	$=$	975	(5)
$(2) - (1)$			1	3	7	15	$=$ 42	(6)
$(3) - (2)$			1	5	19	65	$=$ 122	(7)
$(4) - (3)$			1	7	37	175	$=$ 274	(8)
$(5) - (4)$			1	9	61	369	$=$ 522	(9)
$(7) - (6)$				2	12	50	$=$ 80	(10)
$(8) - (7)$				2	18	110	$=$ 152	(11)
$(9) - (8)$				2	24	194	$=$ 248	(12)
$(11) - (10)$					6	60	$=$ 72	(13)
$(12) - (11)$					6	84	$=$ 96	(14)
$(14) - (13)$						24	$=$ 24	(15)
$\frac{1}{24}(15)$						1	$=$ 1	(16)
$\frac{1}{6}[(13) - 60(16)]$					1		$=$ 2	(17)
$\frac{1}{2}\{(10) - [12(17) + 50(16)]\}$				1			$=$ 3	(18)
$(6) - [3(18) + 7(17) + 15(16)]$ . . .			1				$=$ 4	(19)
$(1) - [(19) + (18) + (17) + (16)]$ . .	1						$=$ 5	(20)

An examination of the Register will show how easy it would have been to shorten the process; thus, (10) is (7) − (6), which is (3) + (1) − 2 (2); similarly, (11) is (4) + (2) − 2 (3); therefore, (13) is (4) + 3 (2) − 3 (3) − (1), etc.

Ex. 56.

Solve the following systems of equations :

1. $2x + 3y = 41,$
$3x + 2y = 39.$

2. $5x + 7y = 17,$
$7x - 5y = 9.$

3. $\frac{1}{3}x + \frac{1}{4}y = 6,$
$3x - 4y = 4.$

4. $\frac{1}{2}x - \frac{1}{3}y = 1,$
$\frac{1}{8}x - \frac{2}{3}y + 5 = 0.$

5. $\frac{1}{3}y = \frac{1}{2}x - 1,$
$\frac{1}{4}y = \frac{2}{5}x - 1.$

6. $1.5x - 2y = 1,$
$2.5x - 3y = 6.$

7. $3.5x + 2\frac{1}{3}y = 13 + 4\frac{1}{7}x - 3.5y,$
$2\frac{1}{7}x + 0.8y = 22\frac{1}{2} + 0.7x - 3\frac{1}{3}y.$

8. $\dfrac{1}{x} + \dfrac{1}{y} = \dfrac{5}{6},$
$\dfrac{1}{x} - \dfrac{1}{y} = \dfrac{1}{6}.$

11. $17x - \dfrac{0.3}{y} = 3,$
$16x - \dfrac{0.4}{y} = 2.$

9. $\dfrac{3}{x} + \dfrac{8}{y} = 3,$
$\dfrac{15}{x} - \dfrac{4}{y} = 4.$

12. $\dfrac{x}{3} + \dfrac{5}{y} = 4\frac{1}{3},$
$\dfrac{x}{6} + \dfrac{10}{y} = 2\frac{2}{3}.$

10. $\dfrac{1.6}{x} = \dfrac{2.7}{y} - 1,$
$\dfrac{0.8}{x} + \dfrac{3.6}{y} = 5.$

13. $\dfrac{5x}{0.7} + \dfrac{0.3}{y} = 6,$
$\dfrac{10x}{7} + \dfrac{9}{y} = 31.$

14. $\frac{3}{4}x - \frac{1}{2}(y+1) = 1,$

     $\frac{1}{3}(x+1) + \frac{3}{4}(y-1) = 9.$

15. $\dfrac{5}{x+2y} = \dfrac{7}{2x+y},$

     $\dfrac{7}{3x-2} = \dfrac{5}{6-y}.$

16. $\dfrac{x+3y}{x-y} = 8,$

     $\dfrac{7x-13}{3y-5} = 4.$

17. $\dfrac{x+2y+1}{2x-y+1} = 2,$

     $\dfrac{3x-y+1}{x-y+3} = 5.$

18. $\dfrac{x+3y+13}{0.4x+0.5y-2.5} = 30,$

     $\dfrac{0.8x+0.1y+0.6}{5x+3y-23} = \dfrac{1}{2}.$

19. $\dfrac{x+1}{3} - \dfrac{y+2}{4} = \dfrac{2(x-y)}{5},$

     $\dfrac{x-3}{4} - \dfrac{y-3}{3} = 2y - x.$

20. $\dfrac{2x-y+3}{3} - \dfrac{x-2y+3}{4} = 4,$

     $\dfrac{3x-4y+3}{4} + \dfrac{4x-2y-9}{3} = 4.$

21. $20(x+1) = 15(y+1) = 12(x+y).$

22. $(x-2):(y+1):(x+y-3)::3:4:5.$

23. $(x-5):(y+9):(x+y+4)::1:2:3.$

24. $\dfrac{x+3}{x+1} = \dfrac{y+8}{y+5},$

     $\dfrac{2x-3}{2(y+1)} = \dfrac{5x-6}{5y+7}.$

25. $(x-4)(y+7) = (x-3)(y+4),$

     $(x+5)(y-2) = (x+2)(y-1).$

26. $(x-1)(5y-3) = 3(3x+1),$

     $(x-1)(4y+3) = 3(7x-1).$

27. $(x+1)(2y+1) = 5x + 9y + 1,$

     $(x+2)(3y+1) = 9x + 13y + 2.$

28. $(3x-2)(5y+1)=(5x-1)(y+2),$
$(3x-1)(y+5)=(x+5)(7y-1).$

29. $x+y=37,$
$y+z=25,$
$z+x=22.$

37. $x+y+z=3,$
$2x+4y+8z=13,$
$3x+9y+27z=34.$

30. $2x+2y=7,$
$7x+9z=29,$
$y+8z=17.$

38. $x+2y+3z=32,$
$2x+3y+z=42,$
$3x+y+2z=40.$

31. $1.3x-1.9y=1,$
$1.7y-1.1z=2,$
$2.9z-2.1x=3.$

39. $x+y+2z=34,$
$x+2y+z=33,$
$2x+y+z=32.$

32. $5x+3y+2z=217,$
$5x-3y=39,$
$3y-2z=20.$

40. $3x+3y+z=17,$
$3x+y+3z=15,$
$x+3y+3z=13.$

33. $\frac{1}{5}x-\frac{1}{2}y=0,$
$\frac{1}{3}x-\frac{1}{2}z=1,$
$\frac{1}{2}z-\frac{1}{3}y=2.$

41. $x+2y-z=4.6,$
$y+2z-x=10.1,$
$z+2x-y=5.7.$

34. $1\frac{1}{3}x+1\frac{1}{2}y=10,$
$2\frac{2}{3}x+2\frac{2}{5}z=20,$
$3\frac{1}{4}y+3\frac{2}{5}z=30.$

42. $x+2y-0.7z=21,$
$3x+0.2y-z=24,$
$0.9x+7y-2z=27.$

35. $x+y-z=17,$
$y+z-x=13,$
$z+x-y=7.$

43. $x+y=1\frac{1}{2}z+8,$
$y+z=2\frac{2}{3}y-14,$
$z+x=3\frac{3}{4}x-32.$

36. $x+y+z=9,$
$x+2y+4z=15,$
$x+3y+9z=23.$

44. $\frac{1}{2}x+\frac{1}{3}y+\frac{1}{4}z=36\frac{1}{2},$
$\frac{1}{3}x+\frac{1}{4}y+\frac{1}{5}z=27,$
$\frac{1}{5}x+\frac{1}{6}y+\frac{1}{7}z=18.$

45. $\dfrac{x+1}{y+1}=2,$

$\dfrac{y+2}{z+1}=4,$

$\dfrac{z+3}{x+1}=\dfrac{1}{2}.$

46. $\dfrac{3x+y}{z+1}=2,$

$\dfrac{3y+z}{x+1}=2,$

$\dfrac{3z+x}{y+1}=2.$

47. $\dfrac{x+y}{y-z}=10,$

$\dfrac{x+z}{x-y}=9,$

$\dfrac{y+z}{x+5}=1.$

48. $\dfrac{x+3}{y+z}=2,$

$\dfrac{y+3}{x+z}=1,$

$\dfrac{z+3}{x+y}=\dfrac{1}{2}.$

49. $\dfrac{4}{x}-\dfrac{3}{y}=1,$

$\dfrac{2}{x}+\dfrac{3}{z}=4.$

$\dfrac{3}{y}-\dfrac{1}{z}=0.$

50. $\dfrac{6}{x}+\dfrac{4}{y}+\dfrac{5}{z}=4,$

$\dfrac{3}{x}+\dfrac{8}{y}+\dfrac{5}{z}=4,$

$\dfrac{9}{x}+\dfrac{12}{y}-\dfrac{10}{z}=4.$

51. $\dfrac{xy}{x+y}=\dfrac{1}{5},$

$\dfrac{yz}{x+z}=\dfrac{1}{6},$

$\dfrac{zx}{z+x}=\dfrac{1}{7}.$

52. $\dfrac{xy}{4y-3x}=20,$

$\dfrac{xz}{2x-3z}=15,$

$\dfrac{yz}{4y-5z}=12.$

53. $(x+2)(2y+1)=(2x+7)y,$
$(x-2)(3z+1)=(x+3)(3z-1),$
$(y+1)(z+2)=(y+3)(z+1).$

54. $(2x-1)(y+1)=2(x+1)(y-1),$
$(x+4)(z+1)=(x+2)(z+2),$
$(y-2)(z+3)=(y-1)(z+1).$

55. $(x+1)(5y-3) = (7x+1)(2y-3),$
$(4x-1)(z+1) = (x+1)(2z-1),$
$(y+3)(z+2) = (3y-6)(3z-1).$

56. $21x+31y+42z = 115,$
$6(2x+y) = 3(3x+z) = 2(y+z).$

57. $15(x-2y) = 5(2x-3z) = 3(y+z),$
$21x+31y+41z = 135.$

58. $6x(y+z) = 4y(z+x) = 3z(x+y),$
$\dfrac{1}{x}+\dfrac{1}{y}+\dfrac{1}{z} = 9.$

59. $3x+y+z = 20,$ **60.** $x+z+8y = 30,$
$3u+x+4y = 30,$ $5u+y+z = 10,$
$3u+6x+z = 40,$ $4u+x+z = 10,$
$5u+8y+3z = 50.$ $3u+x+y = 10.$

§ 45. The principle of symmetry is often of use in the solution of symmetrical equations. For, from one relation which may be found to exist between two or more of the letters involved, other relations may be derived by symmetry; also, when the value of one of the unknown quantities has been determined, the values of the others can be at once written down, etc.

EXAMPLES.

1. $(x+y)(x+z) = a,$
$(x+y)(y+z) = b,$
$(x+z)(y+z) = c.$

Multiply the equations together and extract the square root.

$\therefore (x+y)(y+z)(z+x) = \sqrt{(abc)}.$

Divide this equation by the third.

$$\therefore x + y = \frac{\sqrt{(abc)}}{c} \; ; \text{ and, therefore, by symmetry,}$$

$$y + z = \frac{\sqrt{(abc)}}{a},$$

$$\therefore z + x = \frac{\sqrt{(abc)}}{b}.$$

Hence, we get

$$x = \frac{ab - bc + ca}{2\sqrt{(abc)}},$$

whence y and z may be derived by symmetry.

2. $x + y + z = 0,$ (1)
    $ax + by + cz = 0,$ (2)
    $bcx + cay + abz + (a - b)(b - c)(c - a) = 0.$ (3)

$c \times (1) - (2)$ gives $(c - a)x + (c - b)y = 0.$

Hence, $y = \dfrac{(c - a)x}{b - c}$, and similarly,

$$z = \frac{(a - b)x}{b - c}.$$

Substitute in (3) these values of y and z, and reduce;
then, $x(a - b)(c - a) = (a - b)(b - c)(c - a),$
or, $x = b - c.$
Hence, $y = c - a,$ $z = a - b.$

3. $a(yz - zx - xy) = b(zx - xy - yz) = c(xy - yz - zx) = xyz.$

Divide the first and last equations by $axyz$; then,

$$\frac{1}{a} = \frac{1}{x} - \frac{1}{y} - \frac{1}{z}, \text{ and hence, by symmetry,}$$

$$\frac{1}{b} = \frac{1}{y} - \frac{1}{z} - \frac{1}{x},$$

$$\frac{1}{c} = \frac{1}{z} - \frac{1}{x} - \frac{1}{y}.$$

Therefore, $\dfrac{1}{b}+\dfrac{1}{c}=-\dfrac{2}{x}$, and, by symmetry,

$$\dfrac{1}{c}+\dfrac{1}{a}=-\dfrac{2}{y},$$

$$\dfrac{1}{a}+\dfrac{1}{b}=-\dfrac{2}{z}.$$

4. $ax+by+cz=1$,   (1)
$a^2x+b^2y+c^2z=1$,   (2)
$a^3x+b^3y+c^3z=1$.   (3)

$c\times(1)-(2)$ gives $a(c-a)x+b(c-b)y=c-1$. (4)
$c\times(2)-(3)$ gives $a^2(c-a)x+b^2(c-b)y=c-1$. (5)
$b\times(4)-(5)$ gives $ab(c-a)x-a^2(c-a)x=b(c-1)-(c-1)$,
  or $a(a-b)(a-c)x=(c-1)(b-1)$.

Therefore, $x=\dfrac{(1-b)(1-c)}{a(a-b)(a-c)}$;

whence y and z may be derived by symmetry.

5. Eliminate x, y, z, u (which are supposed all different) from the following equations:

$$x=by+cz+du,$$
$$y=cz+du+ax,$$
$$z=du+ax+by,$$
$$u=ax+by+cz.$$

Subtracting the second equation from the first,

$x-y=by-ax$, or $(1+a)x=(1+b)y$,

which, by symmetry,

$$=(1+c)z=(1+d)u.$$

These relations may also be obtained by adding ax to both members of the first equation by, to both members of the second equation, etc.

Now divide the first equation by these equals.

$$\therefore \frac{1}{1+a} = \frac{b}{1+b} + \frac{c}{1+c} + \frac{d}{1+d}.$$

And since $\dfrac{1}{1+a} = 1 - \dfrac{a}{1+a}$, we have

$$1 = \frac{a}{1+a} + \frac{b}{1+b} + \frac{c}{1+c} + \frac{d}{1+d}.$$

Ex. 57.

1. Given $ax + by = c$,   and that $x = \dfrac{b'c - bc'}{b'a - ba'}$,

  $a'x + b'y = c'$,   derive the value of y.

2. Given $bx = ay$,   and that $x = \dfrac{a(dm - cn)}{bc - ad}$,

  $dx + md = cy + nd$,   derive the value of y.

3. Given $ax + by + cz = d$,   and that $x = \dfrac{a(d-b)(d-c)}{a(a-b)(a-c)}$,

  $a^2x + b^2y + c^2z = d^2$,   write down the values of

  $a^3x + b^3y + c^3z = d^3$, y and z.

4. There is a set of equations in x, y, z, u, and w, with corresponding coefficients (a to x, etc.), a, b, c, d, and e; one of the equations is $x = by + cz + du + ew$, write down the others.

Solve the following equations:

5. $\dfrac{x}{m} + \dfrac{y}{n} = a$, $\dfrac{y}{n} + \dfrac{z}{p} = b$, $\dfrac{x}{m} + \dfrac{z}{p} = c$.

6. $x + ay + bz = m$, $y + az + bx = n$, $z + ax + by = p$.

7. $x + ay = l$, $y + bz = m$, $z + cu = n$, $u + dw = p$,

  $w + ex = r$.

8. Eliminate x, y, z (supposed to be all different), from the following equations: $x = by + cz$, $y = cz + ax$, $z = ax + by$.

9. Eliminate x, y, z from

$$\frac{x}{y+z} = a, \quad \frac{y}{z+x} = b, \quad \frac{z}{x+y} = c.$$

10. Having given

$$x = by + cz + du + ew,$$
$$y = cz + du + ew + ax,$$
$$z = du + ew + ax + by,$$
$$u = ew + ax + by + cz,$$
$$w = ax + by + cz + du,$$

show that $\dfrac{a}{1+a} + \dfrac{b}{1+b} + \dfrac{c}{1+c} + \dfrac{d}{1+d} + \dfrac{e}{1+e} = 1.$

§ 46. RESOLUTION OF PARTICULAR SYSTEMS OF LINEAR EQUATIONS.

EXAMPLES.

1.

$$x + y + z = a, \tag{1}$$
$$y + z + u = b, \tag{2}$$
$$z + u + x = c, \tag{3}$$
$$u + x + y = d. \tag{4}$$

(1)+(2)+(3)+(4) $3(u + x + y + z) = a + b + c + d,$ (5')

$3(1)$ $3(x + y + z) = 3a,$ (6')

$\frac{1}{3}[(5') - (6')]$ $u = \frac{1}{3}(-2a + b + c + d).$

The values of x, y, and z may now be written down by symmetry.

The following is a variation of the above method, applicable to a much more general system.

Assume the auxiliary equation

$$u + x + y + z = s. \tag{5}$$

Hence,   (1) becomes $s - u = a$,   (6)
              (2) becomes $s - x = b$,   (7)
              (3) becomes $s - y = c$,   (8)
              (4) becomes $s - z = d$.   (9)

$(5) + (6) + (7) + (8) + (9)$, $4s = s + a + b + c + d$.

Therefore, $s = \frac{1}{3}(a + b + c + d)$.

s is now a known quantity, and may be treated as such,

in (6) giving $u = s - a$,
in (7) giving $x = s - b$,
in (8) giving $y = s - c$,
in (9) giving $z = s - d$.

2. $yz = a(y + z)$,   (1)
                  $zx = b(z + x)$,   (2)
                  $xy = c(x + y)$.   (3)

$(1) \div ayz$, $\dfrac{1}{y} + \dfrac{1}{z} = \dfrac{1}{a}$,

$(2) \div bzx$, $\dfrac{1}{z} + \dfrac{1}{x} = \dfrac{1}{b}$,

$(3) \div cxy$, $\dfrac{1}{x} + \dfrac{1}{y} = \dfrac{1}{c}$.

This may now be solved like Exam. 1, using the reciprocals of a, b, c, x, y, and z, instead of these quantities themselves.

3. $a_1 u + b_1(x + y + z) = c_1$,   (1)
                  $a_2 x + b_2(y + z + u) = c_2$,   (2)
                  $a_3 y + b_3(z + u + x) = c_3$,   (3)
                  $a_4 z + b_4(u + x + y) = c_4$.   (4)

Assume the auxiliary equation,

                  $u + x + y + z = s$.   (5)

(1) becomes $b_1 s - (b_1 - a_1) u = c_1$.

Therefore, $\dfrac{b_1}{b_1 - a_1} s - u = \dfrac{c_1}{b_1 - a_1}$.   (6)

Similarly, from (2), $\dfrac{b_2}{b_2-a_2}s-x=\dfrac{c_2}{b_2-a_2}.$ （7）

Similarly, from (3), $\dfrac{b_3}{b_3-a_3}s-y=\dfrac{c_3}{b_3-a_3}.$ （8）

Similarly, from (4), $\dfrac{b_4}{b_4-a_4}s-z=\dfrac{c_4}{b_4-a_4}.$ （9）

$(5)+(6)+(7)+(8)+(9),$

$$\left(\frac{b_1}{b_1-a_1}+\frac{b_2}{b_2-a_2}+\frac{b_3}{b_3-a_3}+\frac{b_4}{b_4-a_4}\right)s-s$$

$$=\frac{c_1}{b_1-a_1}+\frac{c_2}{b_2-a_2}+\frac{c_3}{b_3-a_3}+\frac{c_4}{b_4-a_4}. \tag{10}$$

From (10) we can at once get the value of s, which may therefore be treated as a known quantity in (6), giving

$$u=\frac{b_1s-c_1}{b_1-a_1},$$

and the values of x, y, and z may be obtained from (7), (8), and (9), or they may be written down by symmetry.

4.

$$ax+b(y+z)=c, \tag{1}$$
$$ay+b(z+u)=d, \tag{2}$$
$$az+b(u+x)=e, \tag{3}$$
$$au+b(x+y)=f. \tag{4}$$

Assume　　　$u+x+y+z=s,$ 　　　(5)

$(1)+(2)+(3)+(4),\ (a+2b)s=c+d+e+f.$ 　(6)

Hence, s is a known quantity, and may be treated as such.

From (1) and (5), $bs-bu+(a-b)x=c.$

Therefore, 　　　$bu-(a-b)x=bs-c.$ 　　　(7)

Similarly, from (2) and (5),

$$bx-(a-b)y=bs-d. \tag{8}$$

From (3) and (5), $by - (a - b)z = bs - e.$ (9)

From (4) and (5), $bz - (a - b)u = bs - f.$ (10)

$b(7) + (a - b)(8),$

$$b^2u - (a - b)^2y = abs - bc - (a - b)d.$$ (11)

$b(9) + (a - b)(10),$

$$b^2y - (a - b)^2u = abs - be - (a - b)f.$$ (12)

$b^2(11) + (a - b)^2(12),$

$$[b^4 - (a-b)^4]u = abs[b^2 + (a - b)^2] - a[b^2d + (a - b)^2f]$$
$$- b[b^2(c - d) + (a - b)^2(e - f)].$$ (13)

The values of x, y, and z may now be written down by symmetry.

5.
$$a^3 + a^2x + ay + z = 0,$$
$$b^3 + b^2x + by + z = 0,$$
$$c^3 + c^2x + cy + z = 0.$$

The polynome $t^3 + xt^2 + yt + z$ vanishes for $t = a$, $t = b$, $t = c$.

Therefore, by Th. II., page 58, for *all* values of t,

$$t^3 + xt^2 + yt + z = (t - a)(t - b)(t - c)$$
$$= t^3 - (a + b + c)t^2 + (ab + bc + ca)t - abc.$$

Therefore, by Th. III., page 67,

$$x = -(a + b + c),$$
$$y = ab + bc + ca,$$
$$z = -abc.$$

6.
$$x + y + z + u = 1,$$ (1)
$$ax + by + cz + du = 0,$$ (2)
$$a^2x + b^2y + c^2z + d^2u = 0,$$ (3)
$$a^3x + b^3y + c^3z + d^3u = 0.$$ (4)

Employing the method of arbitrary multipliers,

$(4) + l(3) + m(2) + n(1),$

$$\begin{aligned} a^3 x + \;& b^3 \;y + \; c^3 \;z + d^3 \;u = n \\ + la^2 \;& \quad + lb^2 \quad + lc^2 \quad + ld^2 \\ + ma \;& \quad + mb \quad + mc \quad + md \\ + n \;& \quad + n \quad\;\;\; + n \quad\;\; + n \end{aligned}$$ (5)

To determine x, assume

$$b^3 + lb^2 + mb + n = 0, \qquad (6)$$
$$c^3 + lc^2 + mc + n = 0, \qquad (7)$$
$$d^3 + ld^2 + md + n = 0. \qquad (8)$$

Therefore, $\qquad x = \dfrac{n}{a^3 + la^2 + ma + n}. \qquad (9)$

But the system (6), (7), (8) has been solved in Exam. 5, from which it is seen that

$$l = -(b + c + d), \quad m = bc + cd + db, \quad n = -bcd,$$

and $a^3 + a^2l + am + n = (a - b)(a - c)(a - d)$.

Hence, using these values in (9),

$$x = \frac{-bcd}{(a-b)(a-c)(a-d)}.$$

The values of y, z, and u may now be written down by symmetry.

7.
$$\frac{x}{m-a} + \frac{y}{m-b} + \frac{z}{m-c} = 1, \qquad (1)$$

$$\frac{x}{n-a} + \frac{y}{n-b} + \frac{z}{n-c} = 1, \qquad (2)$$

$$\frac{x}{p-a} + \frac{y}{p-b} + \frac{z}{p-c} = 1. \qquad (3)$$

Assume
$$1 - \frac{x}{t-a} - \frac{y}{t-b} - \frac{z}{t-c}$$

$$= \frac{t^3 + Bt^2 + Ct + D}{(t-a)(t-b)(t-c)}. \qquad (4)$$

But in virtue of equations (1), (2), and (3), the first member of (4) vanishes for $t = m$, $t = n$, and $t = p$; and hence, $t^3 + Bt^2 + Ct + D$ vanishes for the same values of t; and therefore, by Th. II., page 58,

$$t^3 + Bt^2 + Ct + D = (t - m)(t - n)(t - p).$$

Therefore, (4) becomes

$$1 - \frac{x}{t-a} - \frac{y}{t-b} - \frac{z}{t-c} = \frac{(t-m)(t-n)(t-p)}{(t-a)(t-b)(t-c)}.$$

To obtain the value of x, multiply both sides of this equation by $(t-a)$.

$$t - a - x - \frac{y(t-a)}{t-b} - \frac{z(t-a)}{t-c} = \frac{(t-m)(t-n)(t-p)}{(t-b)(t-c)}.$$

Now t may have any value in this equation;

let $t = a$.

Hence, $x = \dfrac{(a-m)(a-n)(a-p)}{(a-b)(a-c)}.$

The values of y and z may now be written down by symmetry.

8. $\dfrac{x+a}{p} = \dfrac{y+b}{q} = \dfrac{z+c}{r}.$ (1)

 $lx + my + nz = s^2.$ (2)

By § 37,

$$\frac{x+a}{p} = \frac{y+b}{q} = \frac{z+c}{r} = \frac{lx+my+nz+la+mb+nc}{lp+mq+nr}.$$

(2) $= \dfrac{s^2+la+mb+nc}{lp+mq+nr} = R,$ say ;

therefore, $x = pR - a,\ y = qR - b,\ z = rR - c.$

9. $yz + zx + xy = (a+b+c)xyz.$ (1)

 $\dfrac{yz+zx}{a} = \dfrac{zx+xy}{b} = \dfrac{xy+yz}{c}.$ (2)

(1) $\div xyz,$ $\dfrac{1}{x} + \dfrac{1}{y} + \dfrac{1}{z} = a + b + c.$ (3)

(2) $\div xyz,$ $\dfrac{\frac{1}{x}+\frac{1}{y}}{a} = \dfrac{\frac{1}{y}+\frac{1}{z}}{b} = \dfrac{\frac{1}{z}+\frac{1}{x}}{c}.$ (4)

§ 37 and (3), $\dfrac{\dfrac{2}{x}+\dfrac{2}{y}+\dfrac{2}{z}}{a+b+c}=2.$ (5)

(4) and (5),

$\therefore \dfrac{1}{x}+\dfrac{1}{y}=2a,\ \dfrac{1}{y}+\dfrac{1}{z}=2b,\ \dfrac{1}{z}+\dfrac{1}{x}=2c.$ (6)

(3) − (6), $\dfrac{1}{x}=a-b+c,\ \dfrac{1}{y}=a+b-c,\ \dfrac{1}{z}=-a+b+c.$

10. $\dfrac{x+c}{a+b}+\dfrac{y+b}{a+c}=2,$ (1)

$\dfrac{x-b}{a-c}+\dfrac{y-c}{a-b}=2.$ (2)

(1), $\therefore \dfrac{x+c}{a+b}-1=1-\dfrac{y+b}{a+c}.$

$\therefore \dfrac{x-a-b+c}{a+b}=\dfrac{a+c-b-y}{a+c}.$ (3)

Similarly, from (2),

$\dfrac{x-a-b+c}{a-c}=\dfrac{a-b+c-y}{a-b}.$ (4)

(3) and (4),

$\therefore x-a-b+c=\dfrac{a+b}{a+c}(a-b+c-y)$

$=\dfrac{a-c}{a-b}(a-b+c-y).$

But, unless $\dfrac{a+b}{a+c}=\dfrac{a-c}{a-b},$

this cannot be the case except for

$$a-b+c-y=0,$$

in which case $x-a-b+c=0$ also,

giving $x=a+b-c$ and $y=a-b+c.$ (5)

If $\dfrac{a+b}{a+c}=\dfrac{a-c}{a-b},\ \therefore a^2-b^2=a^2-c^2.$ (6)

$b^2-c^2=0,$ or $(b+c)(b-c)=0.$

Therefore, $b=c$ or $b=-c.$

But, if $b = +c$ or $-c$, (1) and (2) are one and the same equation ; hence, if (1) and (2) are indepen-dent, (6) cannot be true, thus leaving only the alternative (5).

11.
$$2ax = (b+c-a)(y+z), \tag{1}$$
$$2by = (c+a-b)(z+x), \tag{2}$$
$$(x+y+z)^2 + x^2 + y^2 + z^2 = 4(a^2+b^2+c^2). \tag{3}$$

(1) and page 155, (5),

$$\frac{x}{b+c-a} = \frac{y+z}{2a} = \frac{x+y+z}{b+c+a}. \tag{4}$$

(2) and page 155, (5),

$$\frac{y}{c+a-b} = \frac{x+z}{2b} = \frac{x+y+z}{c+a+b}. \tag{5}$$

(4), (5), and page 155, (5),

$$\therefore \frac{x+y+z}{a+b+c} = \frac{x}{b+c-a} = \frac{y}{c+a-b} = \frac{z}{a+b-c}.$$

$$\therefore \frac{x^2}{(b+c-a)^2}$$
$$= \frac{(x+y+z)^2 + x^2 + y^2 + z^2}{(a+b+c)^2 + (b+c-a)^2 + (c+a-b)^2 + (a+b-c)^2}.$$

Reduction and (3) $= \dfrac{(x+y+z)^2 + x^2 + y^2 + z^2}{4(a^2+b^2+c^2)} = 1.$

Therefore, $x^2 = (b+c-a)^2.$

Ex. 58.

1. $ax + by = c,$
  $mx + ny = d.$

2. $ax + by = c,$
  $mx - ny = d.$

3. $ax + by = c,$
  $mx + ny = c.$

4. $\dfrac{x}{a} + \dfrac{y}{b} = 1,$
  $x + y = c.$

5. $\dfrac{x}{a} + \dfrac{y}{b} = 1,$

$\dfrac{x}{b} + \dfrac{y}{a} = 1.$

6. $\dfrac{x}{a} + \dfrac{y}{b} = 1,$

$\dfrac{x}{b} = \dfrac{y}{a}.$

7. $ax + bc = by + ac,$

$x + y = c.$

8. $\dfrac{a}{x} + \dfrac{b}{y} = m,$

$\dfrac{b}{x} + \dfrac{a}{y} = n.$

9. $(a+c)x - (a-c)y = 2ab,$

$(a+b)y - (a-b)x = 2ac.$

10. $\dfrac{x-c}{y-c} = \dfrac{a}{b},$

$x - y = a - b.$

11. $\dfrac{x}{y} = \dfrac{a}{b},$

$\dfrac{x+m}{y+n} = \dfrac{c}{d}.$

12. $\dfrac{x+y}{y+1} = \dfrac{a+b+c}{a-b+c},$

$\dfrac{y-1}{x+1} = \dfrac{a-b-c}{a+b-c}.$

13. $\dfrac{x-a+c}{y-a+b} = \dfrac{b}{c},$

$\dfrac{y+b}{x+c} = \dfrac{c+a}{b+a}.$

14. $\dfrac{x+c}{a+b} + \dfrac{y+b}{a+c} = 2,$

$\dfrac{x-b}{a-c} + \dfrac{y-c}{a-b} = 2.$

15. $\dfrac{x}{m-a} + \dfrac{y}{m-b} = 1,$

$\dfrac{x}{n-a} + \dfrac{y}{n-b} = 1.$

16. $x + y + z = 0,$

$(b+c)x + (a+c)y + (a+b)z = 0,$

$bcx + acy + abz = 1.$

17. $x + y + z = l,$

$ax + by + cz = m,$

$\dfrac{x}{l-a} + \dfrac{y}{l-b} + \dfrac{z}{l-c} = 1.$

18. $\dfrac{x-a}{p}=\dfrac{y-b}{q}=\dfrac{z-c}{r}$,

$l(x-a)+m(y-b)+n(z-c)=1.$

19. $\dfrac{x-a}{p}=\dfrac{y-b}{q}=\dfrac{z-c}{r}$,

$lx+my+nz=1.$

20. $a(x-a)=b(y-b)=c(z-c)$,

$ax+by+cz=m^2.$

21. $x+y+z=a+b+c$,

$bx+cy+az=a^2+b^2+c^2$,

$cx+ay+bx=a^2+b^2+c^2.$

22. $x+y+z=a+b+c$,

$ax+by+cz=ab+bc+ca$,

$(b-c)x+(c-a)y+(a-b)z=0.$

23. $x+y+z=m$,   **24.** $ax+by+cz=r$,

$x:y:z=a:b:c.$ $mx=ny,\ \ qy=pz.$

25. $xy+yz+zx=0,\ \ ayz+bzx+cxy=0$,

$bcyz+acxz+abxy+(a-b)(b-c)(c-a)xyz=0.$

26. $(a+b)x+(b+c)y+(c+a)z=ab+bc+ca$,

$(a+c)x+(a+b)y+(b+c)z=ab+ac+bc$,

$(b+c)x+(a+c)y+(a+b)z=a^2+b^2+c^2.$

27. $mx+ny+pz+qu=r$,

$\dfrac{x}{a}=\dfrac{y}{b}=\dfrac{z}{c}=\dfrac{u}{d}.$

28. $\dfrac{x(y+z)}{a} = \dfrac{y(x+z)}{b} = \dfrac{z(x+y)}{c}$,

$$\frac{1}{x} + \frac{1}{y} + \frac{1}{z} = a + b + c.$$

29. $(a-b)(x+c) - ay + bz = (c-a)(y+b) - cz + ax = 0$,

$x + y + z = 2(a + b + c)$.

30. $ax + by = 1$,

$by + cz = 1$,

$cz + ax = 1$.

31. $ly + mx = n$,

$nx + lz = m$,

$mz + ny = l$.

32. $x + y = a$,

$y + z = b$,

$x + z = c$.

33. $y + z - x = \dfrac{mn}{l}$,

$z + x - y = \dfrac{ln}{m}$,

$x + y - z = \dfrac{lm}{n}$.

34. $\dfrac{1}{y} + \dfrac{1}{z} = 2a$,

$\dfrac{1}{z} + \dfrac{1}{x} = 2b$,

$\dfrac{1}{x} + \dfrac{1}{y} = 2c$.

35. $\dfrac{1}{y} + \dfrac{1}{z} - \dfrac{1}{x} = \dfrac{2}{a}$,

$\dfrac{1}{z} + \dfrac{1}{x} - \dfrac{1}{y} = \dfrac{2}{b}$,

$\dfrac{1}{x} + \dfrac{1}{y} - \dfrac{1}{z} = \dfrac{2}{c}$.

36. $(a+b)x + (a-b)z = 2bc$,

$(b+c)y + (b-c)x = 2ac$,

$(c+a)z + (c-a)y = 2ab$.

37. $x + \dfrac{y}{b} - \dfrac{z}{c} = a$,

$y + \dfrac{z}{c} - \dfrac{x}{a} = b$,

$z + \dfrac{x}{a} - \dfrac{y}{b} = c$.

38. $\dfrac{x}{b+c} + \dfrac{y}{c+a} = b - a$,

$\dfrac{x}{b+c} + \dfrac{z}{a+b} = a - c$,

$\dfrac{y}{c+a} + \dfrac{z}{a+b} = c - b$.

39. $x + y - z = a,$
$y + z - v = b,$
$z + v - x = c,$
$v + x - y = d.$

40. $u + v - x = a,$
$v + x - y = b,$
$x + y - z = c,$
$y + x - u = d.$
$z + u - v = e.$

Ex. 59.

Resolve :

1. $(a + b)x + (a - b)y = 2(a^2 + b^2),$
$(a - b)x + (a + b)y = 2(a^2 - b^2).$

2. $x + y = a,$
$x^2 - y^2 = b.$

3. $2x - 3y = m,$
$2x^2 - 3y^2 = n^2 + xy.$

4. $(a-b)x + (a+b)y = a+b,$

$$\frac{x}{a+b} - \frac{y}{a-b} = \frac{1}{a+b}.$$

5. $(a - b)x + y = \dfrac{a + b + 1}{a + b},$

$$x + (a + b)y = \frac{a - b + 1}{a - b}.$$

6. $(a+b-c)x - (a-b+c)y = 4a(b-c),$

$$\frac{x}{y} = \frac{a + b - c}{a - b + c}.$$

7. $\dfrac{x+y}{x-y} = \dfrac{a}{b-c},$

$$\frac{x+c}{a+b} = \frac{y+b}{a+c}.$$

9. $\dfrac{x-y+1}{x-y-1} = a,$

$$\frac{x+y+1}{x+y-1} = b.$$

8. $\dfrac{x-a}{y-a} = \dfrac{a-b}{a+b},$

$$\frac{x}{y} = \frac{a^3 - b^3}{a^3 + b^3}.$$

10. $\dfrac{x+y+1}{x-y+1} = \dfrac{a+1}{a-1},$

$$\frac{x+y+1}{x-y-1} = \frac{1+b}{1-b}.$$

11. $\dfrac{x-y+1}{x+y-1}=a,$

$\dfrac{x+y+1}{x-y-1}=b.$

12. $\dfrac{x}{a+b}+\dfrac{y}{a-b}=a+b,$

$\dfrac{x}{a}+\dfrac{y}{b}=2a.$

13. $(a+c)x+(a-c)y=2ab,$

$(a+b)y-(a-b)x=2ac.$

14. $a^2+ax+y=0,$

$b^2+bx+y=0.$

15. $y+z-x=a,$

$z+x-y=b,$

$x+y-z=c.$

16. $7x+11y+z=a,$

$7y+11z+x=b,$

$7z+11x+y=c.$

17. $\dfrac{a}{x}+\dfrac{b}{y}-\dfrac{c}{z}=2ab,$

$\dfrac{b}{y}+\dfrac{c}{z}-\dfrac{a}{x}=2bc,$

$\dfrac{c}{z}+\dfrac{a}{x}-\dfrac{b}{y}=2ca.$

18. $(a-b)(x+c)-ay+bz=0,$

$(c-a)(y+b)-cz+ax=0,$

$x+y+z=2(a+b+c).$

19. $\dfrac{x}{b+c}+\dfrac{y}{c-a}=a+b,$

$\dfrac{y}{c+a}+\dfrac{z}{a-b}=b+c,$

$\dfrac{z}{a+b}+\dfrac{x}{b-c}=c+a.$

20. $\dfrac{x}{b+c}+\dfrac{y}{c-a}-\dfrac{z}{a-b}=0,$

$\dfrac{x}{b-c}-\dfrac{y}{c-a}+\dfrac{z}{a+b}=0,$

$\dfrac{x}{b+c}+\dfrac{y}{c-a}+\dfrac{z}{a+b}=2a.$

21. $\dfrac{x}{a}+\dfrac{y}{a-1}+\dfrac{z}{a-2}=1,$

$\dfrac{x}{b}+\dfrac{y}{b-1}+\dfrac{z}{b-2}=1,$

$\dfrac{x}{c}+\dfrac{y}{c-1}+\dfrac{z}{c-2}=1.$

22. $\dfrac{xy}{x+y}=a,$

$\dfrac{yz}{y+z}=b,$

$\dfrac{zx}{z+x}=c.$

23. $\dfrac{x}{a}+\dfrac{y}{b}+\dfrac{z}{c}=\dfrac{x}{b}+\dfrac{y}{c}+\dfrac{z}{a}=\dfrac{x}{c}+\dfrac{y}{a}+\dfrac{z}{b}=\dfrac{1}{a}+\dfrac{1}{b}+\dfrac{1}{c}.$

24. $\dfrac{x}{a} = \dfrac{y}{b} = \dfrac{z}{c} = \dfrac{u}{d},$

$\dfrac{x}{m} + \dfrac{y}{n} + \dfrac{z}{p} + \dfrac{u}{q} = \dfrac{1}{r}.$

25. $ax = by = cz = du,$

$y^2 - z^2 = x - u.$

26. $y + z = au,$

$x + z = bu,$

$x + y = cu,$

$\dfrac{1 - x}{1 - y} = \dfrac{a}{b}.$

27. $x + y = m,$

$y + z = n,$

$z + u = a,$

$u - x = b.$

28. $11x + 9y + z - u = a,$

$11y + 9z + u - x = b,$

$11z + 9u + x - y = c,$

$11u + 9x + y - z = d.$

29. $x + ay + a^2z + a^3u + a^4 = 0,$

$x + by + b^2z + b^3u + b^4 = 0,$

$x + cy + c^2z + c^3u + c^4 = 0,$

$x + dy + d^2z + d^3u + d^4 = 0.$

30. $x + y = a,$

$y + z = b,$

$z + u = c,$

$u + v = d,$

$v + x = e.$

31. $x + ly = a,$

$y + mz = b,$

$z + nu = c,$

$u + pv = d,$

$v + qx = e.$

32. $x + y + z = a,$

$y + z + u = b,$

$z + u + v = c,$

$u + v + x = d,$

$v + x + y = e.$

33. $x - y + z = a,$

$y - z + u = b,$

$z - u + v = c,$

$u - v + x = d,$

$v - x + y = e.$

34. $x + y + z - u = a,$

$y + z + u - v = b,$

$z + u + v - x = c,$

$u + v + x - y = d,$

$v + x + y - z = e.$

35. $x + y + z - u - v = a,$

$y + z + u - v - x = b,$

$z + u + v - x - y = c,$

$u + v + x - y - z = d,$

$v + x + y - z - u = e.$

36.
$$2x - y - z + 2u - v = 3a,$$
$$2y - z - u + 2v - x = 3b,$$
$$2z - u - v + 2x - y = 3c,$$
$$2u - v - x + 2y - z = 3d,$$
$$2v - x - y + 2z - u = 3e.$$

37.
$$v - 2x + 3u - 2y + z = a,$$
$$x - 2y + 3v - 2z + u = b,$$
$$y - 2z + 3x - 2u + v = c,$$
$$z - 2u + 3y - 2v + x = d,$$
$$u - 2v + 3z - 2x + y = e.$$

CHAPTER VII.

Pure Quadratics.

§ 47. (A) If an equation reduces to the form

$$(mx + n)^2 = c^2,$$

then $(mx + n)^2 - c^2 = 0.$

Hence, $(mx_1 + n) - c = 0$, and therefore $x_1 = \dfrac{c - n}{n}$,

or $(mx_2 + n) + c = 0$, and therefore $x_2 = \dfrac{-c - n}{m}$.

(B) If an equation reduces to the form

$$\left(\frac{mx + n}{px + q}\right)^2 = \frac{a^2}{b^2},$$

then $x_1 = \dfrac{qa - nb}{mb - pa}$, $x_2 = \dfrac{-qa - nb}{mb + pa}$.

(See Exams. 4 and 5 below.)

Examples.

1. $\dfrac{x + 3(a - b)}{x - 3(a - b)} = \dfrac{a(3x + 9a - 7b)}{b(3x - 7a + 9b)}$.

Apply, if $\dfrac{m}{n} = \dfrac{p}{q}$, therefore $\dfrac{m + n}{m - n} = \dfrac{p + q}{p - q}$.

Hence, $\dfrac{x}{3(a - b)} = \dfrac{3x(a + b) + 9a^2 - 14ab + 9b^2}{3x(a - b) + 9(a^2 - b^2)}$.

Dividing the denominators by $3(a - b)$,

$\cdot$ $x[x + 3(a + b)] = 3x(a + b) + 9a^2 - 14ab + 9b^2.$

Therefore, $x^2 = 9a^2 - 14ab + 9b^2.$

2. $\left(\dfrac{x-2a+4b}{x+4a-2b}\right)^2 = \dfrac{5x-9a+3b}{5x+3a-9b}$.

Apply, if $\dfrac{m}{n} = \dfrac{p}{q}$, therefore $\dfrac{n-m}{n} = \dfrac{q-p}{p}$,

and factor the numerator

$(x+4a-2b)^2 - (x-2a+4b)^2$;

$\therefore \dfrac{12(x+a+b)(a-b)}{(x+4a-2b)^2} = \dfrac{12(a-b)}{5x+3a-9b}$.

$\therefore \dfrac{x+a+b}{x+4a-2b} = \dfrac{x+4a-2b}{5x+3a-9b} = \dfrac{3(a-b)}{4x-a-7b}$,

by taking difference of numerators and difference of denominators.

To the first and third of these fractions apply, if

$\dfrac{m}{n} = \dfrac{p}{q}$, therefore $\dfrac{m}{n-m} = \dfrac{p}{q-p}$.

$\therefore \dfrac{x+a+b}{3(a-b)} = \dfrac{3(a-b)}{4x-4a-4b}$.

$\therefore 4[x^2 - (a+b)^2] = 9(a-b)^2$.

$\therefore x^2 = \frac{1}{4}[4(a+b)^2 + 9(a-b)^2]$.

3. $\dfrac{(x+2)^2}{x^2-2x} = \dfrac{a}{b}$.

$\therefore \dfrac{(x+2)^2}{m(x+2)^2 + n(x^2-2x)} = \dfrac{a}{ma+nb}$. $\qquad$ (1)

But (B) can be applied if m and n are so determined that $m(x+2)^2 + n(x^2-2x)$ is a square.

This requires that $4m(m+n) = (2m-n)^2$.

$\therefore 4m^2 + 4mn = 4m^2 - 4mn + n^2$.

$\therefore 8m = n$.

Assume $m = 1$, then $n = 8$, and (1) becomes, on substitution and reduction,

$$\frac{(x+2)^2}{(3x-2)^2} = \frac{a}{a+8b} = r^2, \text{ say.}$$

$$\therefore x_1 = \frac{2(1+r)}{3r-1}, \quad x_2 = \frac{2(r-1)}{1+3r}.$$

4. $\dfrac{(x+1)^4}{(x^2+1)(x-1)^2} = \dfrac{a}{b}.$

$$\therefore \frac{(x^2+2x+1)^2}{(x^2+1)(x^2-2x+1)} = \frac{a}{b}.$$

For x^2+1 write xz.

$$\therefore \frac{(xz+2x)^2}{xz(xz-2x)} = \frac{a}{b}. \quad \therefore \frac{(z+2)^2}{z(z-2)} = \frac{a}{b}.$$

This equation was solved in Exam. 3, hence z may be treated as known.

But $\dfrac{x^2+1}{x} = z.$ $\therefore \dfrac{x^2+2x+1}{x^2-2x+1} = \dfrac{z+2}{z-2}.$

$$\therefore \left(\frac{x+1}{x-1}\right)^2 = \frac{z+2}{z-2}, \text{ a form solved in } (B).$$

Ex. 60.

1. $(x+a+b)(x-a+b)+(x+a-b)(x-a-b)=0.$

2. $(a+bx)(b-ax)+(b+cx)(c-bx)+(c+ax)(a-cx)=0.$

3. $(a+bx)(ax-b)+(b+cx)(bx-c)+(c+ax)(cx-a)$
     $= \frac{1}{2}(a^2+b^2+c^2).$

4. $(a+x)(b-x)+(1+ax)(1-bx)=(a+b)(1+x^2).$

5. $(a+x)(b+x)(c-x)+(a+x)(b-x)(c+x)$
     $+(a-x)(b+x)(c+x)+(a-x)(b-x)(c+x)$
     $+(a-x)(b+x)(c-x)+(a+x)(b-x)(c-x)$
     $= 5abc.$

6. $(a+x)(b+x)(c+x)+(a+x)(b+x)(c-x)$
$+(a+x)(b-x)(c+x)+(a-x)(b+x)(c+x)$
$+(a+x)(b-x)(c-x)+(a-x)(b+x)(c-x)$
$+(a-x)(b-x)(c+x)+(a-x)(b-x)(c-x)$
$=8x^2.$

7. $(a+5b+x)(5a+b+x)=3(a+b+x)^2.$

8. $(a+17b+x)(17a+b+x)=9(a+b+x)^2.$

9. $(9a-7b+3x)(9b-7a+3x)=(3a+3b+x)^2.$

10. $\dfrac{ab}{a^2-b^2x^2}+\dfrac{cd}{c^2-d^2x^2}=0.$　**15.** $\dfrac{a-x}{1-ax}=\dfrac{b-x}{1-bx}.$

11. $\dfrac{x-a}{x+1}+\dfrac{x+a}{x-1}=2c.$　**16.** $\dfrac{x+a+2b}{x+a-2b}=\dfrac{b-2a+2x}{b+2a-2x}.$

12. $\dfrac{a+x}{a-x}=\dfrac{x+b}{x-b}.$　**17.** $\dfrac{a+4b+x}{a-4b+x}=\dfrac{3b-a+x}{3b+a-x}.$

13. $\dfrac{ax+b}{a+bx}=\dfrac{cx+d}{c+dx}.$　**18.** $\dfrac{x+5a+b}{x-3a+b}=\dfrac{x-a+b}{a-x+3b}.$

14. $\dfrac{a-x}{1-ax}=\dfrac{1-bx}{b-x}.$　**19.** $\dfrac{a-7b+x}{7a-b-x}=\dfrac{a+5b+x}{5a+b+x}.$

20. $\dfrac{3a-b-x}{a-3b+x}=\dfrac{5b-3a+x}{5a-3b+x}.$

21. $\dfrac{3a-2b+3x}{a-2b+x}=\dfrac{x-a+2b}{3x-3a+2b}.$

22. $\dfrac{3a-2b+3x}{a-2b+x}=\dfrac{x-7a+8b}{3x-5a+4b}.$

23. $\dfrac{5a-6b+x}{a+x}=\dfrac{3a-5b+3x}{a+b+x}.$

24. $\dfrac{a+b-x}{3a-b-3x}=\dfrac{3(a-b+x)}{a-5b+x}.$

25. $\dfrac{7a+b-x}{5a+3b-3x}=\dfrac{3(a-b+x)}{a-17b+x}$.

26. $\dfrac{5a-b+x}{2(a+2b-x)}=\dfrac{2(2a-b+x)}{a+11b-x}$

27. $\dfrac{7a-b+x}{7b-a+x}=\dfrac{a(a+5b+x)}{b(5a+b+x)}$.

28. $\dfrac{x+a-b}{x-a+b}=\dfrac{a(x+a+5b)}{b(x+5a+b)}$.

29. $\left(\dfrac{5a-3b+x}{5b-3a+x}\right)^{2}=\dfrac{7a-9b+3x}{7b-9a+3x}$.

30. $\left(\dfrac{a+5b+x}{5a+b+x}\right)^{2}=\dfrac{a+17b+x}{17a+b+x}$.

31. $\left(\dfrac{7a-b+x}{7b-a+x}\right)^{2}=\dfrac{17a+b-x}{17b+a-x}$.

32. $\dfrac{17a+b-x}{a+17b-x}=\dfrac{a^{2}(a+17b+x)}{b^{2}(17a+b+x)}$.

33. $\dfrac{(x+7a+b)(x-a+b)}{(5x+3a-11b)(x-a+17b)}=\dfrac{x-5a+b}{5x+7a-59b}$.

34. $\dfrac{(1+3x+5x^{2})(x^{2}+3x+5)}{(1+2x+3x^{2})(x^{2}+2x+3)}=\dfrac{9}{4}$.

35. $\dfrac{7}{x^{2}-11x+28}+\dfrac{7}{x^{2}-17x+70}=\dfrac{3\frac{1}{2}x^{2}}{x^{2}-14x+40}$.

36. $\dfrac{8}{x^{2}-6x+5}+\dfrac{8}{x^{2}-14x+45}=\dfrac{x^{4}}{x^{2}-11x+10}$.

37. $x^{3}(b-a^{2})+a^{3}(x-b^{2})+b^{3}(a-x^{2})+abx(abx-1)$
$\qquad =(a-x^{2})(b^{2}-a^{4})$.

QUADRATIC EQUATIONS AND EQUATIONS THAT CAN BE
RESOLVED AS QUADRATICS.

§ 48. (C) If an equation appears under the form

$$(a-x)(x-b)=c, \qquad (1)$$

then $\quad x_1=\tfrac{1}{2}(a+b+r), \quad x_2=\tfrac{1}{2}(a+b-r),$

in which $r^2=(a-b)^2-4c.$

From the identity

$$(a-x)+(x-b)=a-b$$

we get $\quad (a-x)^2+2(a-x)(x-b)+(x-b)^2=(a-b)^2, \quad (2)$

$(2)-4(1) \quad (a-x)^2-2(a-x)(x-b)+(x-b)^2$

$$=(a-b)^2-4c=r^2, \text{ say.}$$

Then, $\quad [(a-x)-(x-b)]^2-r^2=0;$

hence, $\quad [(a-x_1)-(x_1-b)]+r=0,$ and $\therefore x_1=\tfrac{1}{2}(a+b+r);$

or, $\quad [(a-x_2)-(x_2-b)]-r=0,$ and $\therefore x_2=\tfrac{1}{2}(a+b-r).$

EXAMPLES.

1. $x^4+(ab+1)^2=(a^2+b^2)(x^2+1)+2(a^2-b^2)x+1.$

$\therefore x^4+a^2b^2=(a^2+b^2)x^2+2(a^2-b^2)x+(a-b)^2.$

$\therefore x^4+2abx^2+a^2b^2=(a+b)^2x^2+2(a^2-b^2)x+(a-b)^2.$

$\therefore x^2+ab=\pm[(a+b)x+(a-b)],$

or $x^2\mp(a+b)x+ab=\pm(a-b).$

$\therefore x^2\mp(a+b)x+\tfrac{1}{4}(a+b)^2=\tfrac{1}{4}(a-b)^2\pm(a-b).$

$\therefore x\mp\tfrac{1}{2}(a+b)=\tfrac{1}{2}\sqrt{[(a-b)^2\pm4(a-b)]}.$

2. $\dfrac{ax+b}{bx+a}=\dfrac{mx-n}{nx-m}.$

Add and subtract numerators and denominators,

$$\dfrac{(a+b)(x+1)}{(a-b)(x-1)}=\dfrac{(m+n)(x-1)}{(m-n)(x+1)}$$

$$\therefore \left(\frac{x+1}{x-1}\right)^2 = \frac{(a-b)(m+n)}{(a+b)(m-n)} = s^2, \text{ say.}$$

$$\therefore x_1 = \frac{s+1}{s-1}, \quad x_2 = \frac{s-1}{s+1}.$$

3. $(a-x)^4 + (b-x)^4 = c.$

In the identity

$$(u+v)^4 = u^4 + v^4 + 4(u+v)^2 uv - 2u^2 v^2$$

let $u = a-x, \ v = x-b.$

$\therefore u+v = a-b$ and $u^4 + v^4 = c.$

$\therefore (a-b)^4 = c + 4(a-b)^2(a-x)(x-b) - 2(a-x)^2(x-b)^2.$

Write z for $(a-x)(x-b).$

$\therefore z^2 - 2(a-b)^2 z + (a-b)^4 = \frac{1}{2}[c + (a-b)^4] = t^2, \text{ say.}$

$\therefore [z - (a-b)^2]^2 = t^2.$

$\therefore$ by (B), $z_1 = (a-b)^2 - t$; $z_2 = (a-b)^2 + t.$

$\therefore z$ is known.

But $(a-x)(x-b) = z.$

$\therefore$ by (C), $x_1 = \frac{1}{2}(a+b+r)$; $x_2 = \frac{1}{2}(a+b-r),$ (1)

  in which $r^2 = (a-b)^2 - 4z$

$$\left. \begin{array}{l} = (a-b)^2 - 4[(a-b)^2 - t] = 4t - 3(a-b)^2 \\ \text{or } (a-b)^2 - 4[(a-b)^2 + t] = -4t - 3(a-b)^2 \end{array} \right\} \quad (2)$$

  and $t^2 = \frac{1}{2}[c + (a-b)^4].$ (3)

Hence, x is expressed in terms of a, b, and r ;

      r is expressed in terms of a, b, and t ;

      t is expressed in terms of a, b, and c ;

and the expressions for r and t are cases of (A).

4. $(a-x)(b+x)^4 + (a-x)^4(b+x) = ab(a^3 + b^3).$

Let $a-x = n-z$ and $b+x = n+z.$

$\therefore n = \frac{1}{2}(a+b).$ (1)

The equation reduces to

$$(n^2 - z^2)[(n+z)^3 + (n-z)^3] = ab(a^3 + b^3).$$
$$\therefore (n^2 - z^2)(2n^3 + 6nz^2) = ab(a^3 + b^3).$$
$$\therefore (n^2 - z^2)(n^2 + 3z^2) = ab(a^2 - ab + b^2).$$

z^2 may now be found by (C), and from (1)

$$x = \tfrac{1}{2}(a - b) + z,$$
$$3z^2 = \tfrac{3}{4}(a - b)^2 \text{ or } \tfrac{1}{4}(10ab - a^2 - b^2).$$
$$\therefore x = 0, \text{ or } a - b, \text{ or } \tfrac{1}{2}(a-b) + \tfrac{1}{6}\sqrt{(30ab - 3a^2 - 3b^2)}.$$

5. $x^4 - 4 = \dfrac{x^2 + 20}{x^2 - 2}$; $\therefore x^6 - 2x^4 - 5x^2 - 12 = 0.$

Find the rational linear factors of the left-hand member by the method of § 27, page 116.

$$\therefore (x - 2)(x + 2)(x^4 + 2x^2 + 3) = 0.$$
$$\therefore x - 2 = 0, \text{ or } x + 2 = 0, \text{ or } x^4 + 2x^2 + 3 = 0.$$

The last of these equations may be solved as a quadratic, giving

$$x^2 = -1 \pm 2\sqrt{-2}.$$
$$\therefore x = \pm 1 \pm \sqrt{-2}.$$
$$\therefore x_1 = 2; \ x_2 = -2; \ x_3 = 1 + \sqrt{-2}; \ x_4 = 1 - \sqrt{-2};$$
$$x_5 = -1 + \sqrt{-2}; \ x_6 = -1 - \sqrt{-2}.$$

NOTE. In solving numerical equations of the higher orders, *the rational linear factors should always be found and separated, as disjunctive equations*, before other methods of reduction are applied. Such separation may always be effected by the methods of §§ 26–29, and, unless it is done, the application of the higher methods may actually fail. Thus, if it be attempted to solve as a cubic the equation

$$x^3 - 9x - 10 = 0,$$

the result is $\quad x = (5 + \sqrt{-2})^{\frac{1}{3}} + (5 - \sqrt{-2})^{\frac{1}{3}},$

which can be reduced only by trial. The left-hand member can, however, be easily factored by the method of § 27, and the equation reduces to

$$(x + 2)(x^2 - 2x - 5) = 0,$$

which gives $\quad x = 2 \text{ or } 1 \pm \sqrt{6}.$

6. $(x-2)^7 - x^7 + 2^7 = 0.$

    Factor (see Exam. 20, page 113), rejecting constant factors,

    $\therefore x(x-2)(x^2 - 2x + 4)^2 = 0.$

    $\therefore x = 0,$ or $x - 2 = 0,$ or $x^2 - 2x + 4 = 0.$

    The last equation gives $x = 1 \pm \sqrt{-3}.$

Ex. 61.

Solve the following equations:

1. $(x + a + b)^3 = x^3 + a^3 + b^3.$

2. $(x + a + b)^5 = x^5 + a^5 + b^5.$

3. $(a - b)x^3 + (b - x)a^3 + (x - a)b^3 = 0.$

4. $(a - b)x^2 + (x - b)a^2 + (x + a)b^2 = 2abx.$

5. $(x - a)^5 + (a - b)^5 + (b - x)^5 = 0.$

6. $(x - a)^7 + (a - b)^7 + (b - x)^7 = 0.$

7. $(a^3 - b)x^4 + (x^3 - a)b^4 + (b^3 - x)a^4 = abx(a^2b^2x^2 - 1).$

8. $(x - a)(x - b)(a - b) + (x - b)(x - c)(b - c)$
      $+ (x - c)(x - a)(c - a) = 0.$

9. $\dfrac{x^5 - 1}{x - 1} = 0.$ **11.** $\dfrac{x^{16} - 1}{x^4 - 1} = 0.$

10. $\dfrac{x^{12} - 1}{x^4 - 1} = 0.$ **12.** $\dfrac{x^{20} - 1}{x^4 - 1} = 0.$

13. $x^4 + 5x^3 - 16x^2 + 20x - 16 = 0.$ (See § 21.)

14. $x^4 - 3x^3 + 5x^2 + 6x + 4 = 0.$

15. $(x - a)^4 + x^4 + a^4 = 0.$ **17.** $x(x - 2)^2(x + 2) = 2.$

16. $2x^3 = (x - 6)^2.$ **18.** $(4x^2 - 17)x + 12 = 0.$

19. $x^4 + (ab + 1)^2 = (a^2 + b^2)(x^2 + 1) + 2(a^2 - b^2)x + 1.$

20. $x^2(x - 169)^2 + 17x = x^2 - 3540.$

21. $6x(x^2 + 1)^2 + (2x^3 + 5)^3 = 150x + 1.$

22. $2x(x-1)^2 + 2 = (x+1)^2.$ **24.** $5x^4 = 12x^3 + 1.$

23. $x^4 = 12x + 5.$ **25.** $(x+4)^3 = 3(2x - 1)^2.$

26. $\dfrac{a}{x} + \dfrac{b}{x-1} + \dfrac{c}{x-2} + \dfrac{c}{x-3} + \dfrac{b}{x-4} + \dfrac{a}{x-5} = 0.$

27. $\dfrac{(x+1)^4}{(x^2+1)(x-1)^2} = \dfrac{m}{n}.$ **30.** $\dfrac{(x^2+1)(x^3+1)}{(x^2-1)(x^3-1)} = \dfrac{m}{n}.$

28. $\dfrac{(x+1)^5}{x(x^3+1)} = \dfrac{m}{n}.$ **31.** $\dfrac{(x^2+1)(x^3+1)}{x^2(x+1)} = \dfrac{m}{n}.$

29. $\dfrac{(x^2+1)(x^3+1)}{(x+1)(x^4+1)} = \dfrac{m}{n}.$ **32.** $\dfrac{(x^3-1)^2}{x(x^2+1)(x-1)^2} = \dfrac{m}{n}.$

33. $\dfrac{x(x+1)^2}{(x^2+1)(x-1)^2} = \dfrac{n(n-m)}{2m(2m-n)}.$

34. $\dfrac{(x^3+1)^2}{x(x^2-1)^2} = \dfrac{4m^2}{m^2 - n^3}.$

35. $\dfrac{(x-1)(x^2+1)^2}{(x^3-1)(x+1)^2} = \dfrac{2(m-n)^2}{mn}.$

36. $\dfrac{x^6 - 1}{(x+1)(x^5 - 1)} = \dfrac{2m}{2m-n}.$

37. $\dfrac{(x^3-1)(x+1)^3}{(x^3+1)(x-1)^3} = \dfrac{m+n}{m-n}.$

38. $\dfrac{(x+1)(x^4+1)}{(x-1)(x^4-1)} = \dfrac{m+n}{m-n}.$

39. $x^3 = \dfrac{ax - b}{bx - a}.$ **41.** $x^5 = \dfrac{ax - b}{bx - a}.$

40. $x^4 = \dfrac{ax - b}{bx - a}.$ **42.** $x^4 = \dfrac{ax^2 + bx + c}{a + bx + cx^2}.$

43. $x^2 = (x-1)^2 (x^2+1)$.

44. $a^2x^2 = (a-x)^2(a^2-x^2)$.

45. $x^2 = (x-a)^2(x^2-1)$.

46. $m(x+m-n)(x-m+7n)^2$
$\qquad = n(x-m+n)(x+7m-n)^2$.

47. $m^2(x+m+17n)(x-m-5n)^2$
$\qquad = n^2(x+17m+n)(x-5m+n)^2$.

48. $m^2(x+m+17n)(x-m+7n)^2$
$\qquad = n^2(x+17m+n)(x+7m-n)^2$.

49. $\dfrac{a}{x} + \dfrac{x}{a} = \dfrac{m}{n}$.

51. $\dfrac{x^2+a^2}{x^2-ax+a^2} = c$.

50. $\dfrac{x^2+ax+a^2}{x^2-ax+a^2} = c$.

52. $\dfrac{x^2+a^2}{(x+a)^2} = c$.

53. $\dfrac{a-x}{x-b} + \dfrac{x-b}{a-x} = \dfrac{m}{n}$.

54. $\dfrac{2a^2+a(a-x)+(a+x)^2}{2a^2+a(a+x)+(a-x)^2} = \dfrac{c+1}{c-1}$.

55. $x^4 + (a-x)^4 = c$.

56. $x^4 + (x-4)^4 = 82$.

57. $(a-x)^5 + (x-b)^5 = c$.

58. $x^5 + (a-x)^5 = a^5$; $\;x^5 + (6-x)^5 = 1056$.

59. $(a-x)^3(x-b)^2 + (a-x)^2(x-b)^3 = a^2b^2(a-b)$.

60. $(x-a+b)^3 - (x-a)^3 + (x-b)^3 - x^3 + a^3 - (a-b)^3 - b^3$
$\qquad = (a-b)c^2$.

61. $\dfrac{(a-x)^4 + (x-b)^4}{(a-x)^2 + (x-b)^2} = \dfrac{a^4+b^4}{a^2+b^2}$.

62. $\dfrac{(a-x)^4 + (x-b)^4}{(a-x)^3 + (x-b)^3} = \dfrac{a^4+b^4}{a^3-b^3}$.

63. $\dfrac{(a-x)^5+(x-b)^5}{(a-x)^3+(x-b)^3}=\dfrac{a^5-b^5}{a^3-b^3}.$

64. $\dfrac{(a-x)^3}{b-x}+\dfrac{(b-x)^3}{a-x}=\dfrac{a^3}{b}+\dfrac{b^3}{a}.$

65. $\dfrac{a-x}{(x-b)^2}+\dfrac{x-b}{(a-x)^2}=\dfrac{a}{b^2}-\dfrac{b}{a^2}.$

66. $\dfrac{(a-x)^4+(x-b)^4}{(a+b-2x)^2}=\dfrac{a^4+b^4}{(a+b)^2}.$

67. $\dfrac{(a-x)^5+(x-b)^5}{(a+b-2x)^2}=\dfrac{a^5-b^5}{(a+b)^2}.$

68. $\dfrac{(a-x)^5+(x-b)^5}{(a-x)^2+(x-b)^2}=(a-b)^3.$

69. $\dfrac{(a-x)^4-(x-b)^4}{(a-x)-(x-b)}=\dfrac{(a-b)c}{(a-x)(x-b)}.$

70. $\dfrac{(a-x)^5+(x-b)^5}{(a-x)^2+(x-b)^2}=c(a-x)(x-b).$

71. $\dfrac{(a-x)^3+(x-b)^3}{(a-x)^4+(x-b)^4}=\dfrac{c}{(a-x)(x-b)}.$

72. $(1+x^2)^3=(x^3-3)^2.$

73. $\dfrac{x^4+1}{2x(x^2+1)}=\dfrac{a}{b}.$

74. $\dfrac{(x+1)^2(x^2+1)}{(x-1)^2(x^2-x+1)}=\dfrac{a}{b}.$

75. $\dfrac{(x-1)^2x}{(x^2-x+1)^2}=\dfrac{a}{b}.$

76. $\dfrac{(x^2+x+1)^2}{(x+1)^2(x^2+1)}=\dfrac{a}{b}.$

77. $\dfrac{(x^2+1)^2}{x(x+1)^2}=\dfrac{a}{b}.$

78. $\dfrac{(x+1)^4}{x(x^2+1)}=\dfrac{a}{b}.$

79. $\dfrac{x(x+1)^2}{(x-1)^4}=\dfrac{a}{b}.$

80. $\left[\dfrac{x^2+x+1}{(x+1)^2}\right]\left[\dfrac{x^2-x+1}{(x-1)^2}\right]=\dfrac{a}{b}.$

81. $\dfrac{x^4-x^2+1}{(x^2-1)^2}=\dfrac{a}{b}.$

82. $\dfrac{x(x^2+1)}{(x^2-1)^2}=\dfrac{a}{b}.$

83. $\dfrac{(x+1)(x^3+1)}{(x-1)(x^3-1)} = \dfrac{a}{b}.$ **85.** $\dfrac{(x+1)^4}{x^4+1} = \dfrac{a}{b}.$

84. $\dfrac{(x+1)(x^5-1)}{(x-1)(x^5+1)} = \dfrac{a}{b}.$ **86.** $\dfrac{(x+1)^5}{x^5+1} = \dfrac{a}{b}.$

§ 49. Quadratic Equations Involving Two or More Variables.

1. $\qquad (x+y)(x^2+y^2) = a, \qquad\qquad\qquad (1)$

$\qquad\qquad x^2 y + x y^2 = c. \qquad\qquad\qquad\qquad (2)$

$(1)+2(2), \quad \therefore (x+y)^3 = a+2c.$

$\qquad\qquad \therefore x+y = \sqrt[3]{(a+2c)}. \qquad\qquad\qquad (3)$

(Any one of the three cube roots.)

$(1) \div (2), \qquad \dfrac{x^2+y^2}{xy} = \dfrac{a}{c}; \quad \therefore \left(\dfrac{x-y}{x+y}\right)^2 = \dfrac{a-2c}{a+2c}.$

By (3), $\qquad x-y = \dfrac{\sqrt{(a-2c)}}{\sqrt[6]{(a+2c)}}.$

Also, $\qquad x+y = \dfrac{\sqrt{(a+2c)}}{\sqrt[6]{(a+2c)}}.$

$\qquad \therefore x = \dfrac{\sqrt{(a+2c)} + \sqrt{(a-2c)}}{2\sqrt[6]{(a+2c)}}.$

$\qquad y = \dfrac{\sqrt{(a+2c)} - \sqrt{(a-2c)}}{2\sqrt[6]{(a+2c)}}.$

Not any one of the six sixth-roots of $a+2c$ may be used indifferently in the denominator, but only any cube-root of whichever square root of $a+2c$ is used in the numerator. Thus, if the radical sign be restricted to denote merely the arithmetical root, if k be defined by the equation $k^2 - k + 1 = 0$, and if m and n indicate any integers whatever, equal or unequal, the value of x may be written,

$$[k^{2m}\sqrt{(a+2c)}+k^{3n-m}\sqrt{(a-2c)}]$$
$$\div 2\sqrt[6]{(a+2c)}.$$

2.
$$8x^2-5xy+3y^2=9(x+y), \qquad (1)$$
$$11x^2-8xy+5y^3=13(x+y), \qquad (2)$$

First Method. Eliminate $(x+y)$.

$$\therefore 104x^2-65xy+39y^2=99x^2-72xy+45y^2.$$
$$\therefore 5x^2+7xy-6y^2=0.$$
$$\therefore (5x-3y)(x+2y)=0.$$
$$\therefore x=\tfrac{3}{5}y \text{ or } -2y.$$

Substitute these values for x in (1),

$$\therefore 72y^2=360y, \text{ or } 45y^2=-9y.$$
$$\therefore y=0, \text{ or } 5, \text{ or } -\tfrac{1}{5},$$
and $\qquad x=0, \text{ or } 3, \text{ or } \tfrac{2}{5}.$

Second Method. Take the sum of the products of (1) and (2) by arbitrary multipliers k and l.

$$k(8x^2-5xy+3y^2)+l(11x^2-8xy+5y^2)$$
$$=(9k+13l)(x+y). \qquad (3)$$

Determine k and l so that the left-hand member of (3) may, like the right-hand member, be a multiple of $x+y$. This may be done by putting $x=-y$ in (3), from which

$$16k+24l=0.$$
$$\therefore 2k=-3l.$$
$$\therefore \text{ if } k=3, \ l=-2.$$

Substituting these values in (3), it becomes

$$2x^2+xy-y^2=x+y.$$
$$\therefore (x+y)(2x-y)=x+y,$$
or $\qquad (x+y)(2x-y-1)=0.$
$$\therefore \text{ either } x+y=0, \text{ or } 2x-y-1=0.$$
$$\therefore y=-x, \text{ or } 2x-1.$$

Substituting these values for x in (1), it becomes

$$16\,x^2 = 0, \text{ or } 10\,x^2 - 7\,x + 3 = 27\,x - 9.$$

$$\therefore x = 0, \text{ or } 3, \text{ or } \tfrac{2}{5},$$

and $$y = 0, \text{ or } 5, \text{ or } -\tfrac{1}{5}.$$

3. $$\frac{x^5 + y^5}{x^3 + y^3} = \frac{a^5 + b^5}{a^3 + b^3},$$ (1)

$$x^2 + xy + y^2 = a^2 + ab + b^2.$$ (2)

$(1) \div (2)$, $\therefore \ \dfrac{x^4 - x^3y + x^2y^2 - xy^3 + y^4}{(x^2 + y^2)^2 + x^2y^2}$

$$= \frac{a^4 - a^3b + a^2b^2 - ab^3 + b^4}{(a^2 + b^2)^2 - a^2b^2}.$$

$$\therefore \frac{x^3y + xy^3}{(x^2 + y^2)^2 - x^2y^2} = \frac{a^3b + ab^3}{(a^2 + b^2)^2 - a^2b^2}.$$ (3)

Write z for $\dfrac{xy}{x^2 + y^2}$, and k for $\dfrac{ab}{a^2 + b^2}.$

(3), $\therefore \dfrac{z}{1 - z^2} = \dfrac{k}{1 - k^2}.$

$$\therefore z = k \text{ or } -\frac{1}{k}.$$

$$\therefore \frac{xy}{x^2 + y^2} = \frac{ab}{a^2 + b^2} \text{ or } \frac{a^2 + b^2}{-ab}.$$

$$\therefore \frac{xy}{x^2 + xy + y^2} = \frac{ab}{a^2 + ab + b^2} \text{ or } \frac{a^2 + b^2}{a^2 - ab + b^2}.$$

(2), $\therefore xy = ab \text{ or } (a^2 + b^2)\dfrac{a^2 + ab + b^2}{a^2 - ab + b^2}.$ (4)

$\sqrt{[(2) + (4)]},$

$$\therefore x + y = \pm (a + b),$$

or $\sqrt{(2\,a^2 - ab + 2\,b^2)}\sqrt{\dfrac{a^2 + ab + b^2}{a^2 - ab + b^2}}.$

$\sqrt{[(2)-3(4)]}$,

and $\qquad x-y=\pm(a-b)$,

or $\qquad i\sqrt{(2a^2+ab+2b^2)}\sqrt{\dfrac{a^2+ab+b^2}{a^2-ab+b^2}}$.

$\therefore x=\pm a,\pm b$,

or $\qquad \frac{1}{2}[\sqrt{(2a^2-ab+b^2)}$

$\qquad +i\sqrt{(2a^2+ab+2b^2)}]\sqrt{\dfrac{a^2+ab+b^2}{a^2-ab+b^2}}$.

$y=\pm b,\pm a$,

or $\qquad \frac{1}{2}[\sqrt{(2a^2-ab+b^2)}$

$\qquad -i\sqrt{(2a^2+ab+2b^2)}]\sqrt{\dfrac{a^2+ab+b^2}{a^2-ab+b^2}}$.

4. $\qquad (x^2+y^2)(x^3+y^3)=a$, $\qquad$ (1)

$\qquad (x+y)(x^4+y^4)=b$. $\qquad$ (2)

Put $\qquad z=\dfrac{xy}{x^2+y^2}$.

$\therefore \dfrac{1-z}{1-2z^2}=\dfrac{a}{b}$.

$\therefore 2az^2-bz-(a-b)=0$.

$\therefore 4az=b\pm\sqrt{(8a^2-8ab+b^2)}=b+r$, say.

$\therefore \dfrac{xy}{x^2+y^2}=\dfrac{b+r}{4a}$. $\qquad$ (3)

$\therefore \dfrac{x+y}{x-y}=\pm\sqrt{\dfrac{2a+b+r}{2a-b-r}}$.

$\therefore \dfrac{x}{y}=\dfrac{\sqrt{(2a+b+r)}+\sqrt{(2a-b-r)}}{\sqrt{(2a+b+r)}-\sqrt{(2a-b-r)}}$

$\qquad =\dfrac{[\sqrt{(2a+b+r)}+\sqrt{(2a-b-r)}]^2}{2(b+r)}$. $\qquad$ (4)

$(1)^2$, $\qquad (x^2+y^2+2xy)(x^2+y^2)^3[(x^2+y^2)-xy]^2=a^2$.

(3),
$$\therefore (xy)^5 \left(\frac{4a+2b+2r}{b+r}\right)\left(\frac{4a}{b+r}\right)^2\left(\frac{4a-b-r}{b+r}\right)^2 = a^2.$$

$$\therefore x^{10}\left(\frac{y}{x}\right)^5\left[\frac{32a^2(2a+b+r)(4a-b-r)^2}{(b+r)^5}\right] = a^2.$$

$$\therefore x^{10} = \left(\frac{x}{y}\right)^5\left[\frac{(b+r)^5}{32(2a+b+r)(4a-b-r)^2}\right]$$

(4),
$$= \frac{[\sqrt{(2a+b+r)}+\sqrt{(2a-b-r)}]^{10}}{1024(2a+b+r)(4a-b-r)^2}.$$

$$\therefore x = \frac{\sqrt{(2a+b+r)}+\sqrt{(2a-b-r)}}{2\sqrt[10]{[(2a+b+r)(4a-b-r)^2]}},$$

in which $r = \pm\sqrt{(8a^2-8ab+b^2)}.$

The value of y may be derived from that of x by the first
  form in (4).

5. $x^4 = ax - by,$ (1)

            $y^4 = ay - bx.$ (2)

$x \times (1) - y \times (2),$

          $x^5 - y^5 = a(x^2 - y^2).$

$y \times (1) - x \times (2),$

           $xy(x^3 - y^3) = b(x^2 - y^2).$

         $\therefore$ either $x - y = 0,$

  from which $x = y = 0,$ or $\sqrt[3]{(a-b)},$ (3)

  or $x^4 - x^3y + x^2y^2 + xy^3 + y^4 = a(x+y),$ (4)

  and $xy(x^2 + xy + y^2) = b(x+y).$ (5)

(4) + (5), $(x+y)^2(x^2+y^2) = (a+b)(x+y).$ (6)

(5), $(x+y)^4 - (x^2+y^2)^2 = 4b(x+y).$ (7)

$\sqrt{[(7)^2 + 4(6)]},$

        $(x+y)^4 + (x^2+y^2)^2 = 2t(x+y),$ (8)

  in which $t = \sqrt{[(a+b)^2 + 4b^2]}.$ (9)

$\frac{1}{2}[(7)+(8)],$

$$\therefore (x+y)^4 = (2b+t)(x+y).$$

$$\therefore (x+y)^3 = 2b+t.$$

$$\therefore (x+y) = \sqrt[3]{(2b+t)}. \tag{10}$$

$(6) \div (10) \quad \therefore x^2 + y^2 = \dfrac{a+b}{\sqrt[3]{(2b+t)}}. \tag{11}$

$2(11) - (10)^2,$

$$\therefore (x-y)^2 = \frac{2(a+b)}{\sqrt[3]{(2b+t)}} - \sqrt[3]{(2b+t)^2}$$

$$= \frac{2a-t}{\sqrt[3]{(2b+t)}}.$$

$$\therefore x - y = \frac{\sqrt{(2a-t)}}{\sqrt[6]{(2b+t)}}.$$

(10) and $\quad x+y = \dfrac{\sqrt{(2b+t)}}{\sqrt[6]{(2b+t)}}.$

$$\therefore x = \frac{\sqrt{(2b+t)} + \sqrt{(2a-t)}}{\sqrt[6]{(2b+t)}},$$

and $\quad y = \dfrac{\sqrt{(2b+t)} - \sqrt{(2a-t)}}{\sqrt[6]{(2b+t)}},$

in which $\quad t = \sqrt{(a^2 + 2ab + 5b^2)}.$

6. $\quad x^4 - c^4 = m(x+y)^4, \tag{1}$

$\qquad\quad y^4 + c^4 = n(x-y)^4. \tag{2}$

Let $\quad z = \dfrac{x+y}{x-y}.$

$$\therefore z+1 = \frac{2x}{x-y} \text{ and } z-1 = \frac{2y}{x-y}. \tag{3}$$

$(1)+(2), \quad x^4 + y^4 = m(x+y)^4 + n(x-y)^4.$

$$\therefore (z+1)^4 + (z-1)^4 = 16(mz^4 + n).$$

$$\therefore (8m-1)z^4 - 6z^2 + (8n-1) = 0.$$

$$\therefore z = \sqrt{\dfrac{3 + \sqrt{[9 - (8m - 1)(8n - 1)]}}{8m - 1}}. \qquad (4)$$

(2) and (3), $\quad (z - 1)^4 (x - y)^4 + 16c^4 = 16n(x - y)^4.$

$$\therefore x - y = \dfrac{2c}{\sqrt[4]{[16n - (z - 1)^4]}}, \qquad (5)$$

and $\qquad x + y = \dfrac{2cz}{\sqrt[4]{[16n - (z - 1)^4]}}.$

$$\therefore x = \dfrac{c(z + 1)}{\sqrt[4]{[16n - (z - 1)^4]}} = \dfrac{c(z + 1)}{\sqrt[4]{[(z + 1)^4 - 16mz^4]}},$$

and $\qquad y = \dfrac{c(z - 1)}{\sqrt[4]{[16n - (z - 1)^4]}},$

and the value of z is given by (4).

7. $\qquad x^2 + y^2 = \tfrac{1}{3}(2m + n^2),$

$\qquad x^3 + y^3 = mn.$

$\therefore (x + y)^2 - 2xy = \tfrac{1}{3}(2m + n^2),$

and $\qquad (x + y)^3 - 3xy(x + y) = mn.$

Let $u = x + y$ and $v = xy$, and the equations become

$\qquad u^2 - 2v = \tfrac{1}{3}(2m + n^2);$

$\qquad u^3 - 3uv = mn.$

Eliminate v,

$\therefore u^3 - (2m + n^2)u + 2mn = 0.$

$\therefore u^4 - (2m + n^2)u^2 + 2mnu = 0.$

$\therefore u^4 - 2mu^2 + m^2 = n^2u^2 - 2mnu + m^2.$

$\therefore u^2 - m = \pm(nu - m).$

$\therefore u = n,$

(the value $u = 0$ was introduced by the multiplication by u),

or $\qquad u^2 + nu - 2m = 0.$

$\therefore u = \tfrac{1}{2}[-n \pm \sqrt{(n^2 + 8m)}].$

$$\therefore v = \tfrac{1}{3}(n^2 - m),$$

or $$\tfrac{1}{6}[n^2 + 8m \mp 3n \sqrt{(n^2 + 8m)}].$$

$\therefore u$ and v are completely determined.

Also, $$x + y = u, \quad x - y = \sqrt{(u^2 - 4v)}.$$

$$\therefore x = \tfrac{1}{2}[u + \sqrt{(u^2 - 4v)}];$$

$$\therefore y = \tfrac{1}{2}[u - \sqrt{(u^2 - 4v)}].$$

If $m = 7$ and $n = 5$, the above equations become

$$x^2 + y^2 = 13, \text{ and } x^3 + y^3 = 35.$$

Solving, as above, gives

$$u = 5, \text{ or } 2, \text{ or } -7;$$

$$2v = 12, \text{ or } -9, \text{ or } 36.$$

$$\therefore x + y = 5, \text{ or } 2, \text{ or } -7;$$

$$x - y = \pm 1, \text{ or } \pm \sqrt{22}, \text{ or } \pm i\sqrt{23}.$$

$$\therefore x = 3, 2, \tfrac{1}{2}(2 \pm \sqrt{22}), \text{ or } \tfrac{1}{2}(-7 \pm i\sqrt{23});$$

$$y = 2, 3, \tfrac{1}{2}(2 \mp \sqrt{22}), \text{ or } \tfrac{1}{2}(-7 \mp i\sqrt{23}).$$

8. $$x^2 + y = \tfrac{17}{16},$$

$$x + y^2 = \tfrac{5}{4}.$$

$$\therefore \tfrac{17}{16} - y = (\tfrac{5}{4} - y^2)^2.$$

$$\therefore y^4 - \tfrac{5}{2}y^2 + y + \tfrac{1}{2} = 0.$$

Testing this for rational linear factors, it is easily reduced to

$$(y - 1)^2(y^2 + 2y + \tfrac{1}{2}) = 0.$$

$$\therefore y = 1, \text{ or } \tfrac{1}{2}(-2 \pm \sqrt{2});$$

$$x = \tfrac{1}{4}, \text{ or } \tfrac{1}{4}(-1 \pm 4\sqrt{2}).$$

9. $$(2x - y + z)(x + y + z) = 9, \qquad (1)$$

$$(x + 2y - z)(x + y + z) = 1, \qquad (2)$$

$$(x + y - 2z)(x + y + z) = 4. \qquad (3)$$

Let $s = x + y + z$, and the equations may be written

$$(s + x - 2y)s = 9, \qquad (4)$$
$$(s + y - 2z)s = 1, \qquad (5)$$
$$(s - 3z)s = 4. \qquad (6)$$

$(4) + 3\,(5), \quad (4s + x + y - 6z)s = 12,$

or $\qquad (5s - 7z)s = 12. \qquad (7)$

$3\,(7) - 7\,(6), \quad [(15s - 21z) - (7s - 21z)]s = 8.$

$$\therefore 8s^2 = 8. \quad \therefore s = \pm 1.$$

Substituting in (1), (2), and (3), they become

$$2x - y + z = \pm 9, \quad x + 2y - z = \pm 1,$$
$$x + y - 2z = \pm 4.$$
$$\therefore x = \pm 4, \; y = \mp 2, \; z = \mp 1.$$

10.
$$x^2 + y^2 = a,$$
$$u^2 + v^2 = b,$$
$$xy + uv = c,$$
$$xu + yv = e.$$

Let $t = xy - uv.$

$$\therefore (x + y)^2 = a + c + t,$$
$$(x - y)^2 = a - c - t.$$
$$\therefore x = \tfrac{1}{2}[\sqrt{(a + c + t)} + \sqrt{(a + c - t)}],$$
$$y = \tfrac{1}{2}[\sqrt{(a + c + t)} - \sqrt{(a - c - t)}].$$
$$(u + v)^2 = b + c - t,$$
$$(u - v)^2 = b - c + t.$$
$$u = \tfrac{1}{2}[\sqrt{(b + c - t)} + \sqrt{(b - c + t)}],$$
$$v = \tfrac{1}{2}[\sqrt{(b + c - t)} - \sqrt{(b - c + t)}].$$

Also, $\quad 2\,(xu + yv) = (x+y)(u+v) + (x-y)(u-v) = 2e.$

$$\therefore \sqrt{[(a + c + t)(b + c - t)]}$$
$$+ \sqrt{[(a - c - t)(b - c + t)]} = 2e.$$
$$\therefore [4e^2 + (a - c - t)(b - c + t) - (a + c + t)(b + c - t)]^2$$
$$= 16e^2(a - c - t)(b - c + t).$$

$$\therefore [(a-b)^2 + 4e^2]t^2 - 2(a^2 - b^2)ct$$
$$+ (a+b)^2 c^2 - 4e^2(ab+c^2) + 4e^4 = 0.$$

$$\therefore t = \frac{(a^2 - b^2)c \pm 2c\sqrt{\{(ab-e^2)[(a-b)^2 - 4(c^2 - e^2)]\}}}{(a-b)^2 + 4e^2}.$$

11.

$$xy = uv, \tag{1}$$
$$x + y + u + v = a, \tag{2}$$
$$x^3 + y^3 + u^3 + v^3 = b^3, \tag{3}$$
$$x^5 + y^5 + u^5 + v^5 = c^5. \tag{4}$$

Let $x + y = \frac{1}{2}(a+z)$.

$$\therefore u + v = \frac{1}{2}(a-z). \tag{5}$$

Also, let $r = xy = uv.$ $\tag{6}$

$$(x+y)^3 = x^3 + y^3 + 3xy(x+y),$$
$$(u+v)^3 = u^3 + v^3 + 3uv(u+v).$$
$$\therefore a(3z^2 + a^2) = 4(b^3 + 3ar). \tag{7}$$

Also,

$$(x+y)^5 = x^5 + y^5 + 5xy(x^3 + y^3) + 10x^2y^2(x+y),$$
$$(u+v)^5 = u^5 + v^5 + 5uv(u^3 + v^3) + 10u^2v^2(u+v).$$
$$\therefore a(5z^4 + 10a^2z^2 + a^4) = 16(c^5 + 5b^3r + 10ar^2). \tag{8}$$

Eliminating r between (7) and (8),

$$45a^2z^4 - 30a(a^3 + 2b^3)z^2 + a^6 - 20a^3b^3$$
$$- 80b^6 + 144ac^5 = 0.$$

$$\therefore 15az^2 - 5(a^3 + 2b^3)$$
$$= \pm 2\sqrt{[5(a^3 + 5b^3)^2 - 180ac^5]}. \tag{9}$$

$$\therefore z = \sqrt{\frac{a^3 + 2b^3 \pm 2\sqrt{\{\frac{1}{5}[(a^3 + 5b^3)^2 - 36ac^5]\}}}{3a}}. \tag{10}$$

(7) and (9), $12ar = a^3 - 4b^3 + 3az^2$
$$= 2a^3 - 2b^3 \pm 2\sqrt{\{\frac{1}{5}[(a^3 + 5b^3)^2 - 36ac^5]\}}.$$

$$\therefore r = \frac{5(a^3 - b^3) \pm \sqrt{[5(a^3 + 5b^3)^2 - 180ac^5]}}{30a}. \tag{11}$$

(10) and (11) give the values of z and r, which may now be treated as known in (5) and (6).

$$x + y = \tfrac{1}{2}(a + z), \text{ and } xy = r.$$
$$\therefore x - y = \tfrac{1}{2}\sqrt{[(a + z)^3 - 16r]}.$$
$$\therefore x = \tfrac{1}{4}\{a + z \pm \sqrt{[(a + z)^2 - 16r]}\},$$
$$y = \tfrac{1}{4}\{a + z \mp \sqrt{[(a + z)^2 - 16r]}\}.$$

The values of u and v may be obtained from those of x and y, respectively, by changing z into $-z$.

12. $\qquad\qquad ax = by = cz = \dfrac{1}{x} + \dfrac{1}{y} + \dfrac{1}{z}.$ $\qquad$ (1)

$(1) \div xyz, \quad \therefore \dfrac{a}{yz} = \dfrac{b}{zx} = \dfrac{c}{xy}.$

$$\therefore \qquad = \frac{a + b + c}{xy + yz + zx}. \qquad (2)$$

Also from $(1) \div xyz$,

$$\frac{a}{yz} = \frac{1}{xyz}\left(\frac{1}{x} + \frac{1}{y} + \frac{1}{z}\right) = \frac{xy + yz + zx}{x^2 y^2 z^2}. \qquad (3)$$

$(2) \times (3), \qquad \dfrac{a^2}{y^2 z^2} = \dfrac{a + b + c}{x^2 y^2 z^2}.$

$$\therefore a^2 x^2 = a + b + c.$$

13. $\qquad\qquad \dfrac{y + z - x}{a} = \dfrac{z + x - y}{b} = \dfrac{x + y - z}{c},$ $\qquad$ (1)

$$xyz = m^3. \qquad (2)$$

$(1), \qquad\qquad \dfrac{z}{a + b} = \dfrac{x}{b + c} = \dfrac{y}{c + a} = \dfrac{m}{r}, \text{ suppose}; \qquad (3)$

then $\qquad\qquad \dfrac{xyz}{(a + b)(b + c)(c + a)} = \dfrac{m^3}{r^3}.$

$$\therefore r^3 = (a + b)(b + c)(c + a).$$

Hence the value of r is known, and from (3)

$$rx = m(b + c).$$

14.
$$y + z = 2axyz, \tag{1}$$
$$z + x = 2bxyz, \tag{2}$$
$$x + y = 2cxyz. \tag{3}$$

$$\therefore xyz = \frac{y+z}{2a} = \frac{z+x}{2b} = \frac{x+y}{2c} = \frac{x+y+z}{a+b+c}$$

$$= \frac{x}{b+c-a} = \frac{y}{c+a+b} = \frac{z}{a+b-c}. \tag{4}$$

$$\therefore x^3 y^3 z^3 = \frac{xyz}{(b+c-a)(c+a-b)(a+b-c)}.$$

$$\therefore x^2 y^2 z^2 = \frac{1}{(b+c-a)(c+a-b)(a+b-c)}.$$

Hence the value of $x^2 y^2 z^2$ is known, call it $\frac{1}{r^2}$, and substitute

in (4) $\qquad \dfrac{1}{r} = \dfrac{x}{b+c-a}.$

$$\therefore rx = b + c - a,$$

in which $\quad r^2 = (b+c-a)(c+a-b)(a+b-c).$

15.
$$y^2 + z^2 - x(y+z) = a, \tag{1}$$
$$z^2 + x^2 - y(z+x) = b, \tag{2}$$
$$x^2 + y^2 - z(x+y) = c. \tag{3}$$

$(1) + (2) + (3),$
$$2(x^2 + y^2 + z^2 - xy - yz - zx) = a + b + c. \tag{4}$$

(1) may be written
$$x^2 + y^2 + z^2 - x(x+y+z) = a. \tag{5}$$

(2) may be written
$$x^2 + y^2 + z^2 - y(x+y+z) = b. \tag{6}$$

(3) may be written
$$x^2 + y^2 + z^2 - z(x+y+z) = c. \tag{7}$$

$$\therefore x+y+z=\frac{a-b}{y-x}=\frac{b-c}{z-y}=\frac{c-a}{x-z}.$$

$$\therefore (x+y+z)^2=\frac{(a-b)^2+(b-c)^2+(c-a)^2}{(y-x)^2+(z-y)^2+(x-z)^2}$$

$$=\frac{a^2+b^2+c^2-ab-bc-ca}{x^2+y^2+z^2-xy-yz-zx}$$

(4)

$$=\frac{2(a^2+b^2+c^2-ab-bc-ca)}{a+b+c} \qquad (8)$$

$$=\frac{2(a^3+b^3+c^3-3abc)}{(a+b+c)^2}. \qquad (9)$$

Write r^2 for $2(a^3+b^3+c^3-3abc)$.

(9) $\therefore x+y+z=\dfrac{r}{a+b+c}.$ (10)

Returning to (8),

$$(x+y+z)^2=\frac{2(a^2+b^2+c^2-ab-bc-ca)}{a+b+c}. \qquad (8)$$

(4) $2(x^2+y^2+z^2-xy-yz-zx)=\dfrac{(a+b+c)^2}{a+b+c}.$ (11)

$\tfrac{1}{3}[(8)+(11)]$, $x^2+y^2+z^2=\dfrac{a^2+b^2+c^2}{a+b+c}.$ (12)

(5) and (10), $x^2+y^2+z^2-\dfrac{rx}{a+b+c}=a.$

$$\therefore rx=(a+b+c)(x^2+y^2+z^2)-a(a+b+c)$$

(12) $=a^2+b^2+c^2-a(a+b+c)$

$$=b^2+c^2-a(b+c).$$

(5), (6), and (7) are symmetrical with respect to $(xyz|abc)$; (10) shows this substitution does not affect r, and consequently the values of y and z may be written down at once from that of x.

Ex. 62.

1. $6[(7-x)^2+y^2]=13(7-x)y$; $x^2+4y=y^2+4$.

2. $10x^2-9y^2=2x^3$; $8x^2-6y^2=13x$.

3. $xy=(3-x)^2=(2-y)^2$.

4. $x^2+y^2=8x+9y=144$.

5. $x^2+y^2=x+y+12$; $xy+8=2(x+y)$.

6. $x+xy+y=5$; $x^2+xy+y^2=7$.

7. $x^3+y^3=7xy=28(x+y)$.

8. $x^2+xy+y^2=\dfrac{35}{x^2+y^2}=\dfrac{28}{xy}$.

9. $x^4+x^2y^2+y^4=133$; $x^3y+x^2y^2+xy^3=114$.

10. $(x+y)(x^2+y^2)=17xy$; $(x-y)(x^2-y^2)=9xy$.

11. $25(x^3+y^3)=7(x+y)^3=175xy$.

12. $2x^2-y^2=14(x^2-2y^2)=14(x-y)$.

13. $2x^2-3xy=9(x-3y)$; $3(x^2-3y^2)=2(2x^2-3xy)$.

14. $2x^2-xy+5y^2=10(x+y)$; $x^2+4xy+3y^2=14(x+y)$.

15. $(2x-3y)(3x+4y)=39(x-2y)$;
    $(3x+2y)(4x-3y)=99(x-2y)$.

16. $(x+2y)(x+3y)=3(x+y)$; $(2x+y)(3x+y)=28(x+y)$.

17. $x+y=8$; $x^4+y^4=706$.

18. $x+y=5$; $x^5+y^5=275$.

19. $x+y=2$; $13(x^5+y^5)=121(x^3+y^3)$.

20. $x+y=4$; $41(x^5+y^5)=122(x^4+y^4)$.

21. $x^2-5xy+y^2+5=0$; $xy=x+y-1$.

22. $x^2+y=5(x-y)$; $x+y^2=2(x-y)$.

23. $3(x^2 + y) = 3(x + y^2) = 13xy.$

24. $10(x^2 + y) = 10(x + y^2) = 13(x^2 + y^2).$

25. $x^2 + y = \frac{16}{9}; \quad x + y^2 = \frac{22}{9}.$

26. $9(x^2 + y) = 3(x + y^2) = 7.$

27. $x + xy + y = 5; \quad x^3 + xy + y^3 = 17.$

28. $x + y = 2; \quad (x + 1)^5 + (y - 2)^5 = 211.$

29. $3(x-1)(y+1) = 4(x+1)(y-1); \quad \dfrac{x^2+x+1}{y^2+y+1} = \dfrac{31}{39}\left(\dfrac{x^2-x+1}{y^2-y+1}\right).$

30. $x + y = \dfrac{1}{xy}; \quad x - y = xy.$

31. $x + y + 1 = 0; \quad x^6 + y^6 + 2 = 0.$

32. $x + y = 1; \quad 3(x^8 + y^8) = 7.$

33. $4xy^2 = 5(5 - x); \quad 2(x^2 + y^2) = 5.$

34. $27xy = 17; \quad 9(x^3 + y^3) = -8.$

35. $(x^2 + y^2)^2 + 4x^2y^2 = 5 - 12y; \quad y(x^2 + y^2) + 3 = 0.$

36. $x + y = xy; \quad x^2 + y^2 = x^3 + y^3.$

37. $x(y^2 + 3y - 1) = 2y^2 + 2y + 3; \quad y(x^2 + 3x - 1) = 2x^2 + 2x + 3.$

38. $\dfrac{x^3}{a^3} + \dfrac{y^3}{b^3} = 2c^3; \quad \dfrac{x}{a} + \dfrac{y}{b} = e\left(\dfrac{x}{a} - \dfrac{y}{b}\right).$

39. $x^3 + xy^2 = a; \quad y^3 + x^2y = b.$

40. $x + y = a; \quad \dfrac{x}{b - y} + \dfrac{b - y}{x} = c.$

41. $x^2 + ay^2 = \dfrac{a+1}{a-1}; \quad ax^2 + y^2 = (a^2 - 1)y.$

42. $x + y^2 = ax; \quad x^2 + y = by.$

43. $x + y^2 = ay^2; \quad x^2 + y = bx^2.$

44. $x^4 - y^4 = a^2(x - y)^2; \quad x^3 - x^2y + xy^2 - y^3 = b^2(x + y).$

45. $(x+y)(x^2+3y^2)=m$; $(x-y)(x^2+3y^2)=n$.

46. $x^2y^2=y(a-x)^3=x(b-y)^3$.

47. $x^3(b-y)=y^3(a-x)=(a-x)^2(b-y)^2$.

48. $a^2(x^2+e^2)=b^2(x+y)^2$; $a^2(y^2+e^2)=c^2(x+y)^2$.

49. $x^3-y^3=a(x^2-y^2)$; $x^2+y^2=b(x+y)$.

50. $x+y=a$; $x^3+y^3=bxy$.

51. $x+y=xy=x^2+y^2$.

52. $x-y=\dfrac{x}{y}=x^2-y^2$.

53. $x^3(1+y^2)(1+y^4)=a$; $x^3(1-y^2)(1-y^4)=b$.

54. $\dfrac{x^2+xy+y^2}{x^2-xy+y^2}=\dfrac{x^2+y^2}{a}=\dfrac{xy}{b}$.

55. $x^2y+xy^2=\dfrac{a}{x^2+y^2}$; $x^4y+xy^4=b$.

56. $x^2y+xy^2=a(x^2+y^2)$; $x^2y-xy^2=b(x^2-y^2)$.

57. $\left(\dfrac{x}{y}+\dfrac{y}{x}\right)(x+y)=a$; $\dfrac{x^2}{y}+\dfrac{y^2}{x}=b$.

58. $x^2+y^2=ax^2y^2=xy(x+y)$.

59. $abxy=a(x^3+y^3)=b(x+y)^3$.

60. $xy(x+y)=a$; $x^3y^3(x^3+y^3)=b$.

61. $\left(\dfrac{1}{x}+\dfrac{1}{y}\right)(x^3-y^3)=a$; $\left(\dfrac{1}{x}-\dfrac{1}{y}\right)(x^3+y^3)=b$.

62. $x^4+y^4=m(x^2+y^2)$; $x^2+xy+y^2=n$.

63. $ab(x+y)=xy(a+b)$; $x^2+y^2=a^2+b^2$.

64. $x^3+y^3=a(x+y)$; $x^4+y^4=b(x+y)^2$.

65. $x^2+y^2=a$; $x^5+y^5=b(x^3+y^3)$.

66. $xy=a$; $x^5+y^5=b(x^3+y^3)$.

67. $(x-y)(x^3+y^3)=(a-b)(a^3+b^3)$; $x^2-y^2=a^2-b^2$.

68. $x^2-y^2=a$; $x^3+y^3=b(x-y)$.

69. $x+y=a$; $x^4+y^4=b$.

70. $x+y=a$; $x^5+y^5=b$.

71. $x+y=a$; $x^2+y^2=b^2x^2y^2$.

72. $x+y=a+b$; $(a-b)^2(x^4+y^4)=(x-y)^2(a^4+b^4)$.

73. $x+y=a$; $c(x^4+y^4)=xy(x^3+y^3)$.

74. $(x+y)^3=a(x^2+y^2)$; $xy=c(x+y)$.

75. $x^2y+xy^2=a^3$; $c^3(x^3+y^3)=x^3y^3$.

76. $x^3=a(x^2+y^2)-cxy$; $y^3=c(x^2+y^2)-axy$.

77. $x^2-y^2=a^2$; $x^3-y^3=c^4\left(\dfrac{1}{x}-\dfrac{1}{y}\right)$.

78. $x^4-y^4=a^2xy$; $(x^2+y^2)^2=b^2(x^2-y^2)$.

79. $(x+y)x^2y^2=a$; $x^5+y^5=b$.

80. $(x+y)xy=a$; $x^5+y^5=bxy$.

81. $x^4+y^4=a(x+y)^2$; $x^5+y^5=b(x+y)^3$.

82. $x^4+x^2y^2+y^4=a$; $x^2-xy+y^2=1$.

83. $(x^2+y^2)xy=x^2-y^2$; $\dfrac{x^4(1+x^2y^2)}{y^4(1+xy)^2}=\dfrac{a}{b}\left[\dfrac{1+xy}{1-xy}\right]$.

84. $x^2+y^2=a(x+y)$; $x^4+y^4=b(x^3+y^3)$.

85. $x^3+y^3=a$; $(x+y)(x^4+y^4)=b(x^2+y^2)$.

86. $(x^2+y^2)(x^3+y^3)=axy$; $(x+y)(x^4+y^4)=bxy$.

87. $(x+y)^2(x^2+y^2)=a$; $(x^2+y^2)^2(x^4+y^4)=b$.

88. $(x-y)(x^2-y^2)(x^4-y^4)=4axy$;
  $(x+y)(x^2+y^2)(x^3+y^3)=b(x-y)$.

89. $x^4y+xy^4=a(x^3y+xy^3)=b(x^4+y^4)$.

90. $a(x^5 + y^5) = ab(x+y) = bxy(x^3 + y^3).$

91. $\dfrac{x^3 - y^3}{x^2 - y^2} = \dfrac{a^3 - b^3}{a^2 - b^2}$; $\dfrac{x^5 - y^5}{x^4 - y^4} = \dfrac{a^5 - b^5}{a^4 - b^4}.$

92. $\dfrac{x^3 + y^3}{x^2 - y^2} = \dfrac{a^3 + b^3}{a^2 - b^2}$; $\dfrac{x^4 + y^4}{x^3 - y^3} = \dfrac{a^4 + b^4}{a^3 - b^3}.$

93. $x^5 = 2ax - by$; $y^5 = 2ay - bx.$

94. $(x+y)(x^3 + y^3) = a$; $(x-y)(x^3 - y^3) = b.$

95. $\dfrac{(x+y)^4(x^2 + xy + y^2)}{(x^2 + y^2)(x^2 - xy + y^2)} = 3m^2$;

$\dfrac{(x-y)^4(x^2 - xy + y^2)}{(x^2 + y^2)(x^2 + xy + y^2)} = 3n^2.$

96. $(x+y)(x^3 + y^3) = axy$; $(x-y)(x^3 - y^3) = bxy.$

97. $(x+y)(x^3 + y^3) = a(x^2 + y^2)$; $(x-y)(x^3 - y^3) = b(x^2 + y^2).$

98. $\dfrac{(x+y)^3(x^3 + y^3)}{(x^2 + xy + y^2)(x^2 + y^2)} = a^2$; $\dfrac{(x-y)^3(x^3 - y^3)}{(x^2 - xy + y^2)(x^2 + y^2)} = b^2.$

99. $\dfrac{(x^3 + y^3)(x+y)^3}{x^2 + xy + y^2} = 2a^2$; $\dfrac{(x^3 - y^3)(x-y)^3}{x^2 - xy + y^2} = 2b^2.$

100. $\dfrac{(x^3 + y^3)(x+y)^5}{(x^2 + xy + y^2)^2} = 8a^2$; $\dfrac{(x^3 - y^3)(x-y)^5}{(x^2 - xy + y^2)^2} = 8b^2.$

101. $xy(x+y)(x^3 + y^3) = a$; $xy(x-y)(x^3 - y^3) = b.$

102. $x(x+y)(x+2y)(x+3y) = a^2$; $(x+y)^2 + (x+2y)^2 = b.$

103. $(x+1)(y-1) = a(x-1)(y+1)$;

$(x^5 + 1)(y-1)^5 = b^3(x-1)^5(y^5 + 1).$

104. $x + y = a(1 + xy)$; $(x+y)^4 = b^4(1 + x^4y^4).$

105. $x + y = a(1 + xy)$; $x^5 + y^5 = b^3(1 + x^5y^5).$

106. $(x+y)(y-1) = a(x-1)(y+1)$;

$(x^5 - 1)(y-1) = b^2(y^5 - 1)(x-1).$

107. $\dfrac{(1+x)(1+y)}{(1-x)(1-y)}=a$; $\dfrac{(1+x)^3(1+y)^3}{(1-x^3)(1-y^3)}=b.$

108. $\dfrac{(c+x)(c+y)}{(c-x)(c-y)}=a$; $\dfrac{(c^4+x^4)(c^4+y^4)}{(c^4-x^4)(c^4-y^4)}=b.$

109. $\dfrac{(x+m)(y+n)}{(x-m)(y-n)}=a$; $\dfrac{(x^4+m^4)(y^4+n^4)}{(x-m)^4(y-n)^4}=b.$

110. $\dfrac{(x+1)(y+1)}{(x-1)(y-1)}=\dfrac{a}{c}$; $\dfrac{(x^5+1)(y^5+1)}{(x^5-1)(y^5-1)}=\dfrac{b}{c}.$

111. $\dfrac{y(1+x^2)}{x(1+y^2)}=a$; $\dfrac{y^2(1+x^4)}{x^2(1+y^4)}=b.$

112. $\dfrac{1+x}{1+y}=a\sqrt{\dfrac{x}{y}}$; $\dfrac{y(1+x+x^2)}{x(1+y+y^2)}=b.$

113. $\dfrac{y(1+x^2)}{x(1+y^2)}=a$; $\dfrac{y^3(1+x^6)}{x^3(1+y^6)}=b.$

114. $\dfrac{y(1+x^2)}{x(1+y^2)}=a$; $\dfrac{y^4(1+x^8)}{x^4(1+y^8)}=b.$

115. $\dfrac{y(1+x^2)}{x(1+y^2)}=a$; $\dfrac{y^5(1+x^{10})}{x^5(1+y^{10})}=b.$

116. $\dfrac{(x+y)(xy+1)}{(x-y)(xy-1)}=\dfrac{a^2+b^2}{2ab}$; $\dfrac{x(y^2+1)}{y(x^2-1)}=\dfrac{a-b}{a+b}.$

117. $\dfrac{(x+y)(1+xy)}{(x-y)(1-xy)}=a^2(b^2-1)$; $\dfrac{x(1-y^2)}{y(1-x^2)}=b.$

118. $\dfrac{(x+y)(1+xy)}{(x-y)(1-xy)}=a$; $\dfrac{(x^2+y^2)(1+x^2y^2)}{(x^2-y^2)(1-x^2y^2)}=b.$

119. $\dfrac{(x+y)(1+xy)}{(x-y)(1-xy)}=a$; $\dfrac{(x^3+y^3)(1+x^3y^3)}{(x^3-y^3)(1-x^3y^3)}=b.$

120. $\dfrac{(x^4+y^4)(1+x^4y^4)}{(x^4-y^4)(1-x^4y^4)}=a$; $\dfrac{(x+y)(1+xy)}{(x-y)(1-xy)}=b.$

121. $\dfrac{(x+y)(1+xy)}{(x-y)(1-xy)} = a \; ; \; \dfrac{(x^5+y^5)(1+x^5y^5)}{(x^5-y^5)(1-x^5y^5)} = b.$

122. $\dfrac{(x^2+xy+y^2)(1+xy+x^2y^2)}{(x+y)^2(1+xy)^2} = a \; ;$

$\dfrac{(x^2-xy+y^2)(1-xy+x^2y^2)}{(x-y)^2(1-xy)^2} = b.$

123. $x^4 - 3a^2y + 5a^3x + y^2 = 0 \; ; \; y^3 - x^2y^2 - 2a^5x = 0.$

124. $2x(y^2 - 2x)^2 = a \; ; \; y(y^2 - 2x)^2\sqrt{(y^2 - 4x)} = b.$
    Hence, deduce the solution of $x^5 - 5x^2 + 2 = 0.$

125. $2xy(x^2+y^2)^2 = a \; ; \; (x^2 - y^2)(x^2 + y^2)^2 = b.$

Ex. 63.

1. $(2x + y - 4z)(x + y + z) = 24,$
   $(x + 2y - 2z)(x + y + z) = 6,$
   $(-2x + 3y + 5z)(x + y + z) = 30.$

2. $x^2 - yz = 1,$
   $y^2 - xz = 2,$
   $z^2 - xy = 3.$

3. $(x + 2y - 3z)(x + y + z) - 2(xy + yz + zx) = -12,$
   $(2x - 3y + z)(x + y + z) + (xy + yz + zx) = 61,$
   $(3x - y + 2z)(x + y + z) - 5(xy + yz + zx) = 5.$

4. $x^2 - yz = 0,$
   $x + y + z = 7,$
   $x^2 + y^2 + z^2 = 21.$

5. $(x^5 + y^5 + z^5)^3 + (x + y)^2 = 31,$
   $(x^5 + y^5 + z^5)^3 + (x + y + z)^3 = 729,$
   $(x + y)^2 + (x + y + z)^3 = 31.$

6. $x + yz = 14,$
   $y + zx = 11,$
   $z + xy = 10.$

8. $x + y = 5z,$
   $x^2 + y^2 = 39z,$
   $x^3 + y^3 = 105z^2.$

7. $x + y = 8z,$
   $x^3 + y^3 = 134z^3,$
   $x^2 + y^2 + z^2 = 134.$

9. $x + y = 7z,$
   $x^2 + y^2 = 25z^2,$
   $x^4 + y^4 = 674z^3.$

10. $x + y = 7z,$
$x^2 + y^2 = 25z^2,$
$x^5 + y^5 = 20{,}272z.$

11. $x+y:y+z:z+x::a:b:c,$
$(a+b+c)xyz = 2.$

12. $x+y:y+z:z+x::a:b:c,$
$(a+b+c)xyz = 2(x+y+z).$

13. $(x + y - z)x = a,$
$(x - y + z)y = b,$
$(-x + y + z)z = c.$

14. $z\left(\dfrac{x}{y} + \dfrac{y}{z}\right) = a,$

$y\left(\dfrac{x}{z} + \dfrac{z}{x}\right) = b,$

$x\left(\dfrac{y}{z} + \dfrac{z}{y}\right) = c.$

15. $(y+z)(2x+y+z) = a,$
$(z+x)(x+2y+z) = b,$
$(x+y)(x+y+2z) = c.$

16. $x(y+z) : y(z+x) : z(x+y) = b+c : c+a : a+b,$
$xy + yz + zx = (a + b + c)(x + y + z).$

17. $(a+b)x+(b+c)y+(c+a)z = (a+b+c)(x+y+z),$
$a(x + y) = c(y + z),$
$(x+y)^2 + (y+z)^2 + (z+x)^2 = 4(a^2 + b^2 + c^2).$

18. $c(x+y) + b(x-z) - a(y+z) = 0,$
$b(x-z) = (a-c)y,$
$x^2 + y^2 + z^2 = a^2 + b^2 + c^2.$

19. $x + y - az = x - by + z = -cx + y + z = xyz.$

20. $(a + b + c)(x - y) + a(x+z) - b(y+z) = 0,$
$(a + b + c)(x - z) + a(x+y) - c(y+z) = 0,$
$\dfrac{ax^2}{(b+c)^2} + \dfrac{by^2}{(c+a)^2} + \dfrac{cz^2}{(a+b)^2} = 1.$

21. $xy + \dfrac{x}{z} = a, \quad yz + \dfrac{y}{x} = b, \quad zx + \dfrac{z}{y} = c.$

22. $y+z:z+x:x+y::b+c:c+a:a+b,$
$(x + y + z)(xyz) = (a + b + c)(xy + yz + zx).$

23. $x^2 - yz = a, \quad y^2 - xz = b, \quad z^2 - xy = c.$

24. $x^2+(y-z)^2=a^2, \quad y^2+(z-x)^2=b^2, \quad z^2+(x-y)^2=c^2.$

25. $x^2 + xy + y^2 = a^2,\ \ y^2 + yz + z^2 = b^2,\ \ z^2 + zx + x^2 = c^2.$

26. $x + y^3 - z^3 + 3xyz = a(x + y - z),$
$\quad x^3 - y^3 + z^3 + 3xyz = b(x - y + z),$
$\quad -x^3 + y^3 + z^3 + 3xyz = c(-x + y + z).$

27. $x + y + 2az = 0,$
$\quad x^2 + y^2 - 2b^2z^2 = 0,$
$\quad x^n + y^n + z^n = c^n.$

34. $(x - y)^2 = az(x + y),$
$\quad x^3 - y^3 = bz(x + y)^3,$
$\quad (x - y)^3 = cz(x^3 + y^3).$

28. $x + y - az = 0,$
$\quad y(x^2 + y^2) = b^2,$
$\quad x^3 + y^3 = c^3.$

35. $x - y = a,$
$\quad u - v = b,$
$\quad xy = uv,$
$\quad x^5 - y^5 + u^5 - v^5 = c(a + b).$

29. $x\,(y - 1)(z - 1) = 2a,$
$\quad x^2(y^2 - 1)(z^2 - 1) = 4bz,$
$\quad x^3(y^3 - 1)(z^3 - 1) = 6cz^2.$

36. $x + y = a,$
$\quad u + v = b,$
$\quad x^2 + u^2 = c^2,$
$\quad y^2 + v^2 = e^2.$

30. $x\,(y - 1) = a(z - 1),$
$\quad x^2(y^2 - 1) = b^2(z^2 - 1),$
$\quad x^3(y^3 - 1) = c^3(z^3 - 1).$

31. $x\,(y - 1) = a(z - 1),$
$\quad x^2(y^2 - 1) = b^2(z^2 - 1),$
$\quad x^4(y^4 - 1) = c^4(z^4 - 1).$

37. $xy = uv = a^2,$
$\quad x + y + u + x = b,$
$\quad x^3 + y^3 + u^3 + v^3 = c^3.$

32. $x\,(y - 1) = a(z - 1),$
$\quad x^2(y^2 - 1) = b^2(z^2 - 1),$
$\quad x^4(y^4 + 1) = c^4(z^4 + 1).$

38. $xy = uv = a^2,$
$\quad x + y + u + v = b,$
$\quad x^4 + y^4 + u^4 + v^4 = c^4.$

33. $x\,(y - 1) = a(z - 1),$
$\quad x^2(y^2 - 1) = b^2(z^2 - 1),$
$\quad x^5(y^5 - 1) = c^5(z^5 - 1).$

39. $xy = uv = a^2,$
$\quad x + y + u + v = b,$
$\quad x^5 + y^5 + u^5 + v^5 = c^5.$

40. $xy = uv = a^2$,

$x + y + u + v = b$,

$(x + u)^3 + (y + v)^3 = c^3$.

41. $xy = uv = a^2$,

$x + y + u + v = b$,

$(x + u)^4 + (y + v)^4 = c^4$.

42. $xy = uv$,

$x + y + u + v = a$,

$x^2 + y^2 + u^2 + v^2 = b^2$,

$x^3 + y^3 + u^3 + v^3 = c^3$.

43. $xy = uv$,

$x + y + u + v = a$,

$x^2 + y^2 + u^2 + v^2 = b^2$,

$x^4 + y^4 + u^4 + v^4 = c^4$.

44. $xy = uv$,

$x + y + u + v = a$,

$x^2 + y^2 + u^2 + v^2 = b^2$,

$x^5 + y^5 + u^5 + v^5 = c^5$.

45. $xy = uv$,

$x + y + u + v = a$,

$x^3 + y^3 + u^3 + v^3 = b^3$,

$x^4 + y^4 + u^4 + v^4 = c^4$.

46. $xy - uv = 0$,

$xu + yv = a^2$,

$x + y + u + v = b$,

$x^3 + y^3 + u^3 + v^3 = c^3$.

47. $x^2 + y^2 = a^2$,

$u^2 + v^2 = m^2$,

$ux + vy = c^2$,

$vx + uy = n^2$.

48. $x + y + u + v = a$,

$xy + uv = b^2$,

$x^2 + y^2 = m^2$,

$u^2 + v^2 = n^2$.

49. $y(1 + x^2) = 2x$,

$u(1 + y^2) = 2y$,

$v(1 + u^2) = 2u$,

$x(1 + v^2) = 2v$.

50. $x + y + u + v = a$,

$(x + y)^2 + (u + v)^2 = b^2$,

$(x + u)^2 + (y + v)^2 = c^2$,

$(x + v)^2 + (y + u)^2 = e^2$.

51. $\dfrac{x}{y + z} = \dfrac{2a - u}{a - 2u}$,

$\dfrac{y}{z + x} = \dfrac{2b - u}{b - 2u}$,

$\dfrac{z}{x + y} = \dfrac{2c - u}{c - 2u}$,

$x^2 + y^2 + z^2 = e^2$.

52. $\dfrac{1+x+x^2}{1+y+y^2}=a,$

$\dfrac{1+y+x^2}{1+x+y^2}=b.$

56. $\dfrac{x+y}{1+xy}=\dfrac{a}{b+c},$

• $\dfrac{x-y}{1-xy}=\dfrac{b-c}{a}.$

53. $\dfrac{x+1}{y+1}=a\left(\dfrac{x-1}{y-1}\right),$

$\dfrac{x^2+x+1}{y^2+y+1}=b^2\left(\dfrac{x-1}{y-1}\right)^2.$

57. $\dfrac{x+y}{1-xy}=\dfrac{2a\alpha}{a^2-\alpha^2},$

$\dfrac{x-y}{1+xy}=\dfrac{2b\beta}{b^2-\beta^2}.$

54. $\dfrac{(1+x)(1+y)}{(1-x)(1-y)}=\dfrac{1+a}{1-a},$

$\dfrac{(1+x)(1-y)}{(1-x)(1+y)}=\dfrac{1+b}{1-b}.$

58. $\dfrac{1+xy}{x+y}+\dfrac{x+y}{1+xy}=\dfrac{2a}{m},$

$\dfrac{1-xy}{x-y}+\dfrac{x-y}{1-xy}=\dfrac{2b}{n}.$

55. $\dfrac{x+y}{1+xy}=\dfrac{a^2-\alpha^2}{a^2+\alpha^2},$

$\dfrac{x-y}{1-xy}=\dfrac{b^2-\beta^2}{b^2+\beta^2}.$

59. $\dfrac{y(1+x^2)}{x(1+y^2)}=a,$

$\dfrac{y(1-x^2)}{x(1-y^2)}=b.$

60. $2ax=(b+c-a)(y+z),$

$2by=(c+a-b)(x+z),$

$(x+y+z)^2+x^2+y^2+z^2=4(a^2+b^2+c^2).$

61. $\dfrac{x-1}{y-1}=\dfrac{a-1}{b-1},$

$\dfrac{x^3-1}{y^3-1}=\dfrac{a^3-1}{b^3-1}.$

64. $x^3+y^3=\dfrac{a}{x-y},$

$x^2y-xy^2=\dfrac{b}{x+y}.$

62. $\dfrac{x^2+xy+y^2}{x^2-xy+y^2}=\dfrac{x^2+y^2}{a}=\dfrac{xy}{b}.$

65. $xy+\dfrac{x}{y}=a(x^2-y^2),$

$xy-\dfrac{x}{y}=b(x^2+y^2).$

63. $x^4+x^2y^2+y^4=a,$

$x^2+xy+y^2=b.$

66. $x^3=a(x^2+y^2)-bxy,$

$y^3=b(x^2+y^2)-axy.$

67. $4c(x^2+1)=(a+b)(x-y)^2,$
$\qquad 4c(y^2-1)=(a-b)(x-y)^2.$

68. $x^3-n^3=\dfrac{b+c}{2a}(x^2+xy+y^2)(x+y),$

$\qquad y^3-n^3=\dfrac{b-c}{2a}(x^2+xy+y^2)(x+y).$

69. $\dfrac{x+x^2}{y+y^2}=a,$

$\qquad \dfrac{y+x^2}{x+y^2}=b.$

70. $\dfrac{x^2+y^2}{xy}=a,$

$\qquad \dfrac{1+x^2y^2}{xy}=b.$

71. $x(y+z)=a,$
$\qquad y(z+x)=b,$
$\qquad z(x+y)=c.$

72. $(x+y)(x+z)=a,$
$\qquad (y+z)(y+x)=b,$
$\qquad (z+x)(z+y)=c.$

73. $x(x+y+z)=a-yz,$
$\qquad y(x+y+z)=b-zx,$
$\qquad z(x+y+z)=c-xy.$

74. $x^2-(y-z)^2=a,$
$\qquad y^2-(z-x)^2=b,$
$\qquad z^2-(x-y)^2=c.$

75. $x^2+y^2=az,$
$\qquad x+y=bz,$
$\qquad x-y=cz.$

76. $\dfrac{1}{x^2}+\dfrac{1}{y^2}=\dfrac{2a}{z^2},$

$\qquad \dfrac{1}{x^2}-\dfrac{1}{y^2}=\dfrac{2b}{c^2},$

$\qquad \dfrac{1}{x}+\dfrac{1}{y}=\dfrac{1}{c}.$

77. $xy=\dfrac{z-1}{z+1},$

$\qquad (x-y)(z+1)=2a,$

$\qquad (x^2-y^2)(z+1)^2=4bz.$

78. Find the *real* roots of the system of equations,

$\qquad x^2+w^2+v^2=a^2,$ $\qquad vw+u(y+z)=bc,$
$\qquad w^2+y^2+u^2=b^2,$ $\qquad wu+v(z+x)=ca,$
$\qquad v^2+u^2+z^2=c^2,$ $\qquad uv+w(x+y)=ab.$

CHAPTER VIII.

INDICES AND SURDS.

§ **50.** The general Index-laws are :

$$a^{\frac{m}{n}} \times a^{\frac{p}{q}} = a^{\frac{m}{n}+\frac{p}{q}}, \qquad (1)$$

$$a^{\frac{m}{n}} \div a^{\frac{p}{q}} = a^{\frac{m}{n}-\frac{p}{q}}, \qquad (2)$$

$$(ab)^{\frac{m}{n}} = a^{\frac{m}{n}} \times b^{\frac{m}{n}}, \qquad (3)$$

$$(a \div b)^{\frac{m}{n}} = a^{\frac{m}{n}} \div b^{\frac{m}{n}}, \qquad (4)$$

$$(a^{\frac{m}{n}})^{\frac{p}{q}} = a^{\frac{mp}{nq}}. \qquad (5)$$

The law connecting the Index and Surd symbols is

$$a^{\frac{m}{n}} = \sqrt[n]{(a^m)}. \qquad (6)$$

The indices $\frac{1}{2}$, $\frac{1}{3}$, $\frac{1}{4}$, etc., are generally used to denote "either square-root," "any of the cube-roots," "any one of the fourth-roots," etc.

The surd symbols $\sqrt{\ }$, $\sqrt[3]{\ }$, $\sqrt[4]{\ }$, etc., are by some writers restricted to indicate the arithmetical or absolute roots, sometimes called the positive roots. Thus,

$$\sqrt{4} = 2, \text{ but } 4^{\frac{1}{2}} = \pm 2.$$

$$\therefore 4^{\frac{1}{2}} = \pm\sqrt{4}.$$

Also, $\qquad \sqrt{[(-2)^2]} = \sqrt{4} = 2.$

$$\sqrt[3]{27} = 3, \text{ but } 27^{\frac{1}{3}} = 3 \text{ or } 3\left(\frac{-1 \pm i\sqrt{3}}{2}\right).$$

$$\therefore 8^{\frac{1}{3}} = (1^{\frac{1}{3}}) \sqrt[3]{27}.$$

$$\sqrt[4]{16} = 2, \text{ but } 16^{\frac{1}{4}} = \pm 2 \text{ or } \pm 2i.$$

$$\therefore 16^{\frac{1}{4}} = (1^{\frac{1}{4}}) \sqrt[4]{16}.$$

With this restriction, the general connecting formula would be

$$a^{\frac{m}{n}} = (1^{\frac{m}{n}}) \sqrt[n]{(a^m)}.$$

In the following exercises this restriction need not be observed.

Ex. 64.

1. What is the arithmetical value of each of the following:

$36^{\frac{1}{2}}$; $27^{\frac{1}{3}}$; $16^{\frac{1}{4}}$; $32^{\frac{1}{5}}$; $4^{\frac{2}{3}}$; $8^{\frac{2}{3}}$; $27^{\frac{5}{3}}$; $64^{\frac{2}{3}}$; $32^{\frac{2}{5}}$;

$64^{\frac{5}{6}}$; $81^{\frac{3}{4}}$; $(3\frac{3}{8})^{\frac{1}{3}}$; $(5\frac{1}{16})^{\frac{1}{4}}$; $(1\frac{9}{16})^{\frac{3}{2}}$; $(0.25)^{\frac{1}{2}}$;

$(0.027)^{\frac{2}{3}}$; $49^{0.5}$; $32^{0.2}$; $81^{0.75}$?

2. Interpret a^{-2}; a^0; a^{2^2}; $(a^2)^{-2}$; $a^{2^{-2}}$; $a^{-\frac{1}{3}}$; $(a^{-\frac{1}{2}})^{-\frac{1}{3}}$; $a^{\frac{1}{2}}$; $a^{-\frac{1}{3}}$.

3. What is the arithmetical value of $36^{-\frac{1}{2}}$; $27^{-\frac{1}{3}}$; $(0.16)^{-\frac{3}{2}}$; $(0.0016)^{-\frac{3}{4}}$; $(\frac{1}{4})^{-\frac{1}{2}}$; $(\frac{4}{25})^{-\frac{1}{2}}$; $(\frac{9}{16})^{-\frac{3}{2}}$; $(5\frac{1}{16})^{-\frac{1}{3}}$?

4. Prove $(a^m)^n = (a^n)^m$; $(a^m)^{\frac{1}{n}} = (a^{\frac{1}{n}})^m$; $a^{-m} = (a^{-1})^m$; and express these theorems in words.

5. Simplify $a^{\frac{1}{2}} \times a^{\frac{1}{3}}$; $c^{\frac{1}{3}} \times c^{\frac{1}{6}}$; $m^{\frac{1}{2}} \times m^{-\frac{1}{6}}$; $n^{\frac{3}{4}} \times n^{-\frac{1}{12}}$;

$(7\frac{1}{2})^{\frac{1}{3}} \times (2\frac{2}{3})^{\frac{1}{3}} \times (3\frac{1}{5})^{\frac{1}{3}}$;

$\dfrac{a^{\frac{1}{2}}}{a^{-\frac{1}{3}}}$; $\dfrac{c^{\frac{5}{6}}}{c^{-\frac{1}{2}}}$; $\dfrac{d^{\frac{1}{4}}}{d^{\frac{5}{12}}}$; $\dfrac{e^{-\frac{1}{12}}}{e^{\frac{1}{3}}}$; $\dfrac{e^{\frac{1}{6}}}{x^{-\frac{1}{3}}}$; $(2\frac{2}{5})^{\frac{3}{4}} \times (6\frac{2}{3})^{\frac{3}{4}} \div (\frac{1}{4})^{-1\frac{1}{2}}$.

6. Remove the brackets from

$(a^6)^{\frac{1}{2}}$; $(b)^{-\frac{1}{3}}$; $(c^{\frac{3}{4}})^{-\frac{4}{3}}$; $(d^{\frac{2}{3}})^{1\frac{1}{5}}$; $(e^{-\frac{1}{8}})^{\frac{3}{4}}$; $(f^{-\frac{3}{4}})^{-\frac{2}{3}}$; $(a^2b^2)^{\frac{1}{6}}$; $(a^{\frac{3}{4}}b^{\frac{6}{5}})^{\frac{10}{3}}$; $(a^2c^{-1})^{-\frac{1}{2}}$; $(a^{-5}c^{\frac{2}{5}})^{-\frac{1}{2}}$; $(x^{\frac{3}{8}}y^{-\frac{4}{9}})^{-6}$.

7. Remove the brackets and simplify

$$(x^{\frac{1}{5}-\frac{1}{7}})^{\frac{2}{3}}(x^{\frac{1}{7}-\frac{1}{3}})^{\frac{2}{5}}(x^{\frac{1}{3}-\frac{1}{5}})^{\frac{2}{7}}\;;\quad x^{(\frac{1}{5}-\frac{1}{7})\frac{2}{3}}x^{(\frac{1}{7}-\frac{1}{3})\frac{2}{5}}x^{(\frac{1}{3}-\frac{1}{5})\frac{2}{7}}\;;$$

$$x^{(\frac{1}{9}+\frac{1}{16})^{\frac{1}{2}}}x^{(\frac{1}{9}-\frac{1}{25})^{\frac{1}{2}}}\;;\quad x^{(\frac{1}{9}+\frac{1}{16})^{-\frac{1}{2}}}x^{(\frac{1}{9}-\frac{1}{25})^{\frac{1}{2}}}\;;$$

$$(x^{2^{-2}}\div x^{-2^{-2}})(x^{(-2)^2}\div x^{-2^2}).$$

8. Simplify $-x[x^{-\frac{1}{2}}(-x)^{-1}]^{\frac{1}{3}}$; $x[(-x)^{-\frac{2}{3}}(-x)^{-2}]^{-\frac{2}{3}}$;

$(-x)^{-1}(x^{-3}x^{-\frac{3}{4}})^{\frac{3}{4}}.$

9. Determine the commensurable and the surd factors of

$$12^{\frac{1}{2}}\;;\quad 24^{\frac{1}{3}}\;;\quad 18^{-\frac{1}{2}}\;;\quad (-81)^{\frac{1}{3}}\;;\quad 12^{\frac{2}{3}}\;;\quad 64^{\frac{3}{5}}\;;\quad (\tfrac{1}{16})^{\frac{2}{5}}$$

$(6\tfrac{1}{4})^{-\frac{3}{4}}.$

The surd factor must be the incommensurable root of an integer.

10. Simplify $8^{\frac{1}{2}}+18^{\frac{1}{2}}-50^{\frac{1}{2}}$; $72^{\frac{1}{3}}+(\tfrac{24}{125})^{\frac{2}{3}}-(\tfrac{3}{125})^{-\frac{1}{3}}$;

$[(6+2^{\frac{1}{2}})(6-2^{\frac{1}{2}})]^{\frac{1}{5}}$; $(2^{\frac{1}{2}}+3^{\frac{1}{2}})^2+(2^{\frac{1}{2}}-3^{\frac{1}{2}})^2$;

$(2^{\frac{1}{3}}+3^{\frac{1}{3}})(4^{\frac{1}{3}}+9^{\frac{1}{3}}-6^{\frac{1}{3}})$; $(7^{\frac{1}{2}}-3^{\frac{1}{2}})^{\frac{1}{2}}(7^{\frac{1}{2}}+3^{\frac{1}{2}})^{\frac{1}{2}}$;

$\{[(a+x)(x+b)]^{\frac{1}{2}}-[(a-x)(x-b)]^{\frac{1}{2}}\}^2$;

$[a^{\frac{3}{2}}+(a^3-x^3)^{\frac{1}{2}}]^{\frac{1}{3}}\times[a^{\frac{3}{2}}-(a^3-x^3)^{\frac{1}{2}}]^{\frac{1}{3}}.$

Express as surds:

11. $a^{\frac{3}{4}}$; $x^{\frac{5}{8}}$; $p^{3\frac{1}{2}}$; $c^{-\frac{1}{2}}$; $h^{-3\frac{2}{3}}$.

12. $x^{n+\frac{1}{2}}$; $y^{-n+\frac{2}{3}}$; $a^{0.25}$; $b^{-n+\frac{1}{m}}$.

13. $(ax-b)^{\frac{a}{5}}$; $(x^2-4x+1)^{\frac{m-3}{4}}$; $(p-qx)^{n-\frac{2}{3}}$.

Express with indices:

14. $\sqrt[3]{a^2}$; $\sqrt[4]{c^3}$; $\sqrt[n]{x^m}$; $\sqrt[3]{y^{m-n}}$; $\sqrt[5]{(ax)}$; $\sqrt{a^{-b}}$.

15. $\sqrt[3]{(a^3+b^3)}$; $\sqrt[3]{(a^3+b^3)^2}$; $[\sqrt[3]{(a^3+b^3)}]^2$; $\sqrt[x]{[(a-b)x]}$; $\sqrt[n]{(a-bx)^{n-1}}$; $\sqrt[n]{(a^n-b^n)^{m-3n}}$.

16. $(a^{\frac{2}{3}})^{\frac{3}{4}}$; $(b^{-\frac{2}{3}})^{-\frac{3}{4}}$; $(c^{-\frac{2}{3}})^{-\frac{3}{4}}$; $(x^{\frac{3}{5}})^{-\frac{2}{3}}$; $(a^2x)^{-\frac{1}{2}}$; $(a^{-3}x^{-\frac{1}{2}})^{-\frac{1}{2}}$; $(x^{\frac{1}{7}}y^{-\frac{1}{6}})^{14}$.

Simplify the following, expressing the results by both notations:

17. $a \times a^{-\frac{1}{2}}$; $a^0 \times a^{-\frac{1}{2}}$; $a^{\frac{1}{3}} \times a^{-\frac{1}{4}}$; $a \times a^{-\frac{4}{3}}$; $a^{-\frac{1}{2}} \times \sqrt{a}$;

$a^{\frac{2}{3}} \sqrt[3]{a^2}$; $a^{\frac{1}{4}} \sqrt[4]{a}$; $a^{\frac{3}{5}} \sqrt[5]{a^3}$; $a^{\frac{3}{5}} \sqrt[5]{a^{-3}}$; $a^{\frac{5}{3}} \sqrt{a^{-1}}$;

$a^{\frac{2}{3}}b^{\frac{1}{2}}c^{-\frac{1}{4}} \times a^{\frac{1}{3}}b^{-\frac{1}{2}}c^2 d$; $a^{\frac{1}{3}}b^{\frac{2}{3}}c^{\frac{1}{6}} \times a^{-\frac{2}{3}}b^{-\frac{1}{3}}c^{-\frac{1}{6}}$.

18. $\dfrac{a^{\frac{1}{3}}}{\sqrt{a}}$; $\dfrac{a^{\frac{2}{3}}}{\sqrt[3]{a}}$; $\dfrac{\sqrt[6]{c^5}}{\sqrt[3]{c}}$; $\dfrac{\sqrt[4]{x^3}}{x^{\frac{1}{20}}}$; $\dfrac{\sqrt[3]{y^{-2}}}{y^{-\frac{1}{2}}}$; $\dfrac{\sqrt[3]{(24\,a^{-2})}}{2\sqrt[6]{a}}$;

$\dfrac{c(ab)^{\frac{1}{2}} - ac}{bc - c(ab)^{\frac{1}{2}}}$.

19. $\dfrac{a^{\frac{1}{2}}+a^{-\frac{1}{2}}}{a^{\frac{1}{2}}-a^{-\frac{1}{2}}}$; $\dfrac{a^{\frac{3}{2}}-a^{-\frac{3}{2}}}{a^{\frac{1}{2}}-a^{-\frac{1}{2}}}$; $\dfrac{a^{-\frac{3n}{2}}-a^{\frac{3n}{2}}}{a^{-\frac{3n}{2}}+a^{\frac{3n}{2}}}$; $\dfrac{a^2+1+a^{-2}}{a+1+a^{-1}}$.

20. Divide $x-y$ by $x^{\frac{1}{n}}-y^{\frac{1}{n}}$;

$x^{\frac{4}{3}}+a^{\frac{2}{3}}x^{\frac{2}{3}}+a^{\frac{4}{3}}$ by $x^{\frac{2}{3}}+a^{\frac{1}{3}}x^{\frac{1}{3}}+a^{\frac{2}{3}}$;

$x+y+z-3x^{\frac{1}{3}}y^{\frac{1}{3}}z^{\frac{1}{3}}$ by $x^{\frac{1}{3}}+y^{\frac{1}{3}}+z^{\frac{1}{3}}$;

$2ab+2bc+2ca-a^2-b^2-c^2$ by $a^{\frac{1}{2}}+b^{\frac{1}{2}}+c^{\frac{1}{2}}$.

Ex. 65.

1. Express the following quantities (i.) as quadratic surds; (ii.) as cubic surds; (iii.) as quartic surds:

a; $3a$; $2a^2$; a^2x; x^n; $y^{\frac{1}{5}}$; a^{-m}; $\dfrac{x}{y}$; $mx^{-\frac{n}{p}}$; 0.1; 0.01; $1.1x^2$.

2. Reduce to entire surds :

$x\sqrt{x}$; $a\sqrt[3]{a}$; $b^2\sqrt[3]{b^2}$; $3\sqrt[3]{3}$; $4\sqrt[3]{2}$; $\frac{1}{2}\sqrt{2}$; $\frac{1}{3}\sqrt[3]{4}$; $\frac{1}{3}\sqrt[3]{9}$; $3\sqrt[3]{\frac{1}{3}}$;

$a\sqrt{\left(\frac{b}{a}\right)}$; $\frac{a}{b}\sqrt{b}$; $\frac{a}{b}\sqrt{\left(\frac{b}{a}\right)}$; $\frac{a}{b}\sqrt{\left(\frac{a}{b}\right)}$;

$\frac{x}{y}\sqrt[3]{\left(\frac{y}{x}\right)}$; $\frac{x}{y}\sqrt[3]{\left(\frac{y}{x}\right)^2}$; $\frac{x^2}{y}\sqrt[3]{(x^2y^{-1})}$;

$a\sqrt[n]{b}$; $a\sqrt[n]{(a^m)}$; $(a+x)\sqrt[n]{(a+x)^m}$; $(a+x)\sqrt[n]{(a-x)^{n+1}}$;

$\frac{x+y}{m}\sqrt{\left(\frac{x+y}{m}\right)}$; $(x+y)\sqrt{\left(\frac{x-y}{x+y}\right)}$; $\frac{a-b}{a+b}\sqrt[3]{\left(\frac{a+b}{a-b}\right)^2}$;

$(x-y)^{-2}\sqrt[5]{(x^2+2xy+y^2)^{-4}}$; $(x-x^{-1})\sqrt[3]{(x^2+1)^2}$.

3. Reduce to their simplest form :

$\sqrt{12}$; $\sqrt{8}$; $\sqrt{50}$; $\sqrt[3]{16}$; $4\sqrt[3]{0.250}$; $\sqrt{\frac{1}{2}}$; $\sqrt[3]{\frac{1}{4}}$; $\sqrt{\frac{8}{27}}$;

$5\sqrt[3]{(-320)}$; $\sqrt[4]{(1-\frac{1}{81})}$; $\sqrt{a^3}$; $\sqrt{(a^3b^7)}$; $\sqrt[3]{a^5}$;

$3\frac{1}{3}\sqrt[3]{(54x^9)}$; $\sqrt[4]{(x^5y^7z^9)}$; $\sqrt{[a^3(1-x^2)]}$; $\sqrt[3]{[a^2(a^2-1)^4]}$;

$\sqrt{(ab)}$; $\sqrt[n]{a^{n+1}}$; $\sqrt[n]{a^{m+n}}$; $\sqrt[n]{a^{2n+3}}$; $\sqrt[n]{a^{3m-2}}$; $\sqrt{(a^2x+a^3)}$;

$\sqrt[3]{(a^3+2a^4x+a^5x^2)}$; $\sqrt{[(x-1)(x^2-1)]}$;

$\sqrt[3]{[(a^2+2ax+x^2)(a^3+x^3)]}$; $\sqrt[3]{[(x^2-a^2)^2(x-a)]}$;

$\sqrt{(4x^3-8x^2+4x)}$; $\sqrt{(8x^2-16x+8)}$;

$\sqrt[3]{[(x^2-2+x^{-2})(x^4-2x^2+1)]}$;

$\sqrt{\left(\frac{2x-2+2x^{-1}}{x+2+x^{-1}}\right)}$; $\sqrt{\left(\frac{3x^3-6x^2+3x}{27x^2+18x+3}\right)}$; $\sqrt{\left(\frac{(a^2-ab)^2+4a^3b}{a-b}\right)}$.

4. Compare the following quantities by reducing them to the same surd index :

$2:\sqrt{3}$; $2:\sqrt[3]{9}$; $\sqrt{2}:\sqrt[3]{3}$; $\sqrt{10}:\sqrt[3]{30}$; $2\sqrt{2}:\sqrt[3]{22}$;

$a^2:\sqrt{a^3}$; $\sqrt[6]{x}:\sqrt[8]{y}$; $\sqrt[m]{x}:\sqrt[n]{y}$; $\sqrt[m]{x^n}:\sqrt[n]{x^m}$; $\sqrt[4]{a}:\sqrt[6]{b}:\sqrt[8]{c}$;

$\sqrt[m]{a}:\sqrt[n]{b}:\sqrt[p]{c}$; $\sqrt[m]{a^n}:\sqrt[n]{b^p}:\sqrt[p]{c^m}$.

5. Reduce to simple surds with lowest integral surd index:

$$\sqrt{(\sqrt[3]{a})}; \ \sqrt[3]{(\sqrt[4]{b})}; \ \sqrt[3]{(\sqrt{c})}; \ \sqrt[3]{(\sqrt{x^3})}; \ \sqrt[4]{(\sqrt[3]{x^2})}; \ \sqrt[5]{(\sqrt[3]{x^{10}})};$$

$$\sqrt[3]{(\sqrt[4]{x^{15}})}; \ \sqrt[3]{(\sqrt{27})}; \ \sqrt{(\sqrt[3]{81})}; \ \sqrt[4]{(\sqrt[3]{81})}; \ \sqrt{(a\sqrt{a})};$$

$$\sqrt[3]{(a\sqrt{a})}; \ \sqrt{(x\sqrt[3]{x})}; \ \sqrt[3]{(x^2\sqrt[4]{x})}; \ \sqrt[3]{(5\sqrt{5})}; \ \sqrt{(3\sqrt[3]{3})};$$

$$\sqrt[4]{(3\sqrt[3]{3})}; \ \sqrt[6]{(x\sqrt[5]{x})}; \ \sqrt[n]{[a\sqrt[n]{(b\sqrt[n]{c})}]}; \ x\sqrt{(x^{-1}\sqrt{x^{-1}})};$$

$$y\sqrt[3]{(y^{-2}\sqrt[3]{y^{-2}})}; \ z\sqrt[4]{(z^{-3}\sqrt[3]{z^{-2}})}; \ \sqrt[m]{[x\sqrt[n]{(y\sqrt[p]{z})}]}; \ x^{-1}\sqrt[3]{(x^2\sqrt{x^3})}.$$

6. In the following quantities, combine the terms involving the same radical:

$$3\sqrt{2}+5\sqrt{2}-7\sqrt{2}; \ \sqrt{8}-\sqrt{2}; \ \sqrt[3]{16}+3\sqrt[3]{2};$$

$$\sqrt[3]{16}+\sqrt{2}; \ a\sqrt{x}-\sqrt{x}; \ a\sqrt[3]{x}-b\sqrt[3]{x};$$

$$8\sqrt{a}+5\sqrt{x}-7\sqrt{a}+\sqrt{(4a)}-3\sqrt{(4x)}+4\sqrt{(9x)};$$

$$\sqrt{x}+3\sqrt{(2x)}-2\sqrt{(3x)}+\sqrt{(4x)}-\sqrt{(8x)}+\sqrt{(12x)};$$

$$7x-3\sqrt{x}+5\sqrt[3]{x}-2\sqrt[4]{x^2}+\sqrt[6]{x^2};$$

$$4\sqrt{(a^2x)}+2\sqrt{(b^2x)}-3\sqrt{[(a+b)^2x]};$$

$$\sqrt{[(a-b)^2x]}+\sqrt{[(a+b)^2x]}-\sqrt{(a^2x)}+\sqrt{[(1-a)^2x]}-\sqrt{x};$$

$$\sqrt{(a-b)}+\sqrt{(16a-16b)}+\sqrt{(ax^2-bx^2)}-\sqrt{[9(a-b)]};$$

$$\sqrt{(a^3+a^2b)}-\sqrt{(b^3+ab^2)};$$

$$\sqrt{(a^3+2a^2b+ab^2)}-\sqrt{(a^3-2a^2b+ab^2)}-\sqrt{(4ab^2)}.$$

7. In the following quantities, perform, as far as possible, the indicated multiplications and divisions, expressing the results in their simplest forms:

$$\sqrt{2}\times\sqrt{6}; \ \sqrt{3}\times\sqrt{12}; \ \sqrt{14}\times\sqrt{35}\times\sqrt{10}; \ \sqrt{a}\times\sqrt{(3a)};$$

$$\sqrt{c}\times\sqrt{(12c)}; \ \sqrt{(6x)}\times\sqrt{(8x)}; \ \sqrt{y^3}\times\sqrt{y^3}; \ \sqrt[3]{y^5}\times\sqrt[3]{y^7};$$

$$\sqrt[3]{a}\times\sqrt[3]{a^2}\times\sqrt{b}; \ \sqrt{a}\times\sqrt{\left(\frac{x}{a}\right)}; \ \sqrt{a}\times\sqrt{\left(\frac{a}{x}\right)};$$

$\sqrt{3}a \times \sqrt{\left(\dfrac{5c}{6a}\right)}$; $\sqrt{a^{n+1}} \times \sqrt{a^{n+1}}$; $\sqrt[3]{b^{n+1}} \times \sqrt[3]{b^{2n+1}}$;

$\sqrt{12} \div \sqrt{3}$; $\sqrt{(6x)} \div \sqrt{(2x)}$; $a \div \sqrt[3]{a}$; $a^2 \div \sqrt[4]{a^3}$;

$a \div \sqrt[n]{a^{n-1}}$; $a^p \div \sqrt[n]{a^{n-m}}$; $(a+x) \div \sqrt{(a+x)}$;

$(a^2 - x^2) \div \sqrt{(a-x)}$; $(x^2 - 1) \div \sqrt[3]{(x+1)^2}$;

$(3\sqrt{8} - 5\sqrt{2} + \sqrt{18} + \sqrt{32} + \sqrt{72} - 2\sqrt{50}) \times \sqrt{2}$;

$(7\sqrt{2} - 5\sqrt{6} - 3\sqrt{8} + 4\sqrt{20})(\sqrt{18})$;

$(\sqrt{5} + \sqrt{3})(\sqrt{5} - \sqrt{3})$; $(\sqrt{2} + 1)(\sqrt{6} - \sqrt{3})$;

$(3 - \sqrt{2})(2 + 3\sqrt{2})$; $(5\sqrt{3} + \sqrt{6})(5\sqrt{2} - 2)$;

$(\sqrt{a} - \sqrt{b})(\sqrt{a} + \sqrt{b})$; $(a\sqrt{b} + b\sqrt{a})(b\sqrt{a} - a\sqrt{b})$;

$[\sqrt{(x+1)} + \sqrt{(x-1)}][\sqrt{(x+1)} - \sqrt{(x-1)}]$;

$[\sqrt{(3a-b)} + \sqrt{(3b-a)}][\sqrt{(3a-b)} - \sqrt{(3b-a)}]$;

$\sqrt{(a + \sqrt{b})} \times \sqrt{(a - \sqrt{b})}$; $\sqrt{(\sqrt{x} + \sqrt{y})} \times \sqrt{(\sqrt{x} - \sqrt{y})}$;

$\sqrt{[a + \sqrt{(a^2 - x^2)}]} \times \sqrt{[a - \sqrt{(a^2 - x^2)}]}$;

$\sqrt[3]{[x - \sqrt{(x^2 - 1)}]} \times \sqrt[3]{[x + \sqrt{(x^2 - 1)}]}$;

$\sqrt[3]{[a\sqrt{a} - \sqrt{(a^3 - x^3)}]} \times \sqrt[3]{[\sqrt{(a^3 - x^3)} + a\sqrt{a}]}$;

$\sqrt[n]{(8 + 3\sqrt{7})} \times \sqrt[n]{(8 - 3\sqrt{7})}$; $(\sqrt{a} + \sqrt{b})^2$; $(\sqrt{a} + \sqrt{b})^3$;

$(a - c\sqrt{x})^2$; $(\sqrt{x} + \sqrt{x^{-1}})^2$;

$[a + \sqrt{(1 - a^2)}]^2$; $[\sqrt{(a + b - x)} - \sqrt{(a - b + x)}]^2$;

$\{[\sqrt{(a+x)(x-b)}] + \sqrt{[(a-x)(x+b)]}\}^2$;

$\left[\sqrt{\left(\dfrac{2a}{3b}\right)} - \sqrt{\left(\dfrac{3b}{2a}\right)}\right]^2$;

$\{\sqrt{[(a+x)(x+b)]} + \sqrt{[(a-x)(x-b)]}\}^2$;

$\left[\sqrt{\left(\dfrac{a-x}{x-b}\right)} - \sqrt{\left(\dfrac{x-b}{a-x}\right)}\right]^2$; $\sqrt{(x^2 - 1)}\sqrt{\left(\dfrac{x+1}{x-1}\right)}$;

$(\sqrt[3]{a} + \sqrt[3]{b})^3$; $[\sqrt{(\sqrt{a} + \sqrt{b})} + \sqrt{(\sqrt{a} - \sqrt{b})}]^2$;

$$[\sqrt{(\sqrt{10}+1)}-\sqrt{(\sqrt{10}-1)}]^2;$$

$$\{\sqrt{[a+\sqrt{(a^2-x^2)}]}+\sqrt{[a-\sqrt{(a^2-x^2)}]}\}^2;$$

$$(\sqrt{x}+\sqrt{y})^4+(\sqrt{x}-\sqrt{y})^4; \quad (a^2+ab\sqrt{2}+b^2)(a^2-ab\sqrt{2}+b^2);$$

$$(\sqrt[3]{a}+\sqrt[3]{c})[\sqrt[3]{a^2}-\sqrt[3]{(ac)}+\sqrt[3]{c^2}];$$

$$(\sqrt{a}+\sqrt{b}+\sqrt{c})(\sqrt{b}+\sqrt{c}-\sqrt{a})(\sqrt{c}+\sqrt{a}-\sqrt{b})$$
$$(\sqrt{a}+\sqrt{b}-\sqrt{c});$$

$$\left(\sqrt{\tfrac{x}{a}}+\sqrt{\tfrac{y}{b}}+1\right)\left(\sqrt{\tfrac{x}{a}}+\sqrt{\tfrac{y}{b}}-1\right)$$
$$\times\left(\sqrt{\tfrac{x}{a}}-\sqrt{\tfrac{y}{b}}+1\right)\left(-\sqrt{\tfrac{x}{a}}+\sqrt{\tfrac{y}{b}}+1\right);$$

$$(\sqrt[3]{a}+\sqrt[3]{b}+\sqrt[3]{c})[\sqrt[3]{a^2}+\sqrt[3]{b^2}-\sqrt[3]{c^2}-2\sqrt[3]{(bc)}$$
$$-2\sqrt[3]{(ca)}-2\sqrt[3]{(ab)}].$$

8. Find rationalizing multipliers for the following expressions, and also the products of multiplication by these:

$$a+\sqrt{b}; \quad \sqrt{a}+b\sqrt{c}; \quad a\sqrt{b}-b\sqrt{a}; \quad a+\sqrt{(a^2-x^2)};$$

$$\sqrt{(a-x)}-\sqrt{(a+x)}; \quad \sqrt{(a^2+\sqrt{c})}+\sqrt{(a^2-\sqrt{c})};$$

$$\sqrt{[8+\sqrt{(24+\sqrt{5})}]}-\sqrt{[8+\sqrt{(24-\sqrt{5})}]}; \quad \sqrt{a}+\sqrt{b}+\sqrt{c};$$

$$3+\sqrt{2}+\sqrt{7}; \quad \sqrt{6}+\sqrt{5}-\sqrt{3}-\sqrt{2}; \quad \sqrt{a}+\sqrt{b}+\sqrt{c}+\sqrt{d};$$

$$\sqrt{(1+a)}-\sqrt{(1-a)}+\sqrt{(1+b)}-\sqrt{(1-c)}; \quad \sqrt[3]{a}+\sqrt[3]{c};$$

$$\sqrt[3]{a^2}-\sqrt[3]{c^2}; \quad \sqrt[4]{a}+\sqrt[4]{c}; \quad \sqrt{a}-\sqrt[3]{b}; \quad \sqrt[4]{a}+\sqrt{a}; \quad \sqrt{x}+\sqrt[4]{y^3};$$

$$\sqrt{x+1}+\sqrt{x^{-1}}; \quad \sqrt{(ab^{-1})}-\sqrt{(a^{-1}b)}; \quad \sqrt[3]{2}+\sqrt[3]{3}-\sqrt[3]{5};$$

$$\sqrt[3]{a}+\sqrt[3]{b}+\sqrt[3]{c}; \quad a+\sqrt{b}+\sqrt[3]{c}.$$

9. Rationalize the divisors and the denominators in the following, and reduce the results to their simplest form:

$$1\div(2-\sqrt{3}); \quad 3\div(3+\sqrt{6}); \quad 5\div(\sqrt{2}+\sqrt{7});$$

$$(\sqrt{3}+\sqrt{2})\div(\sqrt{3}-\sqrt{2}); \quad (7\sqrt{5}+5\sqrt{7})\div(\sqrt{5}+\sqrt{7});$$

$a \div (\sqrt{a} + a); \ (x - a) \div (\sqrt{x} - \sqrt{a});$

$(a^2 + ab + b^2) \div [a + \sqrt{(ab)} + b]; \ (x + a) \div (\sqrt[3]{x} + \sqrt[3]{a});$

$\dfrac{a\sqrt{x} + b\sqrt{y}}{c\sqrt{x} - e\sqrt{y}}; \ \dfrac{2\sqrt{6}}{\sqrt{2} + \sqrt{3} - \sqrt{5}}; \ \dfrac{1 + 3\sqrt{2} - 2\sqrt{3}}{\sqrt{2} + \sqrt{3} + \sqrt{6}};$

$\dfrac{\sqrt{6} - \sqrt{5} - \sqrt{3} + \sqrt{2}}{\sqrt{6} + \sqrt{5} - \sqrt{3} - \sqrt{2}}; \ \dfrac{2}{\sqrt{(a + 1)} - \sqrt{(a - 1)}};$

$\dfrac{2c}{\sqrt{(a + c)} + \sqrt{(a - c)}}; \ \dfrac{a + x + \sqrt{(a^2 + x^2)}}{a + x - \sqrt{(a^2 + x^2)}};$

$\dfrac{\sqrt{(a + x)} + \sqrt{(a - x)}}{\sqrt{(a + x)} - \sqrt{(a - x)}}; \ \dfrac{1}{a\sqrt{(1 + b^2)} + b\sqrt{(1 + a^2)}};$

$\dfrac{a\sqrt{(1 - b^2)} - b\sqrt{(1 - a^2)}}{\sqrt{(1 - b^2)} + \sqrt{(1 - a^2)}}; \ \dfrac{a\sqrt{(1 - a^2)} + c\sqrt{(1 - c^2)}}{a\sqrt{(1 - c^2)} + c\sqrt{(1 - a^2)}};$

$\dfrac{\sqrt{[(1 + a)(1 + b)]} - \sqrt{[(1 - a)(1 - b)]}}{\sqrt{[(1 + a)(1 + b)]} + \sqrt{[(1 - a)(1 - b)]}};$

$\dfrac{(a - x)\sqrt{(b^2 + y^2)} - (b - y)\sqrt{(a^2 + x^2)}}{(a + x)\sqrt{(b^2 + y^2)} + (b + y)\sqrt{(a^2 + x^2)}};$

$\dfrac{\sqrt{(1 + a)} - \sqrt{(1 - a)} + \sqrt{(1 + b)} - \sqrt{(1 - b)}}{\sqrt{(1 + a)} + \sqrt{(1 - a)} + \sqrt{(1 + b)} + \sqrt{(1 - b)}};$

$\dfrac{\sqrt{(x + a)} - \sqrt{(x - a)} - \sqrt{(x + b)} + \sqrt{(x - b)}}{\sqrt{(x + a)} + \sqrt{(x - a)} + \sqrt{(x + b)} + \sqrt{(x - b)}};$

$\dfrac{\sqrt{a}}{\sqrt{b}} + \dfrac{\sqrt{b}}{\sqrt{a}}; \ \sqrt{\left(\dfrac{a + x}{a - x}\right)} - \sqrt{\left(\dfrac{a - x}{a + x}\right)}; \ \sqrt{\dfrac{a + \sqrt{x}}{a - \sqrt{x}}};$

$\sqrt{\dfrac{\sqrt{x} - \sqrt{y}}{\sqrt{x} + \sqrt{y}}}; \ \sqrt{\dfrac{a + \sqrt{(a^2 - 1)}}{a - \sqrt{(a^2 - 1)}}};$

$\dfrac{\dfrac{1}{\sqrt{x}} - \dfrac{1}{\sqrt{y}}}{\dfrac{1}{\sqrt{x}} + \dfrac{1}{\sqrt{y}}}; \ \dfrac{\dfrac{\sqrt{a}}{\sqrt{x}} - \dfrac{\sqrt{x}}{\sqrt{a}}}{\dfrac{\sqrt{a}}{\sqrt{x}} + \dfrac{\sqrt{x}}{\sqrt{a}}}.$

10. Find the values of the following expressions for $n = 1, 2, 3, 4, 5$, respectively:

$$\frac{1}{\sqrt{5}}\left[\left(\frac{1+\sqrt{5}}{2}\right)^n - \left(\frac{1-\sqrt{5}}{2}\right)^n\right];$$

$$\frac{1}{2\sqrt{6}}\left[\frac{(2+\sqrt{6})^{n+1}-(2+\sqrt{6})}{1+\sqrt{6}} - \frac{(2-\sqrt{6})^{n+1}-(2-\sqrt{6})}{1-\sqrt{6}}\right].$$

11. Show that

$$\frac{1}{2(x-1)}\{[x+\sqrt{(x^2-1)}]^{4n\pm1} + [x-\sqrt{(x^2-1)}]^{4n\pm1} \mp 2\}$$

is a square for $n = 1, 2,$ or 3, respectively.

12. Extract the square roots of:

$x + y - 2\sqrt{(xy)}$; $\quad a + c + e + 2\sqrt{(ac+ce)}$;

$a + 2c + e + 2\sqrt{[(a+c)(c+e)]}$; $\quad 2a + 2\sqrt{(a^2-c^2)}$;

$2[a^2 + b^2 - \sqrt{(a^4 + a^2b^2 + b^4)}]$; $\quad x - 2 + x^{-1}$;

$\sqrt{x} + 2 + \sqrt{x^{-1}}$; $\quad x + 3x^2 + x^3 + 2x\sqrt{x} + 2x^2\sqrt{x}$;

$x^2 - xy + \frac{1}{4}y^2 + \sqrt{(4x^3y - 8x^2y^2 + xy^3)}$;

$2x + \sqrt{(3x^2 + 2xy - y^2)}$; $\quad 5 - 2\sqrt{6}$; $\quad 10 + 2\sqrt{21}$; $\quad 9 + 4\sqrt{5}$;

$12 - 5\sqrt{6}$; $\quad 70 + 3\sqrt{451}$; $\quad 4 - \sqrt{15}$; $\quad 4 - \sqrt{15}$; $\quad 7 + 4\sqrt{3}$;

$9 + 2\sqrt{6} + 4(\sqrt{} + 2\sqrt{3})$; $\quad 15.25 - 5\sqrt{0.6}$.

13. Find the value of:

$\dfrac{(a+b)xy}{ay^2 + bx^2}$, given $x = \dfrac{a\sqrt{a}}{\sqrt{(a+b)}}$ and $y = \dfrac{b\sqrt{b}}{\sqrt{(a+b)}}$;

$\sqrt{(x^2 + y^2)}$, given $x = \sqrt[3]{(a^2c)},\ y = \sqrt[3]{(a^2e)}$;

$\dfrac{x + \sqrt{(x^2+1)}}{x - \sqrt{(x^2+1)}}$, given $x = \dfrac{1}{2}\left(\sqrt{\dfrac{a}{c}} - \sqrt{\dfrac{c}{a}}\right)$;

$\dfrac{\sqrt{(1+x)} - \sqrt{(1-x)}}{\sqrt{(1+x)} + \sqrt{(1-x)}}$, given $x = \dfrac{2ab}{a^2 + b^2}$;

$\dfrac{2a\sqrt{(1+x^2)}}{x+\sqrt{(1+x^2)}}$, given $x=\dfrac{1}{2}\left(\sqrt{\dfrac{a}{e}}-\sqrt{\dfrac{e}{a}}\right)$.

14. If $\sqrt{(x+a+b)}+\sqrt{(x+c+d)}$
$=\sqrt{(x+a-c)}+\sqrt{(x-b+d)}$, $\therefore b+c=0$.

15. Simplify $\dfrac{\frac{1}{2}(1+\sqrt{5})x-2}{x^2-\frac{1}{2}(1+\sqrt{5})x+1}+\dfrac{\frac{1}{2}(1-\sqrt{5})x-2}{x^2-\frac{1}{2}(1-\sqrt{5})x+1}$.

COMPLEX QUANTITIES.

Quantities of the form $a+b\sqrt{-1}$, in which neither a nor b involves $\sqrt{-1}$, are called **Complex Quantities.** The letter i (or j) is frequently used as the symbol of the ditensive unit $\sqrt{-1}$, so that $a+b\sqrt{-1}$ would be written $a+bi$. So also $\sqrt{-x}=i\sqrt{x}$, $\sqrt{-x}\times\sqrt{-y}=i^2\sqrt{(xy)}=-\sqrt{xy}$, and $i^3=-i$.

Ex. 66.

Simplify the following, writing i for $\sqrt{-1}$ in any result in which the latter occurs:

1. $\sqrt{-4}$; $\sqrt{-36}$; $\sqrt{-81}$; $\sqrt{-8}$; $\sqrt{-12}$; $\sqrt{-72}$; $\sqrt[3]{-8}$; $\sqrt{-5}\times\sqrt{-6}$; $\sqrt{-6}\times\sqrt{-8}$; $\sqrt{-8}\times\sqrt{12}$; $\sqrt{-8}\times\sqrt[3]{-8}$; $\sqrt{-5}\times\sqrt{-20}$.

2. $\sqrt{-x}$; $\sqrt{-x^2}$; $\sqrt{-a^3}$; $\sqrt{-a^{2n}}$; $\sqrt{(-a)^2}$; $\sqrt{(-a)^3}$; $\sqrt{(-3ax^3)}$; $\sqrt{-a}\times\sqrt{a^3}$; $\sqrt{-x^2}\times\sqrt{-y^3}$; $\sqrt{-a}\times\sqrt{-1}$; $\sqrt{5}\times\sqrt{-a}$.

3. i^2; i^3; $i^4\cdot$ i^5; i^9; i^{15}; i^{16}; i^{17}; i^{18}; i^{4n}; i^{4n+1}; i^{4n+2}; i^{4n+3}.

4. $ai\times bi$; $i\sqrt{x}\times i\sqrt{y}$; $5i$; $i^2\sqrt{5}$; $i\sqrt{-a}$; $i\sqrt{-a^2}$; $i\sqrt{a}\times\sqrt{-a}$.

5. $\sqrt{-i^2}$; $\sqrt{-i^3}$; $\sqrt{-i^4}$; $\sqrt{-i^5}$; $\sqrt{-i^{2n}}$; $\sqrt{-i^{4n}}$.

6. $\dfrac{\sqrt{-6}}{\sqrt{3}}$; $\dfrac{\sqrt{-6}}{\sqrt{-3}}$; $\dfrac{\sqrt{6}}{\sqrt{-3}}$; $\dfrac{\sqrt{a}}{\sqrt{-b}}$; $\dfrac{\sqrt{-a}}{\sqrt{-b}}$; $\dfrac{1}{\sqrt{-1}}$;

$\dfrac{a}{\sqrt{-a}}$; $\dfrac{a^2}{\sqrt{-a^2}}$; $\dfrac{\sqrt{(-ax)}}{\sqrt{-x}}$; $\dfrac{-\sqrt{-1}}{\sqrt{-a}}$; $\dfrac{a^3}{\sqrt[3]{-a^3}}$;

$\dfrac{-c}{\sqrt{-c^3}}$; $\dfrac{\sqrt{(-a)^{2n+1}}}{\sqrt{(-a)^{2n-1}}}$.

7. $\dfrac{1}{i}$; $\dfrac{1}{i^2}$; $\dfrac{1}{i^3}$; $\dfrac{-1}{i}$; $\dfrac{1}{i^6}$; $\dfrac{1}{i^{4n+1}}$; $\dfrac{-1}{i^{4n+1}}$; $\dfrac{1}{i^{4n-1}}$;

$\dfrac{a^2 i}{\sqrt{-a^3}}$; $\dfrac{x}{i\sqrt{x}}$; $\dfrac{-y}{i\sqrt{-y^2}}$; $-\dfrac{ci^3}{\sqrt{-c^2}}$

8. $\sqrt{(a-b)} \times \sqrt{(b-a)}$; $\sqrt{(3x-4y)} \times \sqrt{(4y-3x)}$;

$(3+5i)(7+4i)$; $(8-9i)(8-7i)$;

$(7-i\sqrt{5})(7+i\sqrt{10})$; $(\sqrt{3}-i\sqrt{6})(\sqrt{2}-i\sqrt{6})$;

$(a+bi)(c+ci)$; $[a+(a-1)i][a+(a+1)i]$;

$(\sqrt{a}+i\sqrt{b})(\sqrt{a}-i\sqrt{c})$; $(a+bi)(a-bi)$;

$(ai+b)(ai-b)$; $(\sqrt{a}+i\sqrt{b})(\sqrt{a}-i\sqrt{b})$;

$(a\sqrt{b}+ci\sqrt{x})(a\sqrt{b}-ci\sqrt{x})$; $\sqrt{(1+i)} \times \sqrt{(1-i)}$;

$\sqrt{(3+4i)} \times \sqrt{(3-4i)}$; $\sqrt{(12+5i)} \times \sqrt{(12-5i)}$;

$(1+i)^2$; $(\sqrt{a}-i\sqrt{b})^2$; $(5-2i\sqrt{6})^2$;

$(a+bi)^2+(a-bi)^2$; $(a+bi)^2-(a-bi)^2$;

$(a+bi)^2+(ai-b)^2$; $[\sqrt{(4+3i)}+\sqrt{(4-3i)}]^2$;

$[\sqrt{(3-4i)}-\sqrt{(3+4i)}]^2$; $[\sqrt{(1+i)}+\sqrt{(1-i)}]^2$;

$(1+i)^3$; $(1+i)^4$; $(a+bi)^4$; $(a+bi)^3+(a-bi)^3$;

$(a+bi)^3-(a-bi)^3$;

$\left(\dfrac{1+i\sqrt{3}}{2}\right)^3$; $\left(\dfrac{-1+i\sqrt{3}}{2}\right)^3$; $\left(\dfrac{-1-i\sqrt{3}}{2}\right)^3$;

$$\left(\frac{1+i}{\sqrt{2}}\right)^4; \quad \left(\frac{1-i}{\sqrt{2}}\right)^4;$$

$$(x+iy)^4 + (x-iy)^4; \quad (x+iy)^4 - (ix+y)^4;$$

$$(1+i\sqrt{5})^4 + (1-i\sqrt{5})^4; \quad (a+bi)^5 + (a-bi)^5;$$

$$(a+bi)^5 - (a-bi)^5; \quad (1+i)^5 + (1-i)^5;$$

$$(1+i\sqrt{2})^5 + (1-i\sqrt{2})^5;$$

$$\left(\frac{i+\sqrt{3}}{2}\right)^6; \quad \left(\frac{i-\sqrt{3}}{2}\right)^6;$$

$$\{\tfrac{1}{8}[\sqrt{(30-6\sqrt{5})}-1-\sqrt{5}]+\tfrac{1}{8}i[\sqrt{15}+\sqrt{3}+\sqrt{(10-2\sqrt{5})}]\}^n$$
for all positive integral values of n.

9. $\dfrac{4}{1+i\sqrt{3}}; \quad \dfrac{64}{1-i\sqrt{7}}; \quad \dfrac{21}{4+3i\sqrt{6}}; \quad \dfrac{5}{\sqrt{2}+i\sqrt{3}};$

$$\frac{1-20i\sqrt{5}}{7-2i\sqrt{5}}; \quad \frac{1-i\sqrt{3}}{1+i\sqrt{3}}; \quad \frac{1+i}{1-i}; \quad \frac{1+i^3}{1+i}; \quad \frac{1-i^3}{1-i}; \quad \frac{1-i^3}{(1+i)^3};$$

$$\frac{1+i^3}{1-i}; \quad \frac{x+yi}{x-yi}; \quad \frac{a+i\sqrt{x}}{a-i\sqrt{x}}; \quad \frac{i\sqrt{a}+\sqrt{-b}}{\sqrt{-a}-i\sqrt{b}}; \quad \frac{a-bi}{ai+b};$$

$$\frac{a+i\sqrt{(1-x^2)}}{a-i\sqrt{(1-x^2)}}; \quad \frac{\sqrt{(x-y)}-\sqrt{(y-x)}}{\sqrt{(y-x)}+\sqrt{(x-y)}}; \quad \frac{1}{1+i}+\frac{1}{1-i};$$

$$\frac{1+i}{1-i}+\frac{1-i}{1+i}; \quad \frac{1}{(1+i)^2}+\frac{1}{(1-i)^2}; \quad \frac{1}{(1+i)^4}-\frac{1}{(1-i)^4};$$

$$\frac{x+yi}{a+bi}+\frac{x-yi}{a-bi}; \quad \frac{x+yi}{a+bi}-\frac{x-yi}{a-bi}; \quad \frac{\sqrt{x}+i\sqrt{y}}{\sqrt{x}-i\sqrt{y}}-\frac{\sqrt{y}+i\sqrt{x}}{\sqrt{y}-i\sqrt{x}};$$

$$\frac{\sqrt{(1+a)}+i\sqrt{(1-a)}}{\sqrt{(1+a)}-i\sqrt{(1-a)}}-\frac{\sqrt{(1-a)}+i\sqrt{(1+a)}}{\sqrt{(1-a)}-i\sqrt{(1+a)}}$$

10. $\sqrt{(3+4i)}+\sqrt{(3-4i)}; \quad \sqrt{(3+4i)}-\sqrt{(3-4i)};$

$$\sqrt{(4+3i)}\pm\sqrt{(4-3i)}; \quad \sqrt{(1+2i\sqrt{6})}\pm\sqrt{(1-2i\sqrt{6})};$$

$$\sqrt{(5+2i\sqrt{6})}\pm\sqrt{(5-2i\sqrt{6})};$$

$$\sqrt{(2\sqrt{15}+30\,i)} \pm \sqrt{(2\sqrt{15}-30\,i)}\,;$$

$$\sqrt{(\sqrt{3}+i\sqrt{105})} \pm \sqrt{(\sqrt{3}-i\sqrt{105})}\,;$$

$$\sqrt{[a+i\sqrt{(x^2-a^2)}]} \pm \sqrt{[a-i\sqrt{(x^2-a^2)}]}\,;$$

$$\sqrt{[a^2+ix\sqrt{(x^2+2a^2)}]} \pm \sqrt{[a^2-ix\sqrt{(x^2+2a^2)}]}.$$

11. Prove that both $\frac{1}{2}(-1+i\sqrt{3})$ and $\frac{1}{2}(-1-i\sqrt{3})$ satisfy the equation $\dfrac{x^3-1}{x-1}=0$;

that $(x+\omega y+\omega^2 z)^3 = x^3+y^3+z^3+3\,(x+\omega y)\,(y+\omega z)\,(z+\omega x)$ and that $(x+y+z)(x+\omega y+\omega^2 z)(x+\omega^2 y+\omega z)=x^3+y^3+z^3-3xyz$, in which ω represents *either* of the preceding complex quantities.

Hence, prove that:

(i.) $[2a-b-c+(b-c)i\sqrt{3}]^3$
$$=[2b-c-a+(c-a)i\sqrt{3}]^3=[2c-a-b+(a-b)i\sqrt{3}]^3;$$

(ii.) $u^3+v^3+w^3-3uvw$
$$=(a^3+b^3+c^3-3abc)(x^3+y^3+z^3-3xyz),$$

if $u=ax+by+cz,\ v=ay+bz+cx,\ w=az+bx+cy,$

or if $u=ax+cy+bz,\ v=cx+by+az,\ w=bx+ay+cz.$

12. Prove that $\frac{1}{4}[\sqrt{5}+1+i\sqrt{(10-2\sqrt{5})}]$ satisfies the equation $\dfrac{x^5+1}{x+1}=0.$

Writing ω for the preceding complex quantity, prove that

$(7+\omega+\omega^2+3\omega^3)(7-\omega^4-\omega^3-3\omega^2)=71,$ and

$(x+y+z)(x+\omega^2 y-\omega^3 z)(x-\omega^3 y-\omega z)(x-\omega z+\omega^4 z)$
$$\times (x+\omega^4 y+\omega^2 z)=x^5+y^5+z^5-5x^3yz+5xy^2z^2.$$

Prove that $[4a+(b-c)(\sqrt{5}-1)+(b+c)i\sqrt{(10+2\sqrt{5})}]^5$
$$=\{(a+b)[-1+i\sqrt{(\sqrt{5}+2)}]$$
$$+(a-b)[\sqrt{5}+i\sqrt{(\sqrt{5}-2)}]\times \sqrt[4]{5}-4c\}^5.$$

13. Solve the following equations, and prove that, if r be a root of any one of them, the other roots of *that* equation will be r^2, r^3, r^4, etc.

(i.) $x^2 + 1 = 0$.

(ii.) $x^4 + 1 = 0$.

(iii.) $x^8 + 1 = 0$.

(iv.) $\dfrac{x^3 - 1}{x - 1} = 0$.

(v.) $\dfrac{x^5 - 1}{x - 1} = 0$.

(vi.) $\dfrac{x^3 + 1}{x + 1} = 0$.

(vii.) $\dfrac{x^5 + 1}{x + 1} = 0$.

(viii.) $\dfrac{x^6 + 1}{x^2 + 1} = 0$.

(ix.) $\dfrac{(x^{15} - 1)(x - 1)}{(x^3 - 1)(x^5 - 1)} = 0$.

§ 51. SURD EQUATIONS.

EXAMPLES.

1. Solve $2 + \sqrt{(4x^2 - 9x + 8)} - 2x = 0$.

Here there is but one surd, and it is convenient to make that surd one side of the equation, and transpose all the rational terms to the other; this gives

$$\sqrt{(4x^2 - 9x + 8)} = 2x - 2;$$

Squaring both sides,

$$4x^2 - 9x + 8 = 4x^2 - 8x + 4. \quad \therefore x = 4.$$

2. Solve $\sqrt{(4x^2 + 19)} + \sqrt{(4x^2 - 19)} = \sqrt{47} + 3$.

We have the identity

$$(4x^2 + 19) - (4x^2 - 19) = 38 = 47 - 9.$$

Now dividing the members of this identity by those of the given equation, we have

$$\sqrt{(4x^2 + 19)} - \sqrt{(4x^2 - 19)} = \sqrt{47} - 3.$$

Adding this to the given equation, then

$$2\sqrt{(4x^2 + 19)} = 2\sqrt{47}.$$

$$\therefore 4x^2 + 19 = 47, \text{ and } x = \pm \sqrt{7}.$$

3. $\dfrac{\sqrt{(3x^2-1)}+\sqrt{(3-x^2)}}{\sqrt{(3x^2-1)}-\sqrt{(3-x^2)}}=\dfrac{a}{b}.$

$\therefore \sqrt{\dfrac{3x^2-1}{3-x^2}}=\dfrac{a+b}{a-b}.$

$\therefore \dfrac{3x^2-1}{3-x^2}=\dfrac{(a+b)^2}{(a-b)^2}.$

$\therefore x^2=\dfrac{3(a+b)^2+(a-b)^2}{(a+b)^2+3(a-b)^2}=\dfrac{a^2+ab+b^2}{a^2-ab+b^2}.$

4. $m\sqrt{(1+x)}-n\sqrt{(1-x)}=\sqrt{(m^2+n^2)}.$ (1)

Square both members and reduce,

$\therefore (m^2-n^2)x-2mn\sqrt{(1-x^2)}=0.$ (2)

Transfer the radical term and square both members,

$\therefore (m^2-n^2)^2x^2=4m^2n^2(1-x^2).$ (3)

$\therefore (m^2+n^2)^2x^2=4m^2n^2.$ (4)

$\therefore x=\dfrac{\pm 2mn}{m^2+n^2}.$ (5)

The above follows the usual mode of solving equations involving radicals; viz., make a radical term the right-hand member; gathering all the other terms into the left-hand member, square each member; repeat, if necessary, until all radicals are rationalized. This method is convenient, but it does not explain the difficulty that only one of the values of x in (4) satisfies (1); viz.,

$$\dfrac{+2mn}{m^2+n^2}$$

The other value, $\dfrac{-2mn}{m^2+n^2}$,   satisfies the equation

$$m\sqrt{(1+x)}+n\sqrt{(1-x)}=\sqrt{(m^2+n^2)}.$$

The explanation is simple. Squaring both members of (1) is really equivalent to substituting for (1) the conjoint equation,

$$[m\sqrt{(1+x)} - n\sqrt{(1-x)} - \sqrt{(m^2+n^2)}]$$
$$[m\sqrt{(1+x)} + n\sqrt{(1-x)} - \sqrt{(m^2+n^2)}] = 0, \quad (6)$$

which reduces to (2) above.

Treating (6) or (2) by transferring and squaring is equivalent to substituting for it the equation

$$[m\sqrt{(1+x)} - n\sqrt{(1-x)} - \sqrt{(m^2+n^2)}]$$
$$\times [m\sqrt{(1+x)} - n\sqrt{(1-x)} + \sqrt{(m^2+n^2)}]$$
$$\times [m\sqrt{(1+x)} + n\sqrt{(1-x)} - \sqrt{(m^2+n^2)}]$$
$$\times [m\sqrt{(1+x)} + n\sqrt{(1-x)} + \sqrt{(m^2+n^2)}] = 0, (7)$$

which reduces to

$$[(m^2 - n^2)x - 2mn\sqrt{(1-x^2)}]$$
$$[(m^2 - n^2)x + 2mn\sqrt{(1-x^2)}] = 0, \quad (8)$$

which further reduces to (3).

Thus the whole process of solving (1) is equivalent to reducing it to an equation of the type $A = 0$ and then multiplying the member A by rationalizing factors. Thus, instead of solving (1) we really solve (7), that is, a conjoint equation equivalent to *four* disjunctive equations. (See page 191, § 42.) Now the values given in (4) will satisfy (7), the positive value making the first factor vanish, the negative value making the third factor vanish, while no values can be found that will make either the second or the fourth factor vanish.

Hence, if one of such a set of disjunctive equations is proposed for solution, the conjoint equation must be solved; and if there be a value of x which satisfies the particular equation proposed, that value must be retained and the others rejected.

This process is the opposite to that given in §§ 42 and 43 : there a conjoint equation is solved by resolving it into its equivalent disjunctive equations. The two processes are related somewhat as involution and evolution are.

Further, it should be noticed that just as there are four factors in (7) while there are only two values in (4), it will in general be possible to form more disjunctive equations than there are values of x that satisfy the conjoint equation, and consequently it will be possible to select disjunctive equations that are not satisfied by any value of x, or, in other words, whose solution is impossible.

This will perhaps be better understood by considering the following problem :

Find a number such that, if it be increased by 4 and also diminished by 4, the difference of the square roots of the results shall be 4.

Reduced to an equation, this is

$$\sqrt{(x+4)} - \sqrt{(x-4)} = 4. \tag{9}$$

Rationalizing, this becomes

$$\begin{aligned}&[4 - \sqrt{(x+4)} + \sqrt{(x-4)}] \\ &\times [4 - \sqrt{(x+4)} - \sqrt{(x-4)}] \\ &\times [4 + \sqrt{(x+4)} + \sqrt{(x-4)}] \\ &\times [4 + \sqrt{(x+4)} - \sqrt{(x-4)}] = 0, \tag{10}\end{aligned}$$

which reduces to

$$[24 - 8\sqrt{(x+4)}][24 + 8\sqrt{(x+4)}] = 0 ;$$

that is, $9 - (x+4) = 0$, or $x = 5$.

Now $x = 5$ satisfies (10) because it makes the factor

$$4 - \sqrt{(x+4)} - \sqrt{(x-4)}$$

vanish, and *it is the only finite value of x that does satisfy* (10), or, in other words, there are no values of x which will make any of the factors

$$4 - \sqrt{(x+4)} + \sqrt{(x-4)},$$
$$4 + \sqrt{(x+4)} + \sqrt{(x-4)},$$
or $\qquad 4 + \sqrt{(x+4)} - \sqrt{(x-4)},$

vanish. There is, therefore, no number that will satisfy the conditions of the problem.

It will be found that as x increases $\sqrt{(x+4)} - \sqrt{(x-4)}$ decreases; hence, as 4 is the least value that can be given to x without involving the square root of a negative, the greatest real value of $\sqrt{(x+4)} - \sqrt{(x-4)}$ is $\sqrt{8}$, which is less than 4. We see by this that our method of solution fails for (9) simply because (9) is impossible.

5. $\sqrt{[(a+x)(b+x)]} - \sqrt{[(a-x)(b-x)]}$
$\qquad = \sqrt{[(a-x)(b+x)]} - \sqrt{[(a+x)(b-x)]}. \qquad (1)$

Collecting the terms involving $\sqrt{(a+x)}$ and $\sqrt{(a-x)}$, respectively the equation becomes

$$[\sqrt{(a+x)} - \sqrt{(a-x)}][\sqrt{(b+x)} + \sqrt{(b-x)}] = 0. \qquad (2)$$

This is satisfied if either

$$\sqrt{(a+x)} - \sqrt{(a-x)} = 0, \qquad (3)$$
or $\sqrt{(b+x)} + \sqrt{(b-x)} = 0. \qquad (4)$

The rational form of (3) is $(a+x) - (a-x) = 0$, which is satisfied by $x = 0$, and this also satisfies (3).

The rational form of (4) is $(b+x) - (b-x) = 0$, which requires $x = 0$; but this does not satisfy (4). Hence, the second factor of the left-hand member of (2) cannot vanish.

Therefore, the only solution of (2), and hence of (1), is $x = 0$, derived from (3).

6. $\sqrt[3]{(a+x)} + \sqrt[3]{(a-x)} = \sqrt[3]{(2a)}.$

Cube by the formula,

$$(u+v)^3 = u^3 + v^3 + 3uv(u+v).$$

$$\therefore (a+x) + (a-x) + 3\sqrt[3]{[2a(a^2-x^2)]} = 2a.$$

$$\therefore 2a(a^2-x^2) = 0.$$

$$\therefore x = \pm a.$$

Both these values belong to the proposed equation.

The rationalizing factors of

$$\sqrt[3]{(a+x)} + \sqrt[3]{(a-x)} - \sqrt[3]{(2a)} = 0$$

are $\sqrt[3]{(a+x)} + \omega\sqrt[3]{(a-x)} - \omega^2\sqrt[3]{(2a)},$

and $\sqrt{(a+x)} + \omega^2\sqrt[3]{(a-x)} - \omega\sqrt[3]{(2a)}.$

See Exam. 11, page 276.

The remarks on Exam. 4 will apply *mutatis mutandis* to equations of this type.

7. $\dfrac{\sqrt[3]{(a+x)^2} + \sqrt[3]{(a^2-x^2)} + \sqrt[3]{(a-x)^2}}{\sqrt[3]{(a+x)^2} - \sqrt[3]{(a^2-x^2)} + \sqrt[3]{(a-x)^2}} = c.$ (1)

Assume $\sqrt[3]{(a+x)} = u$ and $\sqrt[3]{(a-x)} = v.$

$$\therefore u^3 + v^3 = 2a \text{ and } u^3 - v^3 = 2x.$$

$$\therefore \frac{u^3 - v^3}{u^3 + v^3} = \frac{x}{a}.$$ (2)

Also (1) becomes

$$\frac{u^2 + uv + v^2}{u^2 + uv + v^2} = c.$$ (3)

Multiply both numbers by $\dfrac{u-v}{u+v}.$

$$\therefore \frac{u^3 - v^3}{u^3 + v^3} = c\frac{u-v}{u+v}.$$

$\therefore$ by (2),

$$\frac{x}{a} = c\,\frac{u-v}{u+v}. \tag{4}$$

Again, adding and subtracting denominators and numerators in (3),

$$\frac{u^2+v^2}{uv} = \frac{c+1}{c-1}.$$

Adding and subtracting 2 (denominators) and numerators in this,

$$\frac{u^2-2uv+v^2}{u^2+2uv+v^2} = \frac{3-c}{3c-1},$$

or $\quad \left(\dfrac{u-v}{u+v}\right)^2 = \dfrac{3-c}{3c-1}.$

$\therefore$ substituting by (4),

$$\frac{x^2}{a^2} = c^2\,\frac{3-c}{3c-1}.$$

$$\therefore x = ac\sqrt{\frac{3-c}{3c-1}}.$$

8. $[\sqrt[4]{(x+a)} + \sqrt[4]{(x-a)}]^3[\sqrt[4]{(x+a)} - \sqrt[4]{(x-a)}] = 2c.$ (1)

Assume $u = \sqrt[4]{(x+a)}$ and $v = \sqrt[4]{(x-a)}$,

and (1) becomes

$$(u+v)^3(u-v) = 2c,$$

or $\quad (u+v)^2(u^2-v^2) = 2c.$ (2)

Also $\quad u^4 - v^4 = 2a$

or $\quad (u^2+v^2)(u^2-v^2) = 2a,$ (3)

and $\quad u^4 + v^4 = 2x.$ (4)

From (2) and (3),

$$(u-v)^2(u^2-v^2) = 4a - 2c. \tag{5}$$

$\therefore (2) \times (5),$

$$(u^2 - v^2)^2(u^2 - v^2)^2$$

or $\quad (u^2 - v^2)^4 = 4c(2a - c).$ (6)

Also $(3)^2 + (6),$

$$[(u^2 + v^2)^2 + (u^2 - v^2)^2](u^2 - v^2)^2$$
$$= 4(a^2 + 2ac - c^2)$$

or $\quad (u^4 + v^4)(u^2 - v^2)^2 = 2(a^2 + 2ac - c^2).$

Substituting by (4) and (6),

$$2x\sqrt{(2ac - c^2)} = a^2 + 2ac - c^2.$$

9. $\left[\sqrt[4]{(a+x)} + \sqrt[4]{(a-x)}\right]^2\left[\sqrt{(a+x)} + \sqrt{(a-x)}\right] = 2cx.$

Divide the terms of the identity

$$\sqrt[4]{(a+x)^4} - \sqrt[4]{(a-x)^4} = 2x$$

by the corresponding terms of the equation.

$$\therefore \sqrt[4]{\left(\frac{a+x}{a-x}\right)} = \frac{c+1}{c-1}.$$

$$\therefore \frac{a+x}{a-x} = \left(\frac{c+1}{c-1}\right)^4.$$

$$\therefore x = a \times \frac{(c+1)^4 - (c-1)^4}{(c+1)^4 + (c-1)^4}.$$

10. $\sqrt[3]{(a-x)^2} + \sqrt[3]{[(a-x)(b-x)]} + \sqrt[3]{(b-x)^2}$
$= \sqrt[3]{(a^2 + ab + b^2)}.$

Divide the terms of the identity

$$\sqrt[3]{(a-x)^3} - \sqrt[3]{(b+x)^3} = a - b$$

by the corresponding terms of the equation.

$$\therefore \sqrt[3]{(a-x)} - \sqrt[3]{(b-x)} = \frac{a-b}{\sqrt[3]{(a^2 + ab + b^2)}}.$$

Cube, using the form

$$(u - v)^3 = u^3 - v^3 - 3uv(u - v).$$

$$(a - x) - (b - x)$$

$$- 3\sqrt[3]{[(a - x)(b - x)]}\frac{a - b}{\sqrt[3]{(a^2 + ab + b^2)}}$$

$$= \frac{(a - b)^3}{a^2 + ab + b^2} = a - b - \frac{3ab(a - b)}{a^2 + ab + b^2}.$$

$$\therefore \sqrt[3]{[(a - x)(b - x)]} = \frac{ab}{\sqrt[3]{(a^2 + ab + b^2)^2}}.$$

$$\therefore (a - x)(b - x) = \frac{a^3b^3}{(a^2 + ab + b^2)^2},$$

a form solved in (C), page 231.

11. $\dfrac{(a - x)^2\sqrt{(a - x)} + (x - b)^2\sqrt{(x - b)}}{(a - x)\sqrt{(a - x)} + (x - b)\sqrt{(x - b)}} = a - b.$

Write $a - b$ in the form $(a - x) + (x - b)$, and multiply by the denominator of the left-hand member.

$$\therefore (a - x)^2\sqrt{(a - x)} + (x - b)^2\sqrt{(x - b)}$$

$$= (a - x)^2\sqrt{(a - x)} + (x - b)^2\sqrt{(x - b)}$$

$$+ (a - x)(x - b)[\sqrt{(a - x)} + \sqrt{(x - b)}].$$

$$\therefore (a - x)(x - b)[\sqrt{(a - x)} + \sqrt{(x - b)}] = 0$$

$$\therefore (a - x) = 0, \text{ or } x - b = 0,$$

or $\sqrt{(a - x)} + \sqrt{(x - b)} = 0.$

$$x_1 = a, \quad x_2 = b.$$

The equation $\sqrt{(a - x)} + \sqrt{(x - b)} = 0$ has no solution, for the sum of two *positive* square roots cannot vanish.

The solution $x = \frac{1}{2}(a + b)$ belongs to the equation

$$\sqrt{(a - x)} - \sqrt{(x - b)} = 0.$$

12. $\sqrt[3]{\dfrac{a-x}{b+x}} + \sqrt[3]{\dfrac{b+x}{a-x}} = c.$

Square both members, subtract 4, and extract the square root.

$$\therefore \sqrt[3]{\dfrac{a-x}{b+x}} - \sqrt[3]{\dfrac{b+x}{a-x}} = \pm \sqrt{(c^2-4)}.$$

$$\therefore \sqrt[3]{\dfrac{a-x}{b+x}} = \tfrac{1}{2}\left[c \pm \sqrt{(c^2-4)}\right] = e, \text{ say.}$$

$$\therefore \dfrac{a-x}{b+x} = e^3.$$

$$\therefore \dfrac{2x-(a-b)}{a+b} = \dfrac{1-e^3}{1+e^3}.$$

$$\therefore x = \tfrac{1}{2}\left[(a-b) + (a+b)\dfrac{1-e^3}{1+e^3}\right].$$

Or thus, cube both members.

$$\therefore \dfrac{a-x}{b+x} + 3c + \dfrac{b+x}{a-x} = c^3.$$

$$\therefore \dfrac{(a-x)^2 + (b+x)^2}{(a-x)(b+x)} = c^3 - 3c.$$

$$\therefore \left[\dfrac{(b+x)-(a-x)}{(b+x)+(a-x)}\right]^2 = \dfrac{c^3-3c-2}{c^3-3c+2}$$

$$= \dfrac{(c+1)^2(c-2)}{(c-1)^2(c+2)}.$$

$$\therefore \dfrac{2x-(a-b)}{a+b} = \dfrac{c+1}{c-1}\sqrt{\dfrac{c-2}{c+2}}.$$

Prove that

$$\dfrac{1-e^3}{1+e^3} = \dfrac{c+1}{c-1}\sqrt{\dfrac{c-2}{c+2}},$$

if $2e = c \pm \sqrt{(c^2-4)}.$

Ex. 67.

1. $\sqrt{(x+4)} + \sqrt{(x-3)} = 7$.

2. $\sqrt{(3x+1)} + \sqrt{(4x+4)} = 1$.

3. $\sqrt{(2x+10)} + \sqrt{(2x-2)} = 6$.

4. $\sqrt{(mx)} - \sqrt{(nx)} = m - n$.

5. $\sqrt{(bx)} + \sqrt{(ab+bx)} = \sqrt{x}$.

6. $\sqrt{x} + \sqrt{(x+3)} = \dfrac{5}{\sqrt{(x+3)}}$.

7. $\sqrt{(ax+x^2)} = (1+x)$.

8. $\sqrt[3]{(17x-26)} = \dfrac{2}{9}$.

9. $\sqrt{x} - \sqrt{(a+x)} = \sqrt{\dfrac{a}{x}}$.

10. $b + x - \sqrt{(b^2+x^2)} = c^2$.

11. $\sqrt{(8+x)} - \sqrt{x} = 2\sqrt{(1+x)}$.

12. $\sqrt{(2x-27a)} = 9\sqrt{a} - \sqrt{(2x)}$.

13. $\dfrac{5\sqrt{(2x-1)} + 2\sqrt{(3x-3)}}{4\sqrt{(2x-1)} - 2\sqrt{(3x-3)}} = 2\frac{11}{13}$.

14. $\dfrac{\sqrt{2x} + \sqrt{(3-2x)}}{\sqrt{2x} - \sqrt{(3-2x)}} = \dfrac{3}{2}$.

15. $\dfrac{2\sqrt[3]{(3x+3)} + \sqrt[3]{(7x+8)}}{2\sqrt[3]{(3x+3)} - \sqrt[3]{(7x+8)}} = 5$.

16. $33[13 - 2\sqrt{(x-5)}] = 3[13 + 2\sqrt{(x-5)}]$.

17. $(\sqrt{n}+1)[\sqrt{(nx+1)} - \sqrt{nx}]$
    $= (\sqrt{n}-1)[\sqrt{(nx+1)} + \sqrt{nx}]$.

18. $\dfrac{\sqrt{(x+c)} + \sqrt{b}}{\sqrt{(x+c)} - \sqrt{b}} = \dfrac{\sqrt{x} + \sqrt{a}}{\sqrt{x} - \sqrt{a}}$.

19. $\dfrac{\sqrt{x+28}}{\sqrt{x+4}} = \dfrac{\sqrt{x+38}}{\sqrt{x+6}}.$

20. $\dfrac{\sqrt[3]{2x+17}}{\sqrt[3]{2x+9}} = \dfrac{\sqrt[3]{2x+27}}{\sqrt[3]{2x+15}}.$

21. $\dfrac{\sqrt{x+2a}}{\sqrt{x+b}} = \dfrac{\sqrt{x+4a}}{\sqrt{x+3b}}.$

22. $\dfrac{3x-1}{\sqrt{3x+1}} = \dfrac{1+\sqrt{3x}}{2}.$

23. $\dfrac{\sqrt{a} - \sqrt{(a-x)}}{\sqrt{a} + \sqrt{(a-x)}} = a.$

24. $\dfrac{\sqrt{x} + \sqrt{b}}{\sqrt{x} - \sqrt{b}} = \dfrac{a}{b}.$

25. $\dfrac{ax+1 + \sqrt{(a^2x^2-1)}}{ax+1 - \sqrt{(a^2x^2-1)}} = b.$

26. $\dfrac{1 - \sqrt{[1 - \sqrt[3]{(1-x)}]}}{1 + \sqrt{[1 - \sqrt{(1-x)}]}} = a.$

27. $\dfrac{\sqrt[3]{(x+1)} - \sqrt[3]{(x-1)}}{\sqrt[3]{(x-1)} + \sqrt[3]{(x-1)}} = \dfrac{1}{2}.$

28. $\sqrt[3]{(1-x)} + \sqrt[3]{(1+x)} = \sqrt[3]{3}.$

29. $\sqrt[3]{(3+x)} + \sqrt[3]{(3-x)} = \sqrt[3]{7}.$

30. $\sqrt[3]{(x+1)} - \sqrt[3]{(x-1)} = \sqrt[3]{11}.$

31. $\sqrt[3]{(a+x)} + \sqrt[3]{(a-x)} = \sqrt[3]{b}.$

32. $\sqrt[3]{(1 + \sqrt{x})} + \sqrt[3]{(1 - \sqrt{x})} = 2.$

33. $\sqrt{x} - \sqrt{[a - \sqrt{(ax+x^2)}]} = \tfrac{1}{2}\sqrt{a}.$

34. $\sqrt[3]{(25+x)} + \sqrt[3]{(25-x)} = 2.$

35. $x + \sqrt{(a^2+x^2)} = \dfrac{na^2}{\sqrt{(a^2+x^2)}}.$

36. $\sqrt{(1+x)} + \sqrt{[1+x+\sqrt{(1-x)}]} = \sqrt{(1-x)}.$

37. $\sqrt{(x+\sqrt{x})} - \sqrt{(x-\sqrt{x})} = a\sqrt{\dfrac{x}{x+\sqrt{x}}}.$

38. $\sqrt{(1+x+x^2)} + \sqrt{(1-x+x^2)} = mx.$

39. $\sqrt{(a^2-x^2)} + x\sqrt{(a^2-1)} = a^2(1-x^2).$

40. $\dfrac{bx-c^2}{\sqrt{(bx)}+c} = \dfrac{\sqrt{(bx)}+c}{n} - a.$

41. $\sqrt{(2x^2+5)} + \sqrt{(2x^2-5)} = \sqrt{15} + \sqrt{5}.$

42. $\sqrt{(3x^2+10)} + \sqrt{(3x^2-10)} = \sqrt{17} + \sqrt{37}.$

43. $\sqrt{(3x^2+9)} - \sqrt{(3x^2-9)} = \sqrt{34} + 4.$

44. $\sqrt{(3a-3b+x^2)} + \sqrt{(2a-2b+x^2)} = \sqrt{a} + \sqrt{b}.$

45. $\sqrt{(4a^2-3b^2-2x^2)} + \sqrt{(3a^2-3b^2-x^2)} = a+x.$

46. $[\sqrt{(a+x)} + \sqrt{(a-x)}][\sqrt{a} + \sqrt{(a^2+x^2)}] = 2x.$

47. $\sqrt{(x^2+2ax)} + \sqrt{(x^2-2ax)} = \dfrac{nax}{\sqrt{(x^2+2ax)}}.$

48. $\sqrt{\left(\dfrac{\sqrt{}+a}{\sqrt{x}-a}\right)} - \sqrt{\left(\dfrac{\sqrt{x}-a}{\sqrt{x}+a}\right)} = \sqrt{(x-a^2)}.$

49. $\sqrt{[(2a+x)^2+b^2]} + \sqrt{[(2a-x)^2+b^2]} = 2a.$

50. $\dfrac{[\sqrt{(a-x)}+\sqrt{(x-b)}]^2}{\sqrt{(a-x)}-\sqrt{(x-b)}} = \sqrt{c}.$

51. $\dfrac{\sqrt{(1+x^2)}+\sqrt{(1-x^2)}}{\sqrt{(1+x^2)}-\sqrt{(1-x^2)}} = \dfrac{a}{b}.$

52. $\dfrac{\sqrt[3]{(1+x^2)}+\sqrt[3]{(1-x^2)}}{\sqrt[3]{(1+x^2)}-\sqrt[3]{(1-x^2)}} = \dfrac{a}{b}.$

53. $\dfrac{\sqrt[4]{(1+x^2)}+\sqrt[4]{(1-x^2)}}{\sqrt[4]{(1+x^2)}-\sqrt[4]{(1-x^2)}} = \dfrac{a}{b}.$

54. $\dfrac{\sqrt[5]{(1+x^2)}+\sqrt[5]{(1-x^2)}}{\sqrt[5]{(1+x^2)}-\sqrt[5]{(1-x^2)}} = \dfrac{a}{b}.$

55. $\dfrac{\sqrt[6]{(1+x^2)} + \sqrt[6]{(x^2-1)}}{\sqrt[6]{(1+x^2)} - \sqrt[6]{(x^2-1)}} = \dfrac{a}{b}.$

56. $\dfrac{\sqrt[n]{(x^2+1)} + \sqrt[n]{(x^2-1)}}{\sqrt[n]{(x^2+1)} - \sqrt[n]{(x^2-1)}} = \dfrac{a}{b}.$

57. $\sqrt{(4a+b-4x)} - 2\sqrt{(a+b-2x)} = \sqrt{b}.$

58. $\sqrt{(3a-2b+2x)} - \sqrt{(3a-2b-2x)} = 2\sqrt{a}.$

59. $\sqrt{(2a-b+2x)} - \sqrt{(10a-9b-6x)} = 4\sqrt{(a-b)}.$

60. $\sqrt{(3a-4b+5x)} + \sqrt{(x-a)} = 2\sqrt{(x+a)}.$

61. $\sqrt{(3a-4b+5x)} + \sqrt{(x-a)} = 2\sqrt{(2x-2b)}.$

62. $\sqrt{(5x-3a+4b)} + \sqrt{(5x-3a-4b)} = 2\sqrt{(x+a)}.$

63. $\sqrt{(2a+b+2x)} + \sqrt{(10a+9b-6x)} = 2\sqrt{(2a+b-2x)}.$

64. $2\sqrt{(2a+b+2x)} + \sqrt{(10a+b-6x)} = \sqrt{(10a+9b-6x)}.$

65. $\sqrt{(2a-13b+14x)} + \sqrt{[3(b-2a+2x)]} = 2\sqrt{(2a-b+2x)}.$

66. $\sqrt{[3(7a+b+x)]} - \sqrt{(a+7b-x)} = 2\sqrt{(7a+b-x)}.$

67. $\sqrt{[(a+x)(x+b)]} + \sqrt{[(a-x)(x-b)]} = 2\sqrt{(ax)}.$

68. $\sqrt{[(a+x)(x+b)]} - \sqrt{[(a-x)(x-b)]} = 2\sqrt{(bx)}.$

69. $\sqrt{(ax+x^2)} - \sqrt{(ax-x^2)} = \sqrt{(2ax-a^2)}.$

70. $\sqrt{(ax-x^2)} + \sqrt{(ax+x^2)} = \sqrt{(2ax+a^2)}.$

71. $\dfrac{1}{1+\sqrt{(1-x)}} + \dfrac{1}{1-\sqrt{(1-x)}} = \tfrac{2}{9}x.$

72. $\dfrac{x+\sqrt{(ax)}}{a-\sqrt{(ax)}} + \dfrac{a+\sqrt{(ax)}}{x-\sqrt{(ax)}} = \dfrac{x-a}{a}.$

73. $\dfrac{\sqrt{[(a+x)(x+b)]} + \sqrt{[(a-x)(x-b)]}}{\sqrt{[(a+x)(x+b)]} - \sqrt{[(a-x)(x-b)]}} = \sqrt{\dfrac{a}{b}}.$

74. $\sqrt{\dfrac{3a-2b+2x}{3a-2b-2x}} = \dfrac{[\sqrt{a}+\sqrt{(2a-2b)}]^2}{2b-a}.$

75. $\sqrt[3]{(a+x)} + \sqrt[3]{(a-x)} = 2\sqrt[3]{a}$.

76. $\sqrt[3]{(a+x)^2} - \sqrt[3]{(a^2-x^2)} + \sqrt[3]{(a-x)^2} = \sqrt[3]{a^2}$.

77. $\dfrac{\sqrt[3]{(1+x)^2} + \sqrt[3]{(1-x^2)} + \sqrt[3]{(1-x)^2}}{\sqrt[3]{(1+x)^2} - \sqrt[3]{(1-x^2)} + \sqrt[3]{(1-x)^3}} = 2\tfrac{1}{3}$.

78. $\sqrt[3]{(1+x)^2} + \sqrt[3]{(1-x)^2} = 2\tfrac{1}{2}\sqrt[3]{(1-x^2)}$.

79. $\sqrt[3]{(3+x)} + \sqrt[3]{(3-x)} = \sqrt[3]{6}$.

80. $\sqrt[3]{(1+x)^2} + \sqrt[3]{(1-x)^2} = 5[\sqrt[3]{(1+x)} + \sqrt[3]{(1-x)}]^2$.

81. $\sqrt[3]{(14+x)^2} - \sqrt[3]{(196-x^2)} + \sqrt[3]{(14-x)^2} = 7$.

82. $[\sqrt[3]{(9+x)} + \sqrt[3]{(9-x)}]\sqrt[3]{(81-x^2)} = 12$.

83. $[\sqrt[3]{(14+x)^2} - \sqrt[3]{(14-x)^2}][\sqrt[3]{(14+x)} - \sqrt[3]{(14-x)}] = 16$.

84. $[\sqrt[3]{(57+x)^2} + \sqrt[3]{(57-x)^2}][\sqrt[3]{(57-x)} + \sqrt[3]{(57+x)}] = 100$.

85. $5[\sqrt[4]{(41+x)} + \sqrt[4]{(41-x)}]^2 = 8[\sqrt{(41+x)} + \sqrt{(41-x)}]$.

86. $[\sqrt[4]{(x+5)} + \sqrt[4]{(x-5)}]^3[\sqrt[4]{(x+5)} - \sqrt[4]{(x-5)}] = 2$.

87. $[\sqrt[4]{(x+1)} + \sqrt[4]{(x-1)}][\sqrt{(x+1)} + \sqrt{(x-1)}]$
$= 26[\sqrt[4]{(x+1)} - \sqrt[4]{(x-1)}]$.

88. $\sqrt[3]{\dfrac{1+x}{1-x}} + \sqrt[3]{\dfrac{1-x}{1+x}} = a. \quad (y + y^{-1} = a)$.

89. $2[\sqrt[3]{(1+x)^2} + \sqrt[3]{(1-x^2)}] = (c^2+1)[\sqrt[3]{(1+x)} + \sqrt[3]{(1-x)}]^2$.

90. $\sqrt[3]{(a+x)} + \sqrt[3]{(a-x)} = \sqrt[3]{c}$.

91. $[\sqrt[3]{(a+x)} + \sqrt[3]{(a-x)}]\sqrt[3]{(a^2-x^2)} = c$.

92. $\sqrt[3]{(a+x)^2} = \sqrt[3]{(a^2-x^2)} + \sqrt[3]{(a-x)^2} = \sqrt[3]{c^2}$.

93. $[\sqrt[3]{(a+x)^2} - \sqrt[3]{(a-x)^2}][\sqrt[3]{(a+x)} - \sqrt[3]{(a-x)}] = c$.

94. $[\sqrt[3]{(a+x)^2} + \sqrt[3]{(a-x)^2}][\sqrt[3]{(a+x)} + \sqrt[3]{(a-x)}] = c$.

95. $(a+x)\sqrt[3]{(a-x)} - (a-x)\sqrt[3]{(a+x)}$
$= c[\sqrt[3]{(a+x)} - \sqrt[3]{(a-x)}]$.

96. $(a+x)\sqrt[3]{(a+x)} - (a-x)\sqrt[3]{(a-x)}$
$= c[\sqrt[3]{(a+x)} - \sqrt[3]{(a-x)}].$

97. $[\sqrt[3]{(a+x)^2} - \sqrt[3]{(a^2-x^2)} + \sqrt[3]{(a-x)^2}]^2$
$= c[\sqrt[3]{(a+x)} + \sqrt[3]{(a-x)}].$

98. $[\sqrt[4]{(a+x)} + \sqrt[4]{(a-x)}]^2 = (c+1)[\sqrt{(a+x)} + \sqrt{(a-x)}].$

99. $[\sqrt[4]{(x+a)} - \sqrt[4]{(x-a)}][\sqrt{(x+a)} + \sqrt{(x-a)}]^2$
$= c[\sqrt[4]{(x+a)} + \sqrt[4]{(x-a)}].$

100. $\dfrac{\sqrt{(x-a)} + \sqrt{(x-b)}}{\sqrt{(x-a)} - \sqrt{(x-b)}} = \sqrt{\dfrac{x-a}{x-b}}.$

101. $\dfrac{\sqrt{(x-a)} + \sqrt{(x-b)}}{\sqrt{(x-a)} - \sqrt{(x-b)}} = \sqrt{\dfrac{a-x}{x-b}}.$

102. $\sqrt{\dfrac{a-x}{b+x}} - \sqrt{\dfrac{b+x}{a-x}} = c.$

103. $\sqrt{\dfrac{a-x}{b-x}} + \sqrt{\dfrac{b-x}{a-x}} = c.$

104. $\sqrt[3]{\dfrac{a-x}{b+x}} - \sqrt[3]{\dfrac{b+x}{a-x}} = c.$

105. $\sqrt[3]{\left(\dfrac{a-x}{b-x}\right)^2} - \sqrt[3]{\left(\dfrac{b-x}{a-x}\right)^2} = c.$

106. $\sqrt[4]{\dfrac{a-x}{b+x}} + \sqrt[4]{\dfrac{b+x}{a-x}} = c.$ **109.** $\sqrt[5]{\dfrac{a-x}{b-x}} - \sqrt[5]{\dfrac{b-x}{a-x}} = c.$

107. $\sqrt[4]{\dfrac{a-x}{b-x}} - \sqrt[4]{\dfrac{b-x}{a-x}} = c.$ **110.** $\sqrt[6]{\dfrac{a-x}{b+x}} + \sqrt[6]{\dfrac{b+x}{a-x}} = c.$

108. $\sqrt[5]{\dfrac{a-x}{b+x}} + \sqrt[5]{\dfrac{b+x}{a-x}} = c.$ **111.** $\sqrt[6]{\dfrac{a-x}{b-x}} - \sqrt[6]{\dfrac{b-x}{a-x}} = c.$

112. $\dfrac{\sqrt{(a-x)^3} + \sqrt{(b-x)^3}}{\sqrt{(a-x)} + \sqrt{(b-x)}} = c.$

113. $\dfrac{\sqrt{(a-a)^3} + \sqrt{(b-x)^3}}{[\sqrt{(a-x)} + \sqrt{(b-x)}]^3} = c.$

114. $\dfrac{\sqrt{(a-x)^3}+\sqrt{(b-x)^3}}{\sqrt{(a-x)}-\sqrt{(b-x)}}=c.$

115. $\dfrac{[\sqrt{(a-x)}+\sqrt{(b-x)}]^3}{\sqrt{(a-x)}-\sqrt{(b-x)}}=c.$

116. $\dfrac{\sqrt{(a-x)^5}+\sqrt{(x-b)^5}}{\sqrt{(a-x)}+\sqrt{(x-b)}}=c.$

117. $\dfrac{\sqrt{(a-x)^5}-\sqrt{(x-b)^5}}{[\sqrt{(a-x)}-\sqrt{(x-b)}]^5}=c.$

118. $\dfrac{\sqrt{(a-x)^5}+\sqrt{(x+b)^5}}{\sqrt{(a-x)^3}+\sqrt{(x+b)^3}}=\sqrt{[(a-x)(x+b)]}.$

119. $\dfrac{\sqrt{(a-x)^3}+\sqrt{(x+b)^3}}{\sqrt{(a-x)}+\sqrt{(x+b)}}=\dfrac{(a+b)^2}{4\sqrt{[(a-x)(x+b)]}}.$

120. $\dfrac{x^3+(a-x^2)\sqrt{(a-x^2)}}{x+\sqrt{(a-x^2)}}=c.$

121. $\dfrac{x^3+(a^2-x^2)\sqrt{(a^2-x^2)}}{x+\sqrt{(a^2-x^2)}}=cx\sqrt{(a^2-x^2)}.$

122. $\sqrt[3]{(a-x)^2}-\sqrt[3]{[(a-x)(x-b)]}+\sqrt[3]{(x+b)^2}$
$=\sqrt[3]{(a^2-ab+b^2)}.$

123. $\dfrac{b\sqrt{(a-x)}+a\sqrt{(x-b)}}{\sqrt{(a-x)}+\sqrt{(x-b)}}=x.$

124. $\dfrac{a\sqrt{(a-x)}+b\sqrt{(x-b)}}{\sqrt{(a-x)}+\sqrt{(x-b)}}=x.$

125. $\dfrac{\sqrt{(x-a)}+\sqrt{(x+a)}-\sqrt{(2a)}}{\sqrt{(x-a)}-\sqrt{(x+a)}+\sqrt{(2a)}}=\sqrt[4]{\dfrac{x+c}{x-c}}.$

126. $\dfrac{\sqrt{(a-x)}+\sqrt{c}}{\sqrt{(x-b)}+\sqrt{c}}=\sqrt[4]{\dfrac{a-x}{x-b}}.$

127. $\sqrt[3]{(a-x)^2}-\sqrt[3]{[(a-x)(x+b)]}+\sqrt[3]{(x+b)^2}$
$=\sqrt[3]{(a^2-ab+b^2)}.$

128. $\{\sqrt[3]{(a-x)^2} - \sqrt[3]{[(a-x)(x-b)]} + \sqrt[3]{(x-b^2)^2}\}^2$
$\quad = (a-b)[\sqrt[3]{(a-x)} + \sqrt[3]{(x-b)}].$

129. $[\sqrt[3]{(a-x)^2} + \sqrt[3]{(b+x)^2}]^2 = (a+b)[\sqrt[3]{(a-x)} + \sqrt[3]{(b+x)}].$

130. $\sqrt[3]{(a-x)} + \sqrt[3]{(x-b)} = \sqrt[3]{c}.$

131. $\sqrt[3]{(a+x)^2} - \sqrt[3]{(a-x)^2} = \sqrt[3]{(2cx)}.$

132. $\sqrt[3]{(a-x)^2} + \sqrt[3]{[(a-x)(b-x)]} + \sqrt[3]{(b-x)^2} - \sqrt[3]{c^2}.$

133. $\sqrt[3]{(a-x)^2} - \sqrt[3]{[(a-x)(x+b)]} + \sqrt[3]{(x+b)^2}$
$\quad = c[\sqrt[3]{(a-x)} + \sqrt[3]{(x+b)}].$

134. $[\sqrt[3]{(a-x)} + \sqrt[3]{(x+b)}]\sqrt[3]{[(a-x)(x+b)]} = c.$

135. $\sqrt[3]{(a-x)^2} + \sqrt[3]{(x-b)^2} = c[\sqrt[3]{(a-x)} + \sqrt[3]{(x-b)}]^2.$

136. $x + \sqrt[3]{(a^3-x^3)} = \dfrac{c^3}{x\sqrt[3]{(a^3-x^3)}}.$

137. $\dfrac{a^3}{x^3-b^3} = \dfrac{x+\sqrt[3]{(2b^3-x^3)}}{x-\sqrt[3]{(2b^3-x^3)}}.$

138. $(a+x)\sqrt[4]{(a+x)} + (a-x)\sqrt[4]{(a-x)}$
$\quad = a[\sqrt[4]{(a+x)} + \sqrt[4]{(a-x)}].$

139. $(a+x)\sqrt[4]{(a-x)} + (a-x)\sqrt[4]{(a+x)}$
$\quad = a[\sqrt[4]{(a+x)} + \sqrt[4]{(a-x)}].$

140. $\sqrt[4]{(26-x)} + \sqrt[4]{(x-10)} = 2.$

141. $[\sqrt[4]{(a-x)} + \sqrt[4]{(x-b)}]^2 = c[\sqrt{(a-x)} + \sqrt{(x-b)}].$

142. $(a-x)\sqrt[4]{(a-x)} + (x-b)\sqrt[4]{(x-b)}$
$\quad = (a-b)[\sqrt[4]{(a-x)} + \sqrt[4]{(x-b)}].$

143. $[\sqrt[4]{(a-x)} + \sqrt[4]{(x-b)}]^2[\sqrt{(a-x)} + \sqrt{(x-b)}]$
$\quad = c(a+b-2x).$

144. $[\sqrt[4]{(a-x)} + \sqrt[4]{(b-x)}][\sqrt{(a-x)} + \sqrt{(b-x)}]^2$
$\quad = c[\sqrt[4]{(a-x)} - \sqrt[4]{(b-x)}].$

145. $a\sqrt{(1+x^2)} - x\sqrt{(x^2+a^2)} = c.$

146. $(a-x)\sqrt[3]{(x-b)}+(x-b)\sqrt[3]{(a-x)}$
$$=c[\sqrt[3]{(a-x)}+\sqrt[3]{(x-b)}]^4.$$

147. $[\sqrt[3]{(a-x)}+\sqrt[3]{(b+x)}]^5=c[\sqrt[3]{(a-x)^2}+\sqrt[3]{(b+x)^2}].$

148. $[\sqrt[3]{(a-x)}+\sqrt[3]{(b+x)}]^5=c\sqrt[3]{[(a-x)(b+x)]}.$

149. $\sqrt[3]{(a-x)^2}-\sqrt[3]{(b-x)^2}=c\sqrt[3]{(a+b-2x)}.$

150. $\sqrt[4]{(a-x)}+\sqrt[4]{(x-b)}=\sqrt[4]{c}.$

151. $\sqrt[5]{(a-x)}+\sqrt[5]{(x-b)}=\sqrt[5]{c}.$

152. $\dfrac{(a-x)\sqrt[4]{(a-x)}+(x-b)\sqrt[4]{(x-b)}}{(a-x)\sqrt[4]{(x-b)}+(x+b)\sqrt[4]{(a-x)}}=c.$

153. $\dfrac{(a-x)\sqrt[4]{(b-x)}+(b-x)\sqrt[4]{(a-x)}}{\sqrt[4]{(a-x)}-\sqrt[4]{(b-x)}}=c.$

154. $\dfrac{\sqrt[4]{(a-x)}+\sqrt[4]{(x-b)}}{\sqrt[4]{(a-x)}-\sqrt[4]{(x-b)}}=\dfrac{c}{a+b-2x}.$

155. $\dfrac{[\sqrt[4]{(a-x)}+\sqrt[4]{(b-x)}]^5}{\sqrt[4]{(a-x)}-\sqrt[4]{(b-x)}}=c.$

156. $(a-x)\sqrt[5]{(a-x)}-(x-b)\sqrt[5]{(x-b)}$
$$=c[\sqrt[5]{(a-x)}-\sqrt[5]{(x-b)}].$$

157. $(a-x)\sqrt[5]{(x+b)}-(x+b)\sqrt[5]{(x-a)}$
$$=c[\sqrt[5]{(a-x)}-\sqrt[5]{(x+b)}].$$

158. $[\sqrt[5]{(a-x)^3}+\sqrt[5]{(x-b)^3}]\sqrt[5]{[(a-x)(x-b)]}=c.$

159. $[\sqrt[5]{(a-x)}-\sqrt[5]{(x-b)}]^3[\sqrt[5]{(a-x)^2}-\sqrt[5]{(x-b)^2}]=c.$

160. $[\sqrt[5]{(a-x)^2}-\sqrt[5]{(x-b)^2}][\sqrt[5]{(a-x)}+\sqrt[5]{(x-b)}]=c.$

161. $[\sqrt[5]{(a-x)^3}+\sqrt[5]{(x+b)^3}]^2=c[\sqrt[5]{(a-x)}+\sqrt[5]{(x+b)}].$

162. $\sqrt{(x^2-a^2-b^2)}+\sqrt{(x^2-b^2-c^2)}-\sqrt{(x^2-c^2-a^2)}=x.$

163. $\dfrac{\sqrt{(a^2+2x)}+\sqrt{(a^2-2x)}}{\sqrt{(a^2+2x)}-\sqrt{(a^2-2x)}}$
$$=\dfrac{m^2x^2}{a^2}\times\dfrac{\sqrt{(m^2x+2)}+\sqrt{(m^2x-2)}}{\sqrt{(m^2x+2)}-\sqrt{(m^2x-2)}}.$$

164. $\sqrt{(x^2 - a^2)} + \sqrt{(x^2 - b^2)} + \sqrt{(x^2 - c^2)} = x.$

165. $[\sqrt[4]{(a-x)} + \sqrt[4]{(b-x)}][\sqrt[4]{(a-x)} - \sqrt[4]{(b-x)}] = c.$

166. $\dfrac{\sqrt[3]{(a-x)} - \sqrt[3]{(x-b)}}{\sqrt[3]{(a-x)} + \sqrt[3]{(x-b)}} = \dfrac{a+b-2x}{a-b}.$

167. $\sqrt[5]{(a+x)} + \sqrt[5]{(a-x)} = \sqrt[5]{(2a)}.$

168. $\dfrac{[\sqrt[3]{(a-x)^2} + \sqrt[3]{(x-b)^2}]^2}{\sqrt[3]{(a-x)^2} + \sqrt[3]{(x-b)}} = a - b.$

Write u for $\sqrt[3]{(a-x)}$, and v for $\sqrt[3]{(x-b)}$.

169. $\sqrt{(5 - 3x + x^2)} + \sqrt{(5 - 3y + y^2)} = 6 \; ; \; x + y = 3.$

170. $\sqrt{(x - xy)} + \sqrt{(y - xy)} = a \; ; \; x + y = b.$

171. $\sqrt[3]{(x + m)} + \sqrt{(y + n)} = a \; ; \; x + y = b.$

172. $\sqrt[5]{(143 + x)} - \sqrt[5]{(y - 18)} = 1 \; ; \; x - y = 50.$

173. $\sqrt{(xy)} + \sqrt{[(1-x)(1-y)]} = a \; ;$
$\sqrt{[x(1-y)]} + \sqrt{[y(1-x)]} = b.$

174. $xy + \sqrt{[(1-x^2)(1-y^2)]} = ab \; ;$
$x\sqrt{(1-y^2)} + y\sqrt{(1-x^2)} = \dfrac{a^2 - b^2}{2ab}.$

175. $\sqrt{(x - xy)} + \sqrt{(y - xy)} = a \; ;$
$\sqrt{(x - x^2)} + \sqrt{(y - y^2)} = b.$

176. $x^{\frac{3}{2}} + y^{\frac{3}{2}} = 3(x^{\frac{1}{2}} - y^{\frac{1}{2}}) = 3x.$

177. $\left(\dfrac{a^2 - x^2}{y^2 - b^2} + \dfrac{y^2 - b^2}{a^2 - x^2}\right)^{\frac{1}{2}} + \left(\dfrac{a^2 + x^2}{y^2 + b^2} + \dfrac{y^2 + b^2}{a^2 + x^2}\right)^{\frac{1}{2}} = 4 \; ;$
$xy = ab.$

178. $(x + y)^{\frac{1}{3}} + (x - y)^{\frac{1}{3}} = a^{\frac{1}{3}} \; ;$
$(x^2 + y^2)^{\frac{1}{3}} + (x^2 - y^2)^{\frac{1}{3}} = a^{\frac{2}{3}}.$

CHAPTER IX.

Cubic and Quartic Equations.

§ 52. *The Cubic.* Let the general cubic equation be

$$ax^3 + 3bx^2 + 3cx + d = 0. \tag{1}$$

Let $\qquad y = ax + b.$ $\qquad\qquad\qquad\qquad$ (2)

Substitute $\dfrac{y-b}{a}$ for x in (1), and multiply the resulting equation by a^2,

$$y^3 + 3(ac - b^2)y + (a^2d - 3abc + 2b^3) = 0,$$

which may be written

$$y^3 + 3Hy + G = 0, \tag{3}$$

in which $\qquad H = ac - b^2,$ $\qquad\qquad\qquad$ (4)

and $\qquad\quad G = a^2d - 3abc + 2b^3.$ $\qquad\quad$ (5)

Assume $\qquad y = \omega \sqrt[3]{r_1} + \omega^2 \sqrt[3]{r_2},$* $\qquad\quad$ (6)

in which $\qquad \omega^3 = 1.$ $\qquad\qquad\qquad\qquad$ (7)

Cube (6), $\quad y^3 = r_1 + r_2 + 3\sqrt[3]{(r_1 r_2)}(\omega \sqrt[3]{r_1} + \omega^2 \sqrt[3]{r_2})$

$$= r_1 + r_2 + 3\sqrt[3]{(r_1 r_2)}\, y;$$

$$y^3 - 3\sqrt[3]{(r_1 r_2)}\, y - (r_1 + r_2) = 0. \tag{8}$$

Equate coefficients in (3) and (8),

$$r_1 + r_2 = -G, \tag{9}$$

and $\qquad \sqrt[3]{(r_1 r_2)} = -H. \quad \therefore r_1 r_2 = -H^3.$ $\quad$ (10)

---

* Throughout this chapter the symbols $\sqrt{\ }$ and $\sqrt[3]{\ }$ will be used to denote the corresponding arithmetical roots of the quantities they operate on. General roots will be denoted by exponents. See § 50, page 264.

(9) and (10) show that r_1 and r_2 are the roots of the quadratic

$$r^2 + Gr - H^3 = 0, \qquad (11)$$

which may be written

$$(2r + G)^2 - (G^2 + 4H^3) = 0. \qquad (12)$$

By (4) and (5),

$$G^2 + 4H^3 = a^2(a^2d^2 + 4ac^3 - 6abcd + 4b^3d - 3b^2c^2)$$
$$= a^2\Delta, \text{ say.} \qquad (13)$$

$$\therefore (2r + G)^2 - a^2\Delta = 0. \qquad (14)$$

$$\therefore r = -\tfrac{1}{2}G \pm \tfrac{1}{2}a\sqrt{\Delta}.$$

Let $r_1 = -\tfrac{1}{2}G - \tfrac{1}{2}a\sqrt{\Delta}.$ \qquad (15)

$$\therefore r_2 = -\tfrac{1}{2}G + \tfrac{1}{2}a\sqrt{\Delta}. \qquad (16)$$

Substitute these values for r_1 and r_2 in (6).

$$\therefore y = \omega\sqrt[3]{(-\tfrac{1}{2}G - \tfrac{1}{2}a\sqrt{\Delta})} + \omega^2\sqrt[3]{(-\tfrac{1}{2}G + \tfrac{1}{2}a\sqrt{\Delta})}$$
$$= -\omega\sqrt[3]{(\tfrac{1}{2}G + \tfrac{1}{2}a\sqrt{\Delta})} - \omega^2\sqrt[3]{(\tfrac{1}{2}G - \tfrac{1}{2}a\sqrt{\Delta})}. \qquad (17)$$

Substitute this value of y in (2).

$$\therefore x = \frac{1}{a}[-b - \omega\sqrt[3]{(\tfrac{1}{2}G + \tfrac{1}{2}a\sqrt{\Delta})}$$
$$- \omega^2\sqrt[3]{(\tfrac{1}{2}G - \tfrac{1}{2}a\sqrt{\Delta})}]. \qquad (18)$$

Hence, if

$$ax^3 + 3bx^2 + 3cx + d = 0,$$

$$x_1 = \frac{1}{a}[-b - \sqrt[3]{(\tfrac{1}{2}G + \tfrac{1}{2}a\sqrt{\Delta})} - \sqrt[3]{(\tfrac{1}{2}G - \tfrac{1}{2}a\sqrt{\Delta})}],$$

$$x_2 = \frac{1}{a}[-b + \tfrac{1}{2}(1 + i\sqrt{3})\sqrt[3]{(\tfrac{1}{2}G + \tfrac{1}{2}a\sqrt{\Delta})}$$
$$+ \tfrac{1}{2}(1 - i\sqrt{3})\sqrt[3]{(\tfrac{1}{2}G - \tfrac{1}{2}a\sqrt{\Delta})}], \qquad (19)$$

$$x_3 = \frac{1}{a}[-b + \tfrac{1}{2}(1 - i\sqrt{3})\sqrt[3]{(\tfrac{1}{2}G + \tfrac{1}{2}a\sqrt{\Delta})}$$
$$+ \tfrac{1}{2}(1 + i\sqrt{3})\sqrt[3]{(\tfrac{1}{2}G - \tfrac{1}{2}a\sqrt{\Delta})}];$$

in which $a^2\Delta = G^2 + 4H^3$,

$$G = a^2d - 3abc + 2b^3, \qquad (20)$$
$$H = ac - b^2.$$

If the cubic have one or more rational linear factors, the above method of solution should not be attempted; but such factor or factors should be determined, and the cubic resolved into its equivalent disjunctive equations, and these solved. (See *Note*, page 233.) If the cubic have no rational linear factor, $\sqrt[3]{(\frac{1}{2}G + \frac{1}{2}a\sqrt{\Delta})}$ *will not be reducible to the form* $u + \sqrt{v}$; but, if the cubic have such factor, $\sqrt[3]{(\frac{1}{2}G + \frac{1}{2}a\sqrt{\Delta})}$ will be reducible, and *the reduction may be effected by resolving the cubic into its equivalent disjunctive equations, and solving these.*

EXAMPLES.

1. Solve $9x^3 - 9x - 4 = 0$.

Assume $\quad x = \omega\sqrt[3]{r_1} + \omega^2\sqrt[3]{r_2}$.

$\therefore x^3 = r_1 + r_2 + 3\sqrt[3]{(r_1 r_2)}x$.

$\therefore r_1 + r_2 = \frac{4}{9}$,

and $\quad \sqrt[3]{(r_1 r_2)} = \frac{1}{3}$.

$\therefore r_1 r_2 = \frac{1}{27}$.

$\therefore r^2 - \frac{4}{9}r + \frac{1}{27} = 0$.

$\therefore r_1 = \frac{1}{3}$ and $r_2 = \frac{1}{9}$.

$\therefore x_1 = \sqrt[3]{\frac{1}{3}} + \sqrt[3]{\frac{1}{9}} = \frac{1}{3}(\sqrt[3]{9} + \sqrt[3]{3})$.

2. Solve $2x^3 + 6x^2 + 1 = 0$.

First Solution.

Assume $\quad y = x + 1$, and substitute for x.

$\therefore 2y^3 - 6y + 5 = 0$.

Assume $\quad y = \omega\sqrt[3]{r_1} + \omega^2\sqrt[3]{r_2}$,

cube each side, and compare coefficients with those of the equation in y,

$r_1 + r_2 = -\frac{5}{2}$;

$r_1 r_2 = 1$.

$$\therefore r^2 + \tfrac{5}{2}r + 1 = 0.$$
$$\therefore r_1 = -2,$$
and $$r_2 = -\tfrac{1}{2}.$$
$$\therefore y_1 = -\sqrt[3]{2} - \frac{1}{\sqrt[3]{2}}.$$
$$\therefore x_1 = -1 - \sqrt[3]{2} - \frac{1}{\sqrt[3]{2}}.$$

Second Solution.

Let $z = x^{-1}$, and substitute for x.
$$\therefore z^3 + 6z + 2 = 0.$$
Assume $z = \omega \sqrt[3]{r_1} + \omega^2 \sqrt[3]{r_2}.$
$$\therefore r_1 + r_2 = -2;$$
$$r_1 r_2 = \left(\frac{-6}{3}\right)^3 = -8.$$
$$\therefore r_1 = -4,$$
and $r_2 = 2.$
$$\therefore z_1 = \sqrt[3]{2} - \sqrt[3]{4}.$$
$$\therefore x_1 = \frac{1}{\sqrt[3]{2} - \sqrt[3]{4}}$$
$$= \frac{\sqrt[3]{4} + \sqrt[3]{8} + \sqrt[3]{16}}{2 - 4}$$
$$= -\left(1 + \sqrt[3]{2} + \frac{1}{\sqrt[3]{2}}\right).$$

3. Solve $x + y^2 = 7;$
$$x^2 + y = 11.$$
$$\therefore x + (11 - x^2)^2 = 7.$$
$$\therefore x^4 - 22x^2 + x + 114 = 0.$$
$$\therefore (x - 3)(x^3 + 3x^2 - 13y - 38) = 0.$$

Therefore, either

$$x - 3 = 0,$$

or else $\quad x^3 + 3x^2 - 13x - 38 = 0.$

Let $\qquad z = x + 1,$

and substitute for x in the latter of these disjunctive equations.

$$\therefore z^3 - 16z - 23 = 0.$$

Assume $\quad z = \omega \sqrt[3]{r_1} + \omega^2 \sqrt[3]{r_2}.$

$$\therefore r_1 + r_2 = 23,$$

and $\qquad \sqrt[3]{(r_1 r_2)} = \tfrac{16}{3}.$

$$\therefore r_1 r_2 = \tfrac{4096}{27}.$$

$$\therefore r = \tfrac{1}{2}[23 \pm \sqrt{(23^2 - \tfrac{16384}{27})}]$$
$$= \tfrac{1}{2}(23 \pm \tfrac{1}{9} i \sqrt{6303}).$$

$$\therefore x = 3$$

or $\qquad -1 + \omega \sqrt[3]{(11\tfrac{1}{2} + \tfrac{1}{18} i \sqrt{6303})}$
$$+ \omega^2 \sqrt[3]{(11\tfrac{1}{2} - \tfrac{1}{18} i \sqrt{6303})},$$

in which

$$\omega = 1$$

or $\qquad -\tfrac{1}{2}(1 \pm i\sqrt{3}).$

4. Find the cube roots of

$$-10 + 9 i \sqrt{3}.$$

Assume $\quad \tfrac{1}{2}(y + a\sqrt{z}) = (-10 + 9 i \sqrt{3})^{\frac{1}{3}}$
$$= \omega \sqrt[3]{(-10 + 9 i \sqrt{3})}, \qquad (1)$$

in which

$$a^2 = 1,$$

and $\quad \omega = 1$ or $-\tfrac{1}{2}(1 + i\sqrt{3})$ or $-\tfrac{1}{2}(1 - i\sqrt{3}),$

and therefore

$$\omega^2 = 1 \text{ or } -\tfrac{1}{2}(1 - i\sqrt{3}) \text{ or } -\tfrac{1}{2}(1 + i\sqrt{3}).$$

$$\therefore \tfrac{1}{2}(y - a\sqrt{z}) = \omega^2 \sqrt[3]{(-10 - 9 i \sqrt{3})}. \qquad (2)$$

$(1)+(2),\quad y=\omega\sqrt[3]{(-10+9i\sqrt{3})}+\omega^2\sqrt[3]{(-10-9i\sqrt{3})}.$ (3)

$(3)^3,\qquad y^3=-20+21y.$

$\therefore (y-1)(y-4)(y+5)=0.$

$\therefore y_1=1,\quad y_2=4,\quad y_3=-5.$ (4)

$(1)\times(2),\ \tfrac{1}{4}(y^2-z)=7.$

$\therefore z=y^2-28.$ (5)

$\therefore z_1=-27,\quad z_2=-12,\quad z_3=-5.$ (6)

$(1)-(2),\quad a\sqrt{z}=\omega\sqrt[3]{(-10+9i\sqrt{3})}-\omega^2\sqrt[3]{(-10-9i\sqrt{3})},$ (7)

$(7)^3,\qquad az\sqrt{z}=18i\sqrt{3}-21a\sqrt{z}.$

$$\therefore a\sqrt{z}=\frac{18i\sqrt{3}}{21+z}.$$

Substitute for z its values given in (6).

$\therefore a\sqrt{z_1}=-3i\sqrt{3},\ a\sqrt{z_2}=2i\sqrt{3},\ a\sqrt{z_3}=i\sqrt{3}.$ (8)

From (4) and (8),

$$(-10+9i\sqrt{3})^{\frac{1}{3}}=\tfrac{1}{2}(1-3i\sqrt{3}),$$

or $2+i\sqrt{3},$

or $\tfrac{1}{2}(-5+i\sqrt{3}).$

5. Find the cube roots of $4+43i\sqrt{5}.$

Assume $\tfrac{1}{2}(y+a\sqrt{z})=\omega\sqrt[3]{(4+43i\sqrt{5})}.$ (1)

$\therefore \tfrac{1}{2}(y-a\sqrt{z})=\omega^2\sqrt[3]{(4-43i\sqrt{5})}.$ (2)

$(1)+(2),\quad y=\omega\sqrt[3]{(4+43i\sqrt{5})}+\omega^2\sqrt[3]{(4-43i\sqrt{5})}.$ (3)

$(3)^3,\qquad y^3=8+63y.$

$\therefore (y-8)(y^2+8y+1)=0.$ (4)

$\therefore y_1=8,$

$\qquad y_2=-4+\sqrt{15},$ (5)

$\qquad y_3=-4-\sqrt{15}.$

$(1)\times(2),\ \tfrac{1}{4}(y^2-z)=21.$

$\therefore z=y^2-84.$ (6)

$$\therefore z_1 = -20,$$
$$z_2 = -53 - 8\sqrt{15},$$
$$z_3 = -53 + 8\sqrt{15}. \tag{7}$$

$(1)-(2),\quad a\sqrt{z} = \omega\sqrt[3]{(4+43i\sqrt{5})} - \omega^2\sqrt[3]{(4-43i\sqrt{5})}. \tag{8}$

$(8)^3,\qquad az\sqrt{z} = 86i\sqrt{5} - 63a\sqrt{z}.$

$$\therefore a\sqrt{z} = \frac{86i\sqrt{5}}{63+z}.$$

$$\therefore a\sqrt{z_1} = 2i\sqrt{5},$$

$$a\sqrt{z_2} = \frac{86i\sqrt{5}}{10 - 8\sqrt{15}} = \frac{43i\sqrt{5}(5+4\sqrt{15})}{25-240}$$
$$= -i(\sqrt{5}+4\sqrt{3}),$$

$$a\sqrt{z_3} = -i(\sqrt{5}-4\sqrt{3}).$$

Substituting in (1) these values of y and $a\sqrt{z}$,

$$\tfrac{1}{2}(y_1 + a\sqrt{z_1}) = 4 + i\sqrt{5},$$
$$\tfrac{1}{2}(y_2 + a\sqrt{z_2}) = \tfrac{1}{2}(-4+\sqrt{15}-i\sqrt{5}-4i\sqrt{3})$$
$$= -\tfrac{1}{2}(1+i\sqrt{3})(4+i\sqrt{5}),$$
$$\tfrac{1}{2}(y_3 + a\sqrt{z_3}) = -\tfrac{1}{2}(1-i\sqrt{3})(4+i\sqrt{5}).$$

Ex. 68.

Solve the following equations :

1. $x^3 - 3x^2 + 9x - 5 = 0.$

2. $x^3 - 3x^2 + 9x - 9 = 0.$

3. $2x^3 - 6x^2 + 18x + 17 = 0.$

4. $x^3 - 3x^2 - 15x - 13 = 0.$

5. $x^3 - 3x^2 - 15x - 25 = 0.$

6. $x^3 + 9x^2 + 9x + 15 = 0.$

7. $x^3 + 3x^2 + x + 1 = 0.$

8. $3x^3 + 27x^2 - 9x + 41 = 0.$

9. $3x^3 + 27x^2 - 9x + 4 = 0.$

10. $x^3 - 18x - 33 = 0.$

11. $x^3 - 9x - 12 = 0.$

12. $x^3 - 6x^2 + 10x - 1 = 0.$

13. $x^2 + y^2 = 6$, and $x^3 + y^3 = 1.$

14. $x^2 + y^2 = 6mn$, and $x^3 + y^3 = mn(2m-n).$

15. $x^3 - 3x^2 + 9x + (k-8)(1+k^{-1}) = 0.$

16. $x^3 - 3x^2 - 15x + (k+216)(1+k^{-1}) - 200 = 0.$

17. $x^3 + 9x^2 + 9x - 3(k+24)(1+k^{-1}) + 16 = 0.$

18. $x^3 - 2x - 5 = 0.$

Find the cube roots of:

19. $-8 + 6\sqrt{3}.$ **20.** $-55 - 126i\sqrt{3}.$

         21. $5 - 7i\sqrt{5}.$

Solve $ax^3 + 3bx^2 + 3cx + d = 0$, given:

22. $bd = c^2.$ **23.** $2a^2bd = a^2c^2 + b^4.$

24. $a^3bcd = a^2c^2(b^2 + ac) - b^6.$

25. $kG + H^3 = k^2$, k being arbitrary.

26. Show that $H = 0$ is the condition that $ax^2 + 2bx + c$ shall have a square factor; and that $\Delta = 0$ is the condition that $ax^3 + 3bx^2 + 3cx + d$ shall have a square factor.

27. Show how to solve the cubic by assuming
$$ax^3 + 3bx^2 + 3cx + d$$
$$= m(ax + b + t_1)^3 + (1 - m)(ax + b + t_2)^3,$$
and determining m, t_1, and t_2.

28. If $t_1 = (x_1 + \omega x_2 + \omega^2 x_3)^3$ and $t_2 = (x_1 + \omega^2 x_2 + \omega x_3)^3$, where $\omega^2 + \omega + 1 = 0$, find $t_1 + t_2$ and $t_1 t_2$, and apply the result to solve the cubic.

29. If x_1, x_2, and x_3 be the roots of a cubic, express $(x_1 - x_2)^2(x_2 - x_3)^2(x_3 - x_1)^2$ in terms of the coefficients.

30. Prove that if all three roots of a cubic are real and unequal, Δ will be negative; but, that if two of the roots are complex, Δ will be positive.

§ 53. THE QUARTIC.

Let the general quartic equation be

$$ax^4 + 4bx^3 + 6cx^2 + 4dx + e = 0. \tag{21}$$

Assume

$$a(ax^4 + 4bx^3 + 6cx^2 + 4dx + e)$$
$$= (ax^2 + 2bx + c + 2t)^2 - (2\sqrt{r}x + s)^2. \tag{22}$$

Expand and equate coefficients of like powers of x.

$$\therefore r = at - (ac - b^2), \tag{23}$$

$$s\sqrt{r} = 2bt - (ad - bc), \tag{24}$$

$$s^2 = 4t^2 + 4ct - (ae - c^2). \tag{25}$$

$$\therefore [at - (ac - b^2)][4t^2 + 4ct - (ae - c^2)]$$
$$= [2bt - (ad - bc)]^2.$$

$$\therefore 4t^3 - (ae - 4bd + 3c^2)t$$
$$+ (ace + 2bcd - ad^2 - eb^2 - c^3) = 0, \tag{26}$$

which may be written

$$4t^3 - It + J = 0, \tag{27}$$

in which $I = ae - 4bd + 3c^2,$ \hfill (28)

$$J = ace + 2bcd - ad^2 - eb^2 - c^3. \tag{29}$$

Selecting any one of the three values of t determined by the cubic (27), the corresponding value of r may be found by substitution for t in (23), and then that of s by substitution in (24), or if $r = 0$, in (25); and the quartic in (22) may then be resolved into the quadratic factors,

$$ax^2 + 2(b - \sqrt{r})x + c + 2t - s,$$

and $\quad ax^2 + 2(b + \sqrt{r})x + c + 2t + s. \tag{30}$

Each of these factors equated to zero will give a pair of the roots of the quartic equation (21), which will thus be completely resolved. \hfill (31)

The equation (27) is called the Reducing Cubic of the Quartic (21).

EXAMPLES.

1. Solve $x^4 - 12x^3 + 47x^2 - 66x + 27 = 0.$ (See Ex. 41.)

Let $x^4 - 12x^3 + 47x^2 - 66x + 27$

$$= \left(x^2 - 6x + \frac{47}{6} + 2t\right)^2 - (2\sqrt{r}x + s)^2$$

$$= x^4 - 12x^3 + \left(36 + \frac{47}{3} + 4t - 4r\right)x^2$$

$$- (94 + 24t + 4s\sqrt{r})x + \frac{2209}{36} + \tfrac{94}{3}t + 4t^2 - s^2.$$

Equating coefficients of like powers of x,

$$47 = 36 + \frac{47}{3} + 4t - 4r,$$

$$66 = 94 + 24t + 4s\sqrt{r},$$

$$27 = \frac{2209}{36} + \tfrac{94}{3}t + 4t^2 - s^2.$$

$$\therefore 6r = 6t + 7,$$

$$s\sqrt{r} = -(6t + 7),$$

$$36s^2 = 144t^2 + 1128t + 1237.$$

$$\therefore (6t + 7)(144t^2 + 1128t + 1237) = 216(6t + 7)^2.$$

$$\therefore 6t + 7 = 0,$$

or $144t^2 + 1128t + 1237 = 216(6t + 7).$

$$\therefore 144t^2 - 168t - 275 = 0.$$

$$\therefore (12t - 25)(12t + 11) = 0.$$

$$\therefore t_1 = -\frac{7}{6}, \quad t_2 = \frac{25}{12}, \quad t_3 = -\frac{11}{12},$$

and $\therefore r_1 = 0, \quad r_2 = \frac{13}{4}, \quad r_3 = \frac{1}{4},$

$$s_1^2 = \frac{13}{4}, \quad s_2 = -3\sqrt{13}, \quad s_3 = -3.$$

$$\therefore x^4 - 12x^3 + 47x^2 - 66x + 27$$
$$= \left(x^2 - 6x + \frac{11}{2}\right)^2 - \frac{13}{4}$$
$$= (x^2 - 6x + 12)^2 - 13(x-3)^2$$
$$= (x^2 - 6x + 6)^2 - (x-3)^2 = 0.$$

The last gives

$$x^2 - 7x + 9 = 0 \quad \text{or} \quad x^2 - 5x + 3 = 0.$$
$$\therefore x = \tfrac{1}{2}(7 \pm \sqrt{13}) \quad \text{or} \quad x = \tfrac{1}{2}(5 \pm \sqrt{13}).$$

2. Solve $9x^4 - 54x^3 + 60x^2 - 72x + 16 = 0$.

Here $\quad a = 9, \ b = -\dfrac{27}{2}, \ c = 10, \ d = -18, \ e = 16.$

$$\therefore r = 9t - (90 - 182\tfrac{1}{4}) = 2\tfrac{1}{4}(4t + 41),$$
$$s\sqrt{r} = -27t - (-162 + 135) = -27(t-1),$$
$$s^2 = 4t^2 + 40t - (144 - 100)$$
$$= 4(t^2 + 10t - 11) = 4(t+11)(t-1).$$
$$\therefore 9(4t+41)(t+11)(t-1) = 729(t-1).$$
$$\therefore t - 1 = 0,$$

or $\qquad (4t+41)(t+11) = 81(t-1).$

If $\qquad t = 1, \ \therefore r = 101\tfrac{1}{4} \text{ and } s = 0.$

$$\therefore (9x^2 - 27x + 12)^2 - 405x^2 = 0.$$
$$\therefore 3x^2 - (9 + 3\sqrt{5})x + 4 = 0,$$

or $\qquad 3x^2 - (9 - 3\sqrt{5})x + 4 = 0.$

$$\therefore x = \tfrac{1}{6}[9 + 3\sqrt{5} \pm \sqrt{(78 + 54\sqrt{5})}],$$

or $\qquad x = \tfrac{1}{6}[9 - 3\sqrt{5} \pm \sqrt{(78 - 54\sqrt{5})}].$

Ex. 69.

Solve the following equations:

1. $x^4 - 6x^3 - 2x^2 + 36x - 24 = 0.$

2. $x^4 - 2x^3 - 25x^2 + 18x + 24 = 0.$

3. $2x^4 - 5x^3 - 17x^2 + 53x - 28 = 0$. **5.** $x^4 - 12x - 5 = 0$.

4. $x^4 + 14x^2 + 48x + 49 = 0$. **6.** $x^4 - 12x - 17 = 0$.

7. $x^4 - 8x^3 - 12x^2 + 84x - 63 = 0$.

8. $x^4 + 2x^3 - 37x^2 - 38x + 1 = 0$.

9. $121x^4 + 198x^3 - 100x^2 - 36x + 4 = 0$.

10. $x^2 + y = 1\frac{1}{2}$, $x + y^2 = 2\frac{1}{8}$.

§ 54. The cubic in (27) will, in general, give three values of t. Let them be denoted by t_1, t_2, t_3. Also let the corresponding values of r and of s be denoted by r_1, r_2, r_3, and s_1, s_2, s_3, respectively. Let x_1, x_2, x_3, and x_4 denote the roots of the quartic. Then, by (31),

$$ax^2 + 2(b - \sqrt{r_1})x + c + 2t_1 - s_1 = 0 \qquad (32)$$

will furnish a pair of the roots of the quartic (21), say x_1, x_2, and

$$ax^2 + 2(b + \sqrt{r_1})x + c + 2t_1 + s_1 = 0$$

will furnish the complementary pair, x_3, x_4.

So also $ax^2 + 2(b - \sqrt{r_2})x + c + 2t_2 - s_2 = 0 \qquad (33)$

will furnish a pair of roots different from either of the above pairs, say x_1, x_3, and

$$ax^2 + 2(b + \sqrt{r_2})x + c + 2t_2 + s_2 = 0$$

will furnish the complementary pair x_2, x_4.

Finally,

$$ax^2 + 2(b - \sqrt{r_3})x + c + 2t_3 - s_3 = 0$$

will furnish a pair of roots different from any of the preceding pairs. These must therefore be either x_1, x_4, or else x_2, x_3. Then

$$ax^2 + 2(b + \sqrt{r_3})x + c + 2t_3 + s_3 = 0$$

will furnish the complementary pair; that is, either x_2, x_3, or x_1, x_4, as the case may be.

Let $\gamma^2 = 1$, then γ may be so determined that

$$ax^2 + 2(b - \gamma\sqrt{r_3})x + c + 2t_3 - \gamma s_3 = 0 \quad (34)$$

will furnish the pair of roots x_1, x_4, and then

$$ax^2 + 2(b + \gamma\sqrt{r_3})x + c + 2t_3 + \gamma s_3 = 0$$

will furnish the complementary pair x_2, x_3.

By (32), $x_1 + x_2 = -\dfrac{2}{a}(b - \sqrt{r_1})$.

By (33), $x_1 + x_3 = -\dfrac{2}{a}(b - \sqrt{r_2})$.

By (34), $x_1 + x_4 = -\dfrac{2}{a}(b - \gamma\sqrt{r_3})$. $\qquad (35)$

By (21), $x_1 + x_2 + x_3 + x_4 = -\dfrac{4b}{a}$.

$$\therefore ax_1 + b = \sqrt{r_1} + \sqrt{r_2} + \gamma\sqrt{r_3};$$
$$ax_2 + b = \sqrt{r_1} - \sqrt{r_2} - \gamma\sqrt{r_3};$$
$$ax_3 + b = -\sqrt{r_1} + \sqrt{r_2} - \gamma\sqrt{r_3}; \quad (36)$$
$$ax_4 + b = -\sqrt{r_1} - \sqrt{r_2} + \gamma\sqrt{r_3}.$$

Also, from the first three equations of (35),

$$\gamma\sqrt{r_1}\sqrt{r_2}\sqrt{r_3} = [b + \tfrac{1}{2}a(x_1 + x_2)][b + \tfrac{1}{2}a(x_1 + x_3)][b + \tfrac{1}{2}a(x_1 + x_4)]$$
$$= b^3 + \tfrac{1}{2}ab^2(3x_1 + x_2 + x_3 + x_4) + \tfrac{1}{4}a^2b[(x_1 + x_2)(x_1 + x_3)$$
$$+ (x_1 + x_3)(x_1 + x_4) + (x_1 + x_4)(x_1 + x_2)]$$
$$+ \tfrac{1}{8}a^3(x_1 + x_2)(x_1 + x_3)(x_1 + x_4)$$
$$= b^3 + \tfrac{1}{2}ab^2(3x_1 + x_2 + x_3 + x_4)$$
$$+ \tfrac{1}{4}a^2b[3x_1^2 + 2x_1(x_2 + x_3 + x_4) + x_2x_3 + x_2x_4 + x_3x_4]$$
$$+ \tfrac{1}{8}a^3[x_1^3 + x_1^2(x_2 + x_3 + x_4) + x_1(x_2x_3 + x_2x_4 + x_3x_4) + x_2x_3x_4]$$
$$= b^3 + \tfrac{1}{2}ab^2\left(2x_1 - \frac{4b}{a}\right) + \tfrac{1}{4}a^2b\left(2x_1^2 - \frac{4b}{a}x_1 + \frac{6c}{a}\right)$$
$$+ \tfrac{1}{8}a^3\left(-\frac{4b}{a}x_1^2 - \frac{4d}{a}\right) = -b^3 + \tfrac{3}{2}abc - \tfrac{1}{2}a^2d$$
$$= -\tfrac{1}{2}(a^2d - 3abc + 2b^3).$$

That is, $\gamma\sqrt{r_1}\sqrt{r_2}\sqrt{r_3} = -\frac{1}{2}G.$ (37)

[See (5), page 297].

Therefore, $\gamma = +1$ or -1 (38)

according as G is negative or positive.

Hence, by (4), (5), (21), (23), (27), (28), (29), (36), and (38), if $\quad ax^4 + 4bx^3 + 6cx^2 + 4dx + e = 0,$

then shall

$$x_1 = \frac{1}{a}(-b + \sqrt{r_1} + \sqrt{r_2} + \gamma\sqrt{r_3}),$$

$$x_2 = \frac{1}{a}(-b + \sqrt{r_1} - \sqrt{r_2} - \gamma\sqrt{r_3}),$$

$$x_3 = \frac{1}{a}(-b - \sqrt{r_1} + \sqrt{r_2} - \gamma\sqrt{r_3}), \quad\quad (39)$$

$$x_4 = \frac{1}{a}(-b - \sqrt{r_1} - \sqrt{r_2} + \gamma\sqrt{r_3}),$$

in which $\quad r_1 = at_1 - H,$

$$r_2 = at_2 - H, \quad\quad\quad\quad (40)$$

$$r_3 = at_3 - H,$$

t_1, t_2, t_3 being the roots of the equation,

$$4t^3 - It + J = 0, \quad\quad\quad\quad (41)$$

$$H = ac - b^2,$$

$$I = ae - 4bd + 3c^2,$$

$$J = ace + 2bcd - ad^2 - eb^2 - c^3,$$

$$G = a^2d - 3abc + 2b^3,$$

and $\gamma = +1$ or -1 according as G is negative or positive.

§ **55.** The roots given in (39) may also be expressed in terms of any one of the three values of t, as follows:

By (40), $r_1 + r_2 + r_3 = a(t_1 + t_2 + t_3) - 3H.$

But (41), $t_1 + t_2 + t_3 = 0$.

$$\therefore r_1 + r_2 + r_3 = -3H.$$

Now, $\quad \sqrt{r_2} + \gamma\sqrt{r_3} = \sqrt{(r_2 + r_3 + 2\gamma\sqrt{r_2}\sqrt{r_3})}$

$$= \sqrt{\left(-r_1 - 3H - \frac{G}{\sqrt{r_1}}\right)}. \qquad \text{See (37).}$$

In like manner it may be shown that

$$\sqrt{r_2} - \gamma\sqrt{r_3} = \sqrt{\left(-r_1 - 3H + \frac{G}{\sqrt{r_1}}\right)}.$$

Replacing r_1 by $at_1 - H$, see (40), and solving (41), the result becomes

If $\quad ax^4 + 4bx^3 + 6cx^2 + 4dx + e = 0$,

then

$$x_1 = \frac{1}{a}\left\{-b+\sqrt{(at-H)}+\sqrt{\left[-at-2H-\frac{G}{\sqrt{(at-H)}}\right]}\right\},$$

$$x_2 = \frac{1}{a}\left\{-b+\sqrt{(at-H)}-\sqrt{\left[-at-2H-\frac{G}{\sqrt{(at-H)}}\right]}\right\},$$

$$x_3 = \frac{1}{a}\left\{-b-\sqrt{(at-H)}+\sqrt{\left[-at-2H+\frac{G}{\sqrt{(at-H)}}\right]}\right\}, (42)$$

$$x_4 = \frac{1}{a}\left\{-b-\sqrt{(at-H)}-\sqrt{\left[-at-2H+\frac{G}{\sqrt{(at-H)}}\right]}\right\},$$

in which $\quad t = -\frac{1}{2}\sqrt[3]{\left[J+\frac{1}{3\sqrt{3}}\sqrt{(27J^2 - I^3)}\right]}$

$$-\frac{1}{2}\sqrt[3]{\left[J-\frac{1}{3\sqrt{3}}\sqrt{(27J^2 - I^3)}\right]} \qquad (43)$$

$$H = ac - b^2,$$

$$I = ae - 4bd + 3c^2,$$

$$J = ace + 2bcd - ad^2 - eb^2 - c^3,$$

$$G = a^2d - 3abc + 2b^3.$$

Ex. 70.

1. Reduce the quartic $ax^4 + 4bx^3 + 6cx^2 + 4dx + e = 0$ to the form $y^4 + 6Hy^2 + 4Gy + a^2I - 3H^2 = 0$.

2. Show that the two quartics $x^4 + 6Hx^2 \pm 4Gx + K = 0$ have the same reducing cubic.

Solve the quartics:

3. $x^4 - 24x^2 \pm 32x - 132 = 0$.

4. $x^4 - 6x^2 \pm 208x - 321 = 0$.

5. $x^4 - 6x^2 \pm 16x - 33 = 0$.

6. $x^4 - 6x^2 \pm 16x + 39 = 0$.

7. $x^4 - 6x^2 \pm 48x - 117 = 0$.

8. $x^4 + 6x^2 - 60x + 36 = 0$.

Show that, if x_1, x_2, x_3, x_4 be the roots of a quartic,

9. $48H = -\Sigma(x_1 - x_2)^2$.

10. $24a^2I = \Sigma(x_1 - x_2)^2(x_3 - x_4)^2$.

11. $32G = \pm(x_1+x_2-x_3-x_4)(x_1+x_3-x_4-x_2)(x_1+x_4-x_2-x_3)$.

12. $432a^3J = [(x_1 - x_2)(x_3 - x_4) - (x_1 - x_3)(x_4 - x_2)]$
$$\times [(x_1 - x_3)(x_4 - x_2) - (x_1 - x_4)(x_2 - x_3)]$$
$$\times [(x_1 - x_4)(x_2 - x_3) - (x_1 - x_2)(x_3 - x_4)].$$

13. $4H^3 - a^2IH + a^3J + G^2 = 0$.

14. Prove that, if $27J^2 = I^3$, the quartic has a pair of equal roots.

15. Prove that, if $I = J = 0$, the quartic has three equal roots.

16. Prove that, if $a^2I = 12H^2$, and $a^3J = 8H^3$, the quartic has two distinct pairs of equal roots.

Solve the quartic $x^4 + 6 H x^2 + 4 G x + a^2 I - 3 H^2 = 0$:

17. By reducing it to the form
$$(x^2 + 2\sqrt{y}x + z_1)(x^2 - 2\sqrt{y}x + z_2) = 0.$$

18. By reducing it to the form
$$y_2(x^2 + y_1 x + z)^2 - y_1(x^2 + y_2 x)^2 = 0.$$

19. By reducing it to the form
$$(x^2 + y)^2 - (z_1 x + z_2)^2 = 0.$$

20. By assuming the roots to be of the form
$$\alpha\sqrt{r_1} + \beta\sqrt{r_i} + \gamma\sqrt{r_3},$$
in which $\alpha^2 = 1$, $\beta^2 = 1$, $\gamma^2 = 1$.

21. By assuming the roots to be of the form
$$i^n R_1 + i^{2n} R_2 + i^{3n} R_3,$$
in which $i^2 + 1 = 0$, and n is integral.

22. Apply the method of Exam. 20 (Euler's Method) to solve the quartics
$$x^4 - 6 l x^2 \pm 8\sqrt{(l^3 + m^3 + n^3 - 3 lmn)}\, x - 3(4 mn - l^2) = 0.$$

23. Reduce the quartic to the form $y^4 + 6 C y^2 + E = 0$, by assuming $x = \dfrac{z_1 + z_2 y}{1 + y}$, and suitably determining z_1 and z_2.

24. Reduce the quartic to the form
$$y^4 + 4 B y^3 + 6 C y^2 + 4 B y + 1 = 0,$$
by assuming $x = z_1 + z_2 y$, and suitably determining z_1 and z_2.

25. Make the same reduction as in the last question, by assuming $x = z_1 + z_2 y^{-1}$, and suitably determining z_1 and z_2.

26. Eliminate x between $x^4 + 6 H x^2 + 4 G x + a^2 I - 3 H^2 = 0$ and $x^2 + 2 y x + z = 0$, and so determine y that the resulting equation may reduce to the form
$$z^4 + 6 C z^2 + E = 0.$$

If x_1, x_2, x_3, x_4 denote the roots of the quartic

$$x^4 + 6 H x^2 + 4 G x + a^2 I - 3 H^2 = 0,$$

form the cubic whose roots are :

27. $(x_1 + x_2 - x_3 - x_4)^2$, $(x_1 + x_3 - x_4 - x_2)^2$, $(x_1 + x_4 - x_2 - x_3)^2$.

28. $x_1 x_2 + x_3 x_4$, $x_1 x_3 + x_4 x_2$, $x_1 x_4 + x_2 x_3$.

29. $\dfrac{x_1 x_2 - x_3 x_4}{x_1 + x_2 - x_3 - x_4}$, $\dfrac{x_1 x_3 - x_4 x_2}{x_1 + x_3 - x_4 - x_2}$, $\dfrac{x_1 x_4 - x_2 x_3}{x_1 + x_4 - x_2 - x_3}$.

30. $(x_1 x_2 - x_3 x_4)(x_1 + x_2 - x_3 - x_4)$, etc.

31. $(x_1 - x_2)^2 (x_3 - x_4)^2$, $(x_1 - x_3)^2 (x_4 - x_2)^2$, $(x_1 - x_4)^2 (x_2 - x_3)^2$.

32. $(x_1 - x_2)(x_2 - x_3)(x_3 - x_4)(x_4 - x_1)$,
$(x_1 - x_3)(x_3 - x_4)(x_4 - x_2)(x_2 - x_1)$,
$(x_1 - x_4)(x_4 - x_2)(x_2 - x_3)(x_3 - x_1)$.

33. Show how to solve the quartic, knowing the roots of any of the above cubics.

34. Reduce each of the cubics in Exams. 27 to 32 to the standard form $A y^3 + C y + D = 0$.

Form the equation whose roots are :

35. The squares,   **36.** The cubes,

of the roots of $a x^3 + 3 b x^2 + 3 c x + d = 0$.

Form the equation whose roots are :

37. The squares,   **38.** The cubes,

of the roots of $a x^4 + 4 b x^3 + 6 c x^2 + 4 d x + e = 0$.

39. Form the equation whose roots are the squares of the differences of the roots of a cubic.

40. Form the equation whose roots are the squares of the differences of the roots of a quartic.

CHAPTER X.

DETERMINANTS.

I. DEFINITIONS AND NOTATION.

§ 56. The symbol $\begin{vmatrix} a_1 & b_1 \\ a_2 & b_2 \end{vmatrix}$

denotes the expression $a_1 b_2 - a_2 b_1,$

which is called a **Determinant of the Second Order.**

The symbol $\begin{vmatrix} a_1 & b_1 & c_1 \\ a_2 & b_2 & c_2 \\ a_3 & b_3 & c_3 \end{vmatrix}$

denotes the expression

$$a_1 \begin{vmatrix} b_2 & c_2 \\ b_3 & c_3 \end{vmatrix} - a_2 \begin{vmatrix} b_1 & c_1 \\ b_3 & c_3 \end{vmatrix} + a_3 \begin{vmatrix} b_1 & c_1 \\ b_2 & c_2 \end{vmatrix}$$

$$= a_1 b_2 c_3 - a_1 b_3 c_2 - a_2 b_1 c_3 + a_2 b_3 c_1 + a_3 b_1 c_2 - a_3 b_2 c_1,$$

which is called a **Determinant of the Third Order.**

The symbol $\begin{vmatrix} a_1 & b_1 & c_1 & k_1 \\ a_2 & b_2 & c_2 & k_2 \\ a_3 & b_3 & c_3 & k_3 \\ \cdots & \cdots & \cdots & \cdots \\ a_n & b_n & c_n & k_n \end{vmatrix}$

denotes the expression

$$a_1 \begin{vmatrix} b_2 & c_2 & k_2 \\ b_3 & c_3 & \\ \cdots & \cdots & \cdots \\ b_n & c_n & k_n \end{vmatrix} - a_2 \begin{vmatrix} b_1 & c_1 & k_1 \\ b_3 & c_3 & k_3 \\ \cdots & \cdots & \cdots \\ b_n & c_n & k_n \end{vmatrix} + \cdots + (-1)^{n-1} a_n \begin{vmatrix} b_1 & c_1 & k_1 \\ b_2 & c_2 & k_2 \\ \cdots & \cdots & \cdots \\ b_{n-1} & c_{n-1} & k_{n-1} \end{vmatrix}$$

which is called a **Determinant of the nth Order.**

EXAMPLES.

1. $\begin{vmatrix} a+b & a-b \\ a-b & a+b \end{vmatrix} = (a+b)^2 - (a-b)^2 = 4ab.$

2. $\begin{vmatrix} 1 & x & y \\ 1 & x' & y' \\ 1 & x'' & y'' \end{vmatrix} = \begin{vmatrix} x' & y' \\ x'' & y'' \end{vmatrix} - \begin{vmatrix} x & y \\ x'' & y'' \end{vmatrix} + \begin{vmatrix} x & y \\ x' & y' \end{vmatrix}$

$$= x'y'' - x''y' - xy'' + x''y + xy' - x'y.$$

3. $\begin{vmatrix} 1 & 2 & 3 \\ 2 & 3 & 4 \\ 3 & 4 & 5 \end{vmatrix} = 1\begin{vmatrix} 3 & 4 \\ 4 & 5 \end{vmatrix} - 2\begin{vmatrix} 2 & 3 \\ 4 & 5 \end{vmatrix} + 3\begin{vmatrix} 2 & 3 \\ 3 & 4 \end{vmatrix}$

$$= 1(15-16) - 2(10-12) + 3(8-9)$$
$$= -1 + 4 - 3 = 0.$$

4. $\begin{vmatrix} 0 & a & b & c \\ a & 0 & z & y \\ b & z & 0 & x \\ c & y & x & 0 \end{vmatrix} = -a\begin{vmatrix} a & b & c \\ z & 0 & x \\ y & x & 0 \end{vmatrix} + b\begin{vmatrix} a & b & c \\ 0 & z & y \\ y & x & 0 \end{vmatrix} - c\begin{vmatrix} a & b & c \\ 0 & z & y \\ z & 0 & x \end{vmatrix}$

$$
\begin{aligned}
&= + a^2x^2 \quad - abxy \qquad\qquad\qquad\; - cazx \\
& + b^2y^2 \quad - abxy \quad - bcyz \\
& + c^2y^2 \qquad\qquad\qquad - bcyz \quad - cazx \\
\hline
&= a^2x^2 + b^2y^2 + c^2z^2 - 2abxy - 2bcyz - 2cazx
\end{aligned}
$$

Ex. 71.

Expand the following, *i.e.*, write them in ordinary algebraic notation:

1. $\begin{vmatrix} a & y \\ b & x \end{vmatrix}$ **2.** $\begin{vmatrix} a & b \\ y & x \end{vmatrix}$ **3.** $\begin{vmatrix} y & a \\ x & b \end{vmatrix}$ **4.** $\begin{vmatrix} b & a \\ x & y \end{vmatrix}$

5. $\begin{vmatrix} ma & y \\ mb & x \end{vmatrix}$ **6.** $\begin{vmatrix} ma & my \\ b & x \end{vmatrix}$ **7.** $\begin{vmatrix} a & my \\ b & mx \end{vmatrix}$ **8.** $\begin{vmatrix} a & y \\ mb & mx \end{vmatrix}$

9. $\begin{vmatrix} a+my & y \\ b+mx & x \end{vmatrix}$ **10.** $\begin{vmatrix} a+mb & y+mx \\ b & x \end{vmatrix}$ **11.** $\begin{vmatrix} a, & -b \\ a^2, & -b^2 \end{vmatrix}$

12. $\begin{vmatrix} x^2 + a^2 & ab \\ ab & x^2 + b^2 \end{vmatrix}$ 13. $\begin{vmatrix} ab & c^2 \\ -a^2 & -bc \end{vmatrix}$ 14. $\begin{vmatrix} a-b, & -2a \\ -2b, & b-a \end{vmatrix}$

15. $\begin{vmatrix} a_1 & a_2 & a_3 \\ b_1 & b_2 & b_3 \\ c_1 & c_2 & c_3 \end{vmatrix}$ 16. $\begin{vmatrix} a_3 & b_3 & c_3 \\ a_2 & b_2 & c_2 \\ a_1 & b_1 & c_1 \end{vmatrix}$ 17. $\begin{vmatrix} x & y & z \\ z & x & y \\ y & z & x \end{vmatrix}$

18. $\begin{vmatrix} 1 & 1 & 1 \\ x & y & z \\ a & b & c \end{vmatrix}$ 19. $\begin{vmatrix} 1 & 1 & 1 \\ a & b & c \\ a^2 & b^2 & c^2 \end{vmatrix}$ 20. $\begin{vmatrix} a & x & y \\ x & a & z \\ y & z & a \end{vmatrix}$

21. $\begin{vmatrix} a+b & c & c \\ a & b+c & a \\ b & b & c+a \end{vmatrix}$ 22. $\begin{vmatrix} 1 & 1 & 1 \\ 1 & 1+x & 1 \\ 1 & 1 & 1+y \end{vmatrix}$

23. $\begin{vmatrix} ma_1 & mb_1 & mc_1 \\ a_2 & b_2 & c_2 \\ a_3 & b_3 & c_3 \end{vmatrix}$ 24. $\begin{vmatrix} a_1 + ma_3 & b_1 + mb_3 & c_1 + mc_3 \\ a_2 & b_2 & c_2 \\ a_3 & b_3 & c_3 \end{vmatrix}$

25. $\begin{vmatrix} a_1 & b_1 & c_1 & d_1 \\ a_2 & b_2 & c_2 & d_2 \\ a_3 & b_3 & c_3 & d_3 \\ a_4 & b_4 & c_4 & d_4 \end{vmatrix}$ 26. $\begin{vmatrix} a_1 & a_2 & a_3 & a_4 \\ b_1 & b_2 & b_3 & b_4 \\ c_1 & c_2 & c_3 & c_4 \\ d_1 & d_2 & d_3 & d_4 \end{vmatrix}$

27. $\begin{vmatrix} 1 & 1 & 1 & 1 \\ a & b & c & d \\ a^2 & b^2 & c^2 & d^2 \\ a^3 & b^3 & c^3 & d^3 \end{vmatrix}$ 28. $\begin{vmatrix} x & 0 & 0 & a_3 \\ -y & x & 0 & a_2 \\ 0 & -y & x & a_1 \\ 0 & 0 & -y & a_0 \end{vmatrix}$

29. $\begin{vmatrix} 1 & 1 & 1 & 1 \\ 1 & 1+x & 1 & 1 \\ 1 & 1 & 1+y & 1 \\ 1 & 1 & 1 & 1+z \end{vmatrix}$ 30. $\begin{vmatrix} 1+a & 1 & 1 & 1 \\ 1 & 1+b & 1 & 1 \\ 1 & 1 & 1+c & 1 \\ 1 & 1 & 1 & 1+d \end{vmatrix}$

§ 57. The quantities which in the determinant notation stand unconnected, and which are taken as factors to form the terms of the expanded determinant, are called the

Elements of the determinant; *e.g.*, the elements of the determinant

$$\begin{vmatrix} a+b & a & b \\ c & c+d & d \\ e & f & e+f \end{vmatrix}$$

are the nine quantities $a+b$, a, b, c, $c+d$, d, e, f, $e+f$.

The elements standing in any horizontal line constitute a *Row*, and those standing in any vertical line constitute a *Column*. The rows are numbered *first, second, third*, etc., beginning at the top, and the columns are similarly numbered, beginning on the left. The elements of a row are numbered *first, second, third*, etc., beginning on the left, and those of a column are similarly numbered, beginning at the top. Hence the mth element of the nth column, called the (m, n)th element, is the nth element of the mth row; it is often denoted by $a_{m, n}$, the first suffix denoting the row, and the second the column to which the element belongs. Thus, in the above determinant,

$$a_{2, 3} = d, \ \ a_{3, 2} = f, \ \ a_{2, 2} = c + d.$$

A determinant is said to have two diagonals, called, respectively, *principal* and *secondary*. The elements standing in a line from the upper left-hand corner to the lower right-hand corner constitute the principal diagonal; those standing in a line from the upper right-hand corner to the lower left-hand corner constitute the secondary diagonal. Thus, in the determinant given above, the principal diagonal elements are $a+b$, $c+d$, $e+f$, and those of the secondary diagonal are b, $c.+d$, e.

The product of the elements standing in the principal diagonal is called the principal or leading term of the determinant.

Where there is no danger of ambiguity, a determinant is

often denoted by writing only its principal diagonal elements between vertical bars or within parentheses. Thus,

$| a_1 b_2 c_3 |$ denotes $\begin{vmatrix} a_1 & b_1 & c_1 \\ a_2 & b_2 & c_2 \\ a_3 & b_3 & c_3 \end{vmatrix}$

$| a \, x' \, y'' |$ denotes $\begin{vmatrix} a & x & y \\ a' & x' & y' \\ a'' & x'' & y'' \end{vmatrix}$

and $| c_5 g_2 k_0 |$ denotes $\begin{vmatrix} c_5 & g_5 & k_5 \\ c_2 & g_2 & k_2 \\ c_0 & g_0 & k_0 \end{vmatrix}$

If, in any determinant of order n, the pth row and the qth column be erased, and all the rows above the pth and all the columns to the left of the qth be transferred in order over the others, the resulting determinant multiplied by $(-1)^{(n-1)(p+q)}$ is called the complement of the (p, q)th element.

If n be odd, $(-1)^{(n-1)(p+q)} = 1$ for *all* values of $p + q$.

If the elements of any determinant are each denoted by a small letter, and are all different, the complement of any element may be denoted by the corresponding capital letter affected with the suffix or suffixes of the element.

Thus, in $\begin{vmatrix} a_1 & b_1 & c_1 \\ a_2 & b_2 & c_2 \\ a_3 & b_3 & c_3 \end{vmatrix}$

A_1 denotes the complement of a_1, which is $\begin{vmatrix} b_2 & c_2 \\ b_3 & c_3 \end{vmatrix}$

A_2 denotes the complement of a_2, which is $\begin{vmatrix} b_3 & c_3 \\ b_1 & c_1 \end{vmatrix}$

B_3 denotes the complement of b_3, which is $\begin{vmatrix} c_1 & a_1 \\ c_2 & a_2 \end{vmatrix}$

In these $n - 1 = 2$, and $\therefore (-1)^{(n-1)(p+q)} = 1$.

In
$$
\begin{vmatrix}
a_1 & b_1 & c_1 & d_1 \\
a_2 & b_2 & c_2 & d_2 \\
a_3 & b_3 & c_3 & d_3 \\
a_4 & b_4 & c_4 & d_4
\end{vmatrix}
$$

$$
A_2 = -\begin{vmatrix}
b_3 & c_3 & d_3 \\
b_4 & c_4 & d_4 \\
b_1 & c_1 & d_1
\end{vmatrix}
$$
Here $n - 1 = 3$, $p = 1$, $q = 2$, and $\therefore (-1)^{(n-1)(p+q)} = -1$.

and $C_3 = \begin{vmatrix}
d_4 & a_4 & b_4 \\
d_1 & a_1 & b_1 \\
d_2 & a_2 & b_2
\end{vmatrix}$
Here $n - 1 = 3$, $p = 3$, $q = 3$, and $\therefore (-1)^{(n-1)(p+q)} = 1$.

Written in the diagonal notation, the last example would be :

In $|\, a_1\, b_2\, c_3\, d_4\,|$ $A_2 = -\,|\, b_3\, c_4\, d_1\,|$ and $C_3 = |\, d_4\, a_1\, b_2\,|$

To find the complement of a product of two or more elements, find first the complement of one of the elements; then in this complement find the complement of a second element of the product; next, in this second complement find the complement of a third element of the product; and proceed thus through the whole product. The final complement will be the one required.

For example, to find the complement of $a_2 b_3$, first find A_2; then in A_2 find the complement of b_3. Similarly, to find the complement of $a_4 b_1 d_3$, first find A_4; then in A_4 find the complement of b_1; then in this complement find the complement of d_3. The order in which the partial complements are found will affect the form, but not the value of the result. Thus the expansion of the complement of $a_4 b_1 d_3$ will be the same whether found in the order a_4, b_1, d_3, or in the order a_4, d_3, b_1, or again, in the order d_3, b_1, a_4.

The mth element of the nth column is called the conjugate of the nth element of the mth column, and *vice versâ;* i.e., $a_{m,\,n}$ and $a_{n,\,m}$ are each conjugate with respect to the other, or are a pair of conjugates. Each element of the principal diagonal is its own conjugate or is self-conjugate.

A symmetrical determinant is one in which each element is equal to its conjugate.

A skew determinant is one in which the sum of each *pair* of conjugates is zero.

A skew symmetrical determinant is a skew determinant whose principal diagonal elements are all zeros.

Ex. 72.

Write in the "square" notation:

1. $|\, a_1 \; b_3 \; c_5 \,|$ **2.** $|\, x_2 \; y_1 \; z_3 \,|$ **3.** $|\, a_0 \; c_3 \; e_2 \,|$

4. $|\, w_0 \; x_1 \; y_2 \; z_3 \,|$ **5.** $|\, a_k \; b_l \; c_m \; d_n \,|$ **6.** $|\, a_{1,1} \; a_{2,2} \; a_{3,3} \; a_{4,4} \,|$

In $|\, a_1 \; b_2 \; c_3 \; d_4 \; e_5 \,|$ find the complements:

7. A_2.   **8.** B_4.   **9.** B_5.   **10.** D_3.   **11.** C_2.   **12.** E_4.

In the same determinant find the complements of:

13. $a_2 b_3$. **14.** $a_1 c_4$. **15.** $a_4 c_1$. **16.** $a_4 b_2 c_1$. **17.** $b_2 d_3 e_1$. **18.** $a_2 c_3 d_1$.

Prove that:

19. $|\, a_1 \; b_2 \; c_3 \,| = a_1 A_1 + b_1 B_1 + c_1 C_1 = a_1 A_1 + a_2 A_2 + a_3 A_3$.

20. $a_1 A_2 + b_1 B_2 + c_1 C_2 = a_1 A_3 + b_1 B_3 + c_1 C_3$
$$= a_1 B_1 + a_2 B_2 + a_3 B_3 = a_1 C_1 + a_2 C_2 + a_3 C_3 = 0.$$

II. Transformation.

§ 58. Theorem I. *The value of a determinant will not be altered if the columns be written in order as rows, and vice versâ.*

Hence, in any theorem in which the word "row" occurs, the word "column" may be substituted therefor, and *vice versâ;* and, in any theorem in which both " row " and " column " occur, these words may be interchanged without affecting the truth of the theorem.

    THEOREM II. *If any two rows (or two columns) of a determinant be interchanged, the resulting determinant will differ only in sign from the original one.*

      COR. 1. *If a row (or a column) be transferred over* n *other rows (or columns), the determinant will be multiplied by* $(-1)^n$.

      COR. 2. *A transfer of* p *consecutive rows (or columns) over* m $-$ p *other consecutive rows (or columns) multiplies the determinant by* $(-1)^{(m-1)p}$.

      COR. 3. *If* $A_{p,q}$ *denote the complement of the* (p, q)*th element of any determinant* Δ *of order* n, *and* $\Delta_{p,q}$ *denote the determinant formed from* Δ *by striking out the* p*th row and the* q*th column, then will* $A_{p,q} = (-1)^{(p+q)} \Delta_{p,q}$.

EXAMPLES.

1.
$$\begin{vmatrix} a_1 & b_1 & c_1 \\ a_2 & b_2 & c_2 \\ a_3 & b_3 & c_3 \end{vmatrix} = \begin{vmatrix} a_1 & a_2 & a_3 \\ b_1 & b_2 & b_3 \\ c_1 & c_2 & c_3 \end{vmatrix} = -\begin{vmatrix} b_1 & a_1 & c_1 \\ b_2 & a_2 & c_2 \\ b_3 & a_3 & c_3 \end{vmatrix} = -\begin{vmatrix} b_1 & b_2 & b_3 \\ a_1 & a_2 & a_3 \\ c_1 & c_2 & c_3 \end{vmatrix}$$
$$= +\begin{vmatrix} b_2 & b_1 & b_3 \\ a_2 & a_1 & a_3 \\ c_2 & c_1 & c_3 \end{vmatrix}$$

2. Transform $|\, a_1\ b_2\ c_3\,|$ so that $b_3 c_3 a_3$ shall be the first row.
$$\begin{vmatrix} a_1 & b_1 & c_1 \\ a_2 & b_2 & c_2 \\ a_3 & b_3 & c_3 \end{vmatrix} = \begin{vmatrix} a_3 & b_3 & c_3 \\ a_1 & b_1 & c_1 \\ a_2 & b_2 & c_2 \end{vmatrix} = \begin{vmatrix} b_3 & c_3 & a_3 \\ b_1 & c_1 & a_1 \\ b_2 & c_2 & a_2 \end{vmatrix}$$
or $|\, a_1\ b_2\ c_3\,| = |\, a_3\ b_1\ c_2\,| = |\, b_3\ c_1\ a_2\,|$

3. Transform $|\, a_1\ b_2\ c_3\,|$ so that $b_2 a_3 c_1$ shall be the principal diagonal elements in order.
$$|\, a_1\ b_2\ c_3\,| = |\, a_2\ b_3\ c_1\,| = -|\, b_2\ a_3\ c_1\,|$$

4. If $\Delta = |\, a_1\, b_2\, c_3\, d_4\, e_5\, f_6\, |$, find $\Delta_{3,5}$ and $A_{3,5}$.

Here $\quad a_{3,5} = e_3$; $\quad \therefore \Delta_{3,5} = |\, a_1\, b_2\, c_4\, d_5\, f_6\, |$

and $\quad A_{3,5} = (-1)^{(6-1)(3+5)} |\, f_4\, a_5\, b_6\, c_1\, d_2\, |$

Ex. 73.

Transform
$$\begin{vmatrix} a & b & c \\ d & e & f \\ g & h & k \end{vmatrix}$$

into an equivalent determinant having:

1. g, h, k as its first row.
2. b, e, h as its first row.
3. d, f, e as its first row.
4. f, k, c as its first row.
5. c, e, g as its principal diagonal.
6. g, f, b as its secondary diagonal.

Transform
$$\begin{vmatrix} a & b & c & d \\ e & f & g & h \\ k & l & m & n \\ p & q & r & s \end{vmatrix}$$

into an equivalent determinant having:

7. a, d, c, b as its first row, and a, k, e, p as its first column.
8. s, n, h, d as its third row, and m, l, n, k as its second column.
9. e, m, q, d as its principal diagonal.
10. s, m, f, a as its secondary diagonal.

Prove that:

11. $$\begin{vmatrix} a & b & 0 & c \\ d & e & f & 0 \\ 0 & 0 & g & 0 \\ h & k & l & m \end{vmatrix} = \begin{vmatrix} g & 0 & 0 & 0 \\ l & m & h & k \\ 0 & c & a & b \\ f & 0 & d & e \end{vmatrix}$$

12. $\begin{vmatrix} 0 & a & b & 0 \\ c & d & e & f \\ g & h & k & l \\ 0 & m & n & 0 \end{vmatrix} = \begin{vmatrix} d & c & f & e \\ a & 0 & 0 & b \\ m & 0 & 0 & n \\ h & g & l & k \end{vmatrix}$

13. $\begin{vmatrix} d & a & c & b \\ l & 0 & 0 & 0 \\ g & 0 & f & e \\ k & 0 & h & 0 \end{vmatrix} = \begin{vmatrix} a & 0 & 0 & 0 \\ b & e & 0 & 0 \\ c & f & h & 0 \\ d & g & k & l \end{vmatrix}$

Transform $\qquad \begin{vmatrix} a & b & c & d \\ b & c & d & a \\ c & d & a & b \\ d & a & b & c \end{vmatrix}$

so as to have the principal diagonal composed of :

14. The four a's.   15. The four b's.

16. The four c's.   17. The four d's.

Prove that :

18. $|\, a_1 \; b_2 \,| = -\, |\, b_1 \; a_2 \,|$

19. $|\, a_1 \; b_2 \; c_3 \,| = -\, |\, c_1 \; b_2 \; a_3 \,|$

20. $|\, a_1 \; b_2 \; c_3 \; d_4 \,| = |\, d_1 \; c_2 \; b_3 \; a_4 \,|$

21. $|\, a_1 \; b_2 \; c_3 \; d_4 \; e_5 \,| = |\, e_1 \; d_2 \; c_3 \; b_4 \; a_5 \,|$

22. $|\, a_1 \; b_2 \; c_3 \; d_4 \; e_5 \; f_6 \,| = -\, |\, f_1 \; e_2 \; d_3 \; c_4 \; b_5 \; a_6 \,|$

23. If two determinants, Δ and Δ', of the nth degree be such that the first row of the one is the same as the last row of the other, the second row of the one the same as the $(n-1)$th row of the other, the third row of the one the same as the $(n-2)$th row of the other, and so on, then will $\Delta = (-1)^{\frac{1}{2}n(n-1)} \Delta'$.

Transform, by cyclic transposition of the rows and columns, the determinant $|\, a_1 \; b_2 \; c_3 \; d_4 \; e_5 \,|$ into an equal determinant having :

24. c_4 in the first row and first column.

25. d_2 in the first row and first column.

26. c_3 in the second row and fifth column.

27. e_4 in the third row and second column.

Given $\Delta = | a_1 \ b_2 \ c_3 \ d_4 \ e_5 \ f_6 \ g_7 |$, determine :

28. $\Delta_{2,6}$.   **29.** $\Delta_{3,5}$.   **30.** $\Delta_{3,7}$.   **31.** $\Delta_{5,2}$.

32. $A_{2,6}$.   **33.** $A_{3,5}$.   **34.** $A_{3,7}$.   **35.** $A_{5,2}$.

36. Prove that a determinant will not be changed in value by any permutation of the rows and the columns which merely changes the order of the elements of either diagonal, without changing the elements themselves.

§ 59. Theorem III. *If Δ denote any determinant of order* n, $\Delta_{p,q}$ *the determinant formed from Δ by striking out the* p*th row and the* q*th column, and* $a_{p,q}$ *the* (p, q)*th element of Δ, then will*

$$\Delta = (-1)^{q-1}[a_{1,q}\Delta_{1,q} - a_{2,q}\Delta_{2,q} + \cdots\cdots + (-1)^{p-1}a_{p,q}\Delta_{p,q} + \cdots\cdots].$$

Cor. 1. $\Delta = a_{1,q}A_{1,q} + a_{2,q}A_{2,q} + a_{3,q}A_{3,q} + \cdots\cdots + a_{n,q}A_{n,q}$
$$= a_{p,1}A_{p,1} + a_{p,2}A_{p,2} + a_{p,3}A_{p,3} + \cdots\cdots + a_{p,n}A_{p,n}.$$

EXAMPLES.

1. Let $\Delta = | a_1 \ b_2 \ c_3 |$ and $q = 3$.

$$\therefore \Delta = \begin{vmatrix} a_1 & b_1 & c_1 \\ a_2 & b_2 & c_2 \\ a_3 & b_3 & c_3 \end{vmatrix} = (-1)^2 \begin{vmatrix} c_1 & a_1 & b_1 \\ c_2 & a_2 & b_2 \\ c_3 & a_3 & b_3 \end{vmatrix}$$

$$= (-1)^2 \left\{ c_1 \begin{vmatrix} a_2 & b_2 \\ a_3 & b_3 \end{vmatrix} - c_2 \begin{vmatrix} a_1 & b_1 \\ a_3 & b_3 \end{vmatrix} + c_3 \begin{vmatrix} a_1 & b_1 \\ a_2 & b_2 \end{vmatrix} \right\}$$

$$= (-1)^2 (c_1 \Delta_{1,3} - c_2 \Delta_{2,3} + c_3 \Delta_{3,3}).$$

2. Let $\Delta = \mid a_1 \; b_2 \; c_3 \; d_4 \; e_5 \mid$ and $q = 4.$

$$\therefore \Delta = (-1)^3 \mid d_1 \; a_2 \; b_3 \; c_4 \; e_5 \mid$$

$$= (-1)^3 \, (d_1 \mid a_2 \; b_3 \; c_4 \; e_5 \mid - \, d_2 \mid a_1 \; b_3 \; c_4 \; e_5 \mid$$

$$+ \, d_3 \mid a_1 \; b_2 \; c_4 \; e_5 \mid - \, d_4 \mid a_1 \; b_2 \; c_3 \; e_5 \mid$$

$$+ \, d_5 \mid a_1 \; b_2 \; c_3 \; e_4 \mid)$$

$$= (-1)^3 (d_1 \Delta_{1,4} - d_2 \Delta_{2,4} + d_3 \Delta_{3,4} - d_4 \Delta_{4,4} + d_5 \Delta_{5,4}).$$

COR. 2. *If the elements of the* p*th row all vanish except the* q*th, then shall* $\Delta = (-1)^{p+q} a_{p, q} \Delta_{p, q}.$

EXAMPLE.

$$\begin{vmatrix} a & b & 0 & c \\ d & e & f & g \\ h & k & 0 & l \\ m & n & 0 & p \end{vmatrix} = (-1)^{1+2} \begin{vmatrix} f & d & e & g \\ 0 & a & b & c \\ 0 & h & k & l \\ 0 & m & n & p \end{vmatrix} = (-1)^{2+3} f \begin{vmatrix} a & b & c \\ h & k & l \\ m & n & p \end{vmatrix}$$

COR. 3. *If the elements on one side of the principal diagonal of a determinant be all zero, the determinant will be equal to the product of the diagonal elements.*

If the elements on one side of the secondary diagonal be all zero, the determinant will be equal to the product of the secondary diagonal elements multiplied by $(-1)^{\frac{1}{2}n(n-1)}$, n *being the order of the determinant.*

EXAMPLES.

1. $\begin{vmatrix} a & b & c \\ 0 & d & e \\ 0 & 0 & f \end{vmatrix} = a \begin{vmatrix} d & e \\ 0 & f \end{vmatrix} = adf.$

2. $\begin{vmatrix} 0 & 0 & 0 & a \\ 0 & 0 & b & c \\ 0 & d & e & f \\ g & h & k & l \end{vmatrix} = -a \begin{vmatrix} 0 & 0 & b \\ 0 & d & e \\ g & h & k \end{vmatrix} = -ab \begin{vmatrix} 0 & d \\ g & h \end{vmatrix} = abdg.$

COR. 4. *The order of a determinant may be raised without altering its value by prefixing a column of zeros, and superposing a row of elements, the first of which must be unity, but the others may be any finite quantities whatever.*

EXAMPLE.

$$\begin{vmatrix} a_1 - x & b_1 - y \\ a_2 - x & b_2 - y \end{vmatrix} = \begin{vmatrix} 1 & x & y \\ 0 & a_1 - x & b_1 - y \\ 0 & a_2 - x & b_2 - y \end{vmatrix}$$

THEOREM IV. *If each element of a row of a determinant consist of two terms, the determinant may be resolved into the sum of two determinants, the first of which is got from the original determinant by striking out one term of each of the elements in question, and the second, by restoring these and striking out the others.*

Conversely : *The sum of any number of determinants which are alike, except as regards the* m*th row in each, is equal to a determinant which is like the given determinants except that each element of its* m*th row is equal to the sum of the corresponding elements of all the given determinants.*

EXAMPLES.

1. $\begin{vmatrix} a+x & d & g \\ b-y & e & h \\ c+z & f & k \end{vmatrix} = \begin{vmatrix} a & d & g \\ b & e & h \\ c & f & k \end{vmatrix} + \begin{vmatrix} x & d & g \\ -y & e & h \\ z & f & k \end{vmatrix}$

2. $\begin{vmatrix} a & b & c \\ d & e & f \\ g+m & h & k-n \end{vmatrix} = \begin{vmatrix} a & b & c \\ d & e & f \\ g & h & k \end{vmatrix} + \begin{vmatrix} a & b & c \\ d & e & f \\ m & 0 & -n \end{vmatrix}$

3. $|\, a_1 \, b_2 \, c_3 \, d_4 \,| + |\, a_1 \, b_2 \, e_3 \, d_4 \,| + |\, a_1 \, b_2 \, g_3 \, d_4 \,|$

$\qquad = |\, a_1 \, b_2 \, (c_3 + e_3 + g_3) \, d_4 \,|$

THEOREM V. *If each element of any row of a determinant be multiplied (or divided) by the same factor, the determinant will be multiplied (or divided) by the said factor.*

COR. 1. *If all the elements of any row be divisible by a common factor, such common factor may be struck out of these elements and written as a coefficient outside the bars of the resulting determinant.*

COR. 2. *If the sign of every element of a row be changed, the sign of the determinant will be changed.*

EXAMPLES.

1.
$$\begin{vmatrix} 3a & b & c \\ 3d & e & f \\ 3g & h & k \end{vmatrix} = 3 \begin{vmatrix} a & b & c \\ d & e & f \\ g & h & k \end{vmatrix}$$

2.
$$\begin{vmatrix} 2 & 4 & 8 \\ 3 & 6 & 9 \\ 5 & 15 & 20 \end{vmatrix} = 2 \cdot 3 \cdot 5 \begin{vmatrix} 1 & 2 & 4 \\ 1 & 2 & 3 \\ 1 & 3 & 4 \end{vmatrix} = 30 \times 1 = 30.$$

Here the common factor 2 is struck out of the first row of elements, 3 out of the second row, and 5 out of the third row, and their product is written as coefficient of the resulting determinant.

3.
$$\begin{vmatrix} bc & 1 & a \\ ca & 1 & b \\ ab & 1 & c \end{vmatrix} = \frac{1}{abc} \begin{vmatrix} abc & a & a^2 \\ abc & b & b^2 \\ abc & c & c^2 \end{vmatrix} = \begin{vmatrix} 1 & a & a^2 \\ 1 & b & b^2 \\ 1 & c & c^2 \end{vmatrix}$$

4.
$$\begin{vmatrix} 0 & a & b \\ -a & 0 & c \\ -b & -c & 0 \end{vmatrix} = - \begin{vmatrix} 0 & a & b \\ a & 0 & \\ b & -c & 0 \end{vmatrix} = (-1)^2 \begin{vmatrix} 0 & -a & b \\ a & 0 & c \\ b & c & 0 \end{vmatrix}$$

$$= (-1)^3 \begin{vmatrix} 0 & -a & -b \\ a & 0 & -c \\ b & c & 0 \end{vmatrix}$$

But this is the original determinant, say Δ, with its columns written as rows ;

$$\therefore \Delta = (-1)^3\Delta ; \ \therefore \Delta = 0.$$

THEOREM VI. *Any determinant can always be transformed into a determinant of the same order in which the non-zero elements of any one row or one column are all unity.*

EXAMPLES.

1. Let
$$\Delta = \begin{vmatrix} a_1 & b_1 & c_1 \\ a_2 & b_2 & c_2 \\ a_3 & b_3 & c_2 \end{vmatrix}$$

Multiply each element of the first column by b_1c_1, each element of the second column by c_1a_1, and each element of the third column by a_1b_1 ;

$$\therefore a_1^2b_1^2c_1^2\Delta = \begin{vmatrix} a_1b_1c_1 & b_1c_1a_1 & c_1a_1b_1 \\ a_2b_1c_1 & b_2c_1a_1 & c_2a_1b_1 \\ a_3b_1c_1 & b_3c_1a_1 & c_3a_1b_1 \end{vmatrix}$$

$$= a_1b_1c_1 \begin{vmatrix} 1 & 1 & 1 \\ a_2b_1c_1 & b_2c_1a_1 & c_2a_1b_1 \\ a_3b_1c_1 & b_3c_1a_1 & c_3a_1b_1 \end{vmatrix} = a_1b_1c_1\Delta', \text{ say} ;$$

$$\therefore \Delta = \frac{\Delta'}{a_1b_1c_1}.$$

2. Reduce
$$\begin{vmatrix} 3 & -2 & 7 & 4 \\ -5 & 4 & 3 & 7 \\ 6 & 3 & 5 & -2 \\ 4 & 6 & -3 & 5 \end{vmatrix}$$

to an equivalent determinant having the elements of its second column all unity.

The least common multiple of the elements of the second column is 12, and the quotients of 12 by these elements are $-6, 3, 4$, and 2, respectively. Multiply the first *row* by

—6, the second by 3, the third by 4, and the fourth by 2, and divide the determinant by $-6 \times 3 \times 4 \times 2 = -144$. The result is

$$\frac{1}{144}\begin{vmatrix} -18 & 12 & -42 & -24 \\ -15 & 12 & 9 & 21 \\ 24 & 12 & 20 & -8 \\ 8 & 12 & -6 & 10 \end{vmatrix} = -\frac{1}{12}\begin{vmatrix} -18 & 1 & -42 & -24 \\ -15 & 1 & 9 & 21 \\ 24 & 1 & 20 & -8 \\ 8 & 1 & -6 & 10 \end{vmatrix}$$

Ex. 74.

Expand the following determinants:

1. $\begin{vmatrix} a & b & c \\ d & 0 & e \\ g & 0 & h \end{vmatrix}$
  2. $\begin{vmatrix} a & b & c \\ d & e & 0 \\ g & h & 0 \end{vmatrix}$
  3. $\begin{vmatrix} x_1 & y_1 & z_1 \\ 0 & 0 & z_2 \\ x_3 & y_3 & z_3 \end{vmatrix}$

4. $\begin{vmatrix} 1 & 0 & a^2 \\ 1 & 0 & b^2 \\ 1 & c & c^2 \end{vmatrix}$
  5. $\begin{vmatrix} 5 & 0 & 4 \\ 3 & 7 & 3 \\ 4 & 0 & 5 \end{vmatrix}$
  6. $\begin{vmatrix} 5 & 0 & 4 \\ 0 & -2 & 0 \\ 4 & 0 & 3 \end{vmatrix}$

7. $\begin{vmatrix} a_1 & a_2 & a_3 & a_4 \\ 0 & b_2 & 0 & b_4 \\ 0 & c_2 & c_3 & c_4 \\ 0 & e_2 & 0 & 0 \end{vmatrix}$
  8. $\begin{vmatrix} x & y & 0 & 0 \\ 0 & x & y & 0 \\ 0 & 0 & x & y \\ y & 0 & 0 & x \end{vmatrix}$

9. $\begin{vmatrix} a & 0 & e & 0 & x \\ b & 0 & f & x & k \\ c & 0 & x & 0 & 0 \\ d & x & g & h & l \\ x & 0 & 0 & 0 & 0 \end{vmatrix}$
  10. $\begin{vmatrix} a_1 & b_1 & c_1 & d_1 & e_1 \\ 0 & b_2 & c_2 & d_2 & e_2 \\ 0 & b_3 & 0 & d_3 & 0 \\ 0 & b_4 & 0 & d_4 & 0 \\ 0 & b_5 & 0 & d_5 & e_5 \end{vmatrix}$

Show that:

11. $\begin{vmatrix} a_1 & b_1 & 0 & 0 \\ a_2 & b_2 & 0 & 0 \\ a_3 & b_3 & x_1 & y_1 \\ a_4 & b_4 & x_2 & y_2 \end{vmatrix} = \begin{vmatrix} a_1 & b_1 \\ a_2 & b_2 \end{vmatrix} \times \begin{vmatrix} x_1 & y_1 \\ x_2 & y_2 \end{vmatrix}$

12. $\begin{vmatrix} x & y^2 & 0 & 0 \\ -1 & x & y^2 & 0 \\ 0 & -1 & x & y^2 \\ 0 & 0 & -1 & x \end{vmatrix} = \begin{vmatrix} x & y & 0 & 0 \\ -y & x & y & 0 \\ 0 & -y & x & y \\ 0 & 0 & -y & x \end{vmatrix}$

Resolve the following determinants into determinants with monomial elements:

13.
$$\begin{vmatrix} a_1 & b_1+1 & c_1 \\ a_2 & b_2+x & c_2 \\ a_3 & b_3+x^2 & c_3 \end{vmatrix}$$

14.
$$\begin{vmatrix} a_1 & b_1 & c_1+x^3 \\ a_2 & b_2 & c_2+x^2 \\ a_3 & b_3 & c_3+x \end{vmatrix}$$

15.
$$\begin{vmatrix} a_1+x+y & b_1 & c_1 \\ a_2-x+y & b_2 & c_2 \\ a_3+x-y & b_3 & c_3 \end{vmatrix}$$

16.
$$\begin{vmatrix} a_1+x & b_1+y & c_1 \\ a_2-x & b_2+y & c_2 \\ a_3+x & b_3-y & c_3 \end{vmatrix}$$

17.
$$\begin{vmatrix} x_1-a & y_1 & a^2-z_1 \\ x_2-b & y_2 & b^2-z_2 \\ x_3-c & y_3 & c^2-z_3 \end{vmatrix}$$

18.
$$\begin{vmatrix} x+a & d & g \\ b & y+e & h \\ c & f & z+k \end{vmatrix}$$

Combine into a single determinant:

19.
$$\begin{vmatrix} x_1 & y_1 & z_1 \\ x_2 & 0 & z_2 \\ x_3 & y_3 & z_3 \end{vmatrix} + \begin{vmatrix} x_3 & y_3 & z_3 \\ u_2 & u_2 & 0 \\ x_1 & y_1 & z_1 \end{vmatrix}$$

20.
$$\begin{vmatrix} a_1 & b_1 & c_1 \\ a_2 & b_2 & c_2 \\ a_3 & b_3 & c_3 \end{vmatrix} + \begin{vmatrix} a_1 & a_3 & a_4 \\ b_1 & b_3 & b_4 \\ c_1 & c_3 & c_4 \end{vmatrix}$$

21.
$$\begin{vmatrix} x_1-a_1 & y_1 & z_1 \\ x_2-a_2 & y_2 & z_2 \\ x_3-a_3 & y_3 & z_3 \end{vmatrix} + \begin{vmatrix} y_1 & z_1 & a_1-u_1 \\ y_2 & z_2 & a_2-u_2 \\ y_3 & z_3 & a_3-u_3 \end{vmatrix}$$

22.
$$\begin{vmatrix} m+n+p & m+n-p \\ x-y-z & x+y-z \end{vmatrix} + \begin{vmatrix} x+y+z & x-y+z \\ m-n-p & m-n+p \end{vmatrix}$$

23.
$$\begin{vmatrix} 3 & y_1 & z_1 \\ 4 & y_2 & z_2 \\ 5 & y_3 & z_3 \end{vmatrix} + \begin{vmatrix} 7 & y_1 & z_1 \\ 6 & y_2 & z_2 \\ 5 & y_3 & z_3 \end{vmatrix} - \begin{vmatrix} 10 & y_1 & z_1 \\ 10x_1 & x_2 & x_3 \\ 10 & y_3 & z_3 \end{vmatrix}$$
$$- \begin{vmatrix} -10 & y_1 & z_1 \\ -10x_1 & y_2-x_2 & z_2-x_3 \\ 5 & y_3 & z_3 \end{vmatrix}$$

Show that:

24. $\begin{vmatrix} a+u & b+v \\ c+x & d+y \end{vmatrix} + \begin{vmatrix} a-u & b-v \\ c-x & d-y \end{vmatrix} = 2\begin{vmatrix} a & b \\ c & d \end{vmatrix} + 2\begin{vmatrix} u & v \\ x & y \end{vmatrix}$

25. $\begin{vmatrix} a_1 & b_1 & c_1 \\ a_2 & b_2 & c_2 \\ a_3 & b_3 & c_3 \end{vmatrix} = \begin{vmatrix} 0 & b_1 & c_1 \\ a_2 & 0 & c_2 \\ a_3 & b_3 & 0 \end{vmatrix} + a_1\begin{vmatrix} b_2 & c_2 \\ b_3 & c_3 \end{vmatrix} + b_2\begin{vmatrix} 0 & c_1 \\ a_3 & c_3 \end{vmatrix} + c_3\begin{vmatrix} 0 & b_1 \\ a_2 & 0 \end{vmatrix}$

26. $\begin{vmatrix} a_1 & b_1 & c_1 & d_1 \\ a_2 & b_2 & c_2 & d_2 \\ a_3 & b_3 & c_3 & d_3 \\ a_4 & b_4 & c_4 & d_4 \end{vmatrix} = \begin{vmatrix} 0 & b_1 & c_1 & d_1 \\ a_2 & 0 & c_2 & d_2 \\ a_3 & b_3 & 0 & d_3 \\ a_4 & b_4 & c_4 & 0 \end{vmatrix} + a_1\begin{vmatrix} b_2 & c_2 & d_2 \\ b_3 & c_3 & d_3 \\ b_4 & c_4 & d_4 \end{vmatrix}$

$+ b_2\begin{vmatrix} 0 & c_1 & d_1 \\ a_3 & c_3 & d_3 \\ a_4 & c_4 & d_4 \end{vmatrix} + c_3\begin{vmatrix} 0 & b_1 & d_1 \\ a_2 & 0 & d_2 \\ a_4 & b_4 & d_4 \end{vmatrix} + d_4\begin{vmatrix} 0 & b_1 & c_1 \\ a_2 & 0 & c_2 \\ a_3 & b_3 & 0 \end{vmatrix}$

III. EVALUATION.

§ 60. THEOREM VII. *If two rows of a determinant be identical, the determinant will be equal to zero.*

COR. *If the corresponding elements of two rows of a determinant have a constant ratio, the determinant will be equal to zero.*

EXAMPLES.

1. $\begin{vmatrix} mx_1 & y_1 & nx_1 \\ mx_2 & y_2 & nx_2 \\ mx_3 & y_3 & nx_3 \end{vmatrix} = mn\begin{vmatrix} x_1 & y_1 & x_1 \\ x_2 & y_2 & x_2 \\ x_3 & y_3 & x_3 \end{vmatrix} = 0.$

2. $\begin{vmatrix} ma_1+nb_1 & a_1 & b_1 \\ ma_2+nb_2 & a_2 & b_2 \\ ma_3+nb_3 & a_3 & b_3 \end{vmatrix} = \begin{vmatrix} ma_1 & a_1 & b_1 \\ ma_2 & a_2 & b_2 \\ ma_3 & a_3 & b_3 \end{vmatrix} + \begin{vmatrix} nb_1 & a_1 & b_1 \\ nb_2 & a_2 & b_2 \\ nb_3 & a_3 & b_3 \end{vmatrix} = 0.$

THEOREM VIII. *The value of a determinant will not be altered if to the elements of any row there be added equimultiples of the corresponding elements of any other row.*

EXAMPLES.

1. $\begin{vmatrix} a_1 & b_1 & c_1 \\ a_2 & b_2 & c_2 \\ a_3 & b_3 & c_3 \end{vmatrix} = \begin{vmatrix} a_1 & b_1 & c_1 \\ a_2 & b_2 & c_2 \\ a_3 & b_3 & c_3 \end{vmatrix} + \begin{vmatrix} mb_1 & b_1 & c_1 \\ mb_2 & b_2 & c_2 \\ mb_3 & b_3 & c_3 \end{vmatrix} + \begin{vmatrix} nc_1 & b_1 & c_1 \\ nc_2 & b_2 & c_2 \\ nc_3 & b_3 & c_3 \end{vmatrix}$

$= \begin{vmatrix} a_1 + mb_1 + nc_1 & b_1 & c_1 \\ a_2 + mb_2 + nc_2 & b_2 & c_2 \\ a_3 + mb_3 + nc_3 & b_3 & c_3 \end{vmatrix}$

2. Evaluate
$$\begin{vmatrix} 1 & 2 & 3 & 4 \\ 8 & 7 & 6 & 5 \\ 1 & 3 & 6 & 10 \\ 36 & 28 & 21 & 15 \end{vmatrix}$$

From the elements of the fourth column subtract the corresponding elements of the third column, and write the remainders as the corresponding elements of a new fourth column. Do the same with the second column instead of the third, and the third instead of the fourth, and then with the first and second columns instead of the third and fourth. The resulting determinant will be :

$$\begin{vmatrix} 1 & 1 & 1 & 1 \\ 8 & -1 & -1 & -1 \\ 1 & 2 & 3 & 4 \\ 36 & -8 & -7 & -6 \end{vmatrix}$$

In this determinant, add the elements of the first row to the corresponding elements of the second, and also ten times the elements of the first row to the corresponding elements of the fourth row; the result will be :

$$\begin{vmatrix} 1 & 1 & 1 & 1 \\ 9 & 0 & 0 & 0 \\ 1 & 2 & 3 & 4 \\ 46 & 2 & 3 & 4 \end{vmatrix} = -9 \begin{vmatrix} 1 & 1 & 1 \\ 2 & 3 & 4 \\ 2 & 3 & 4 \end{vmatrix} = 0$$

by Theor. III., Cor. 2, and Theor. VII. Hence the given determinant is equal to zero.

3. Evaluate
$$\begin{vmatrix} 1 & 2 & 4 & 8 \\ 2 & 6 & 7 & 10 \\ 5 & 9 & 3 & 1 \\ 7 & 3 & 4 & 0 \end{vmatrix}$$

Take twice the third column from the fourth for a new fourth, twice the second from the third for a new third, and twice the first from the second for a new second; the result is:

$$\begin{vmatrix} 1 & 0 & 0 & 0 \\ 2 & 2 & -5 & -4 \\ 5 & -1 & -15 & -5 \\ 7 & -11 & -2 & 1 \end{vmatrix} = \begin{vmatrix} 2 & -5 & -4 \\ -1 & -15 & -5 \\ -11 & -2 & 1 \end{vmatrix}$$

To the second row add the third, and from the result subtract the first row, and write the remainders as a new second row. To the first row add four times the third, and write the sums as a new first row. The result is:

$$\begin{vmatrix} -42 & -13 & 0 \\ -14 & -12 & 0 \\ -11 & -2 & 1 \end{vmatrix} = \begin{vmatrix} -42 & -13 \\ -14 & -12 \end{vmatrix} = (-)^2 14 \begin{vmatrix} 3 & 13 \\ 1 & 12 \end{vmatrix} = 14 \begin{vmatrix} 2 & 1 \\ 1 & 12 \end{vmatrix}$$

$$= 14(24 - 1) = 322.$$

4.
$$\begin{vmatrix} 1 & 1 & 1 \\ a & b & c \\ a^2 & b^2 & c^2 \end{vmatrix} = \begin{vmatrix} 1 & 0 & 0 \\ a & b-a & c-a \\ a^2 & b^2-a^2 & c^2-a^2 \end{vmatrix}$$

$$= (b-a)(c-a) \begin{vmatrix} 1 & 0 & 0 \\ a & 1 & 1 \\ a^2 & b+a & c+a \end{vmatrix}$$

$$= (b-a)(c-a)[(c+a)-(b+a)]$$

$$= (a-b)(b-c)(c-a).$$

This determinant may also be evaluated thus:

The determinant vanishes for $a = b$; therefore $a - b$ is a factor of it. By symmetry, $b - c$ and $c - a$ are also factors. Now the determinant is of the third degree; there are, therefore, no other literal factors than these three; the determinant therefore $= m(a-b)(b-c)(c-a)$, wherein m is numerical. To determine m : The principal diagonal is bc^2, and the factors give mbc^2, hence $m = +1$, and therefore the determinant is equal to $(a-b)(b-c)(c-a)$, as was otherwise already proved.

5.
$$\Delta = \begin{vmatrix} 0 & x & y & z \\ x & 0 & z & y \\ y & z & 0 & x \\ z & y & x & 0 \end{vmatrix} = \begin{vmatrix} x+y+z & x & y & z \\ x+y+z & 0 & z & y \\ x+y+z & z & 0 & x \\ x+y+z & y & x & 0 \end{vmatrix}$$

which shows that $x+y+z$ is a factor.

Multiply the first and fourth columns, and the second and third rows, each by -1, which is equivalent to multiplying the determinant by $(-1)^4 = 1$;

$$\begin{vmatrix} 0 & x & y & -z \\ x & 0 & -z & y \\ y & -z & 0 & x \\ -z & y & x & 0 \end{vmatrix}$$

and taking the sum of the columns, as before, for a new first column, $x+y-z$ is seen to be a factor. Similarly, $x-y+z$ and $-x+y+z$ may be shown to be factors. The determinant is of the fourth degree, and four linear factors have been found;

$$\therefore \Delta = m(x+y+z)(y+z-x)(z+x-y)(x+y-z).$$

The secondary diagonal is $+z^4$, and the factors give $-mz^4$, $m = -1$.

$$\therefore \Delta = -(x+y+z)(y+z-x)(z+x-y)(x+y-z).$$

6.
$$\begin{vmatrix} a^2+1 & ab & ac & ad \\ ab & b^2+1 & bc & bd \\ ac & bc & c^2+1 & cd \\ ad & bd & cd & d^2+1 \end{vmatrix} = \Delta, \text{ say.}$$

Multiply the first column by a, and then strike out of the first row the common factor a; this will not change the value of the determinant, which denote by Δ.

$$\therefore \Delta = \begin{vmatrix} a^2+1 & b & c & d \\ a^2b & b^2+1 & bc & bd \\ a^2c & bc & c^2+1 & cd \\ a^2d & bd & cd & d^2+1 \end{vmatrix}$$

Similarly, operate with b on the second column and second row, with c on the third column and third row, and with d on the fourth column and fourth row; then

$$\Delta = \begin{vmatrix} a^2+1 & b^2 & c^2 & d^2 \\ a^2 & b^2+1 & c^2 & d^2 \\ a^2 & b^2 & c^2+1 & d^2 \\ a^2 & b^2 & c^2 & d^2+1 \end{vmatrix}$$

Take the sum of the columns for a new first column, and write the common factor outside the bars.

$$\therefore \Delta = (a^2+b^2+c^2+d^2+1) \begin{vmatrix} 1 & b^2 & c^2 & d^2 \\ 1 & b^2+1 & c^2 & d^2 \\ 1 & b^2 & c^2+1 & d^2 \\ 1 & b^2 & c^2 & d^2+1 \end{vmatrix}$$

Subtract the first row from each of the others;

$$\Delta = (a^2+b^2+c^2+d^2+1) \begin{vmatrix} 1 & b^2 & c^2 & d^2 \\ 0 & 1 & 0 & 0 \\ 0 & 0 & 1 & 0 \\ 0 & 0 & 0 & 1 \end{vmatrix}$$

$$= a^2+b^2+c^2+d^2+1. \quad \text{(See Theor. III., Cor 3.)}$$

7.
$$\Delta = \begin{vmatrix} x & y & z \\ z & x & y \\ y & z & x \end{vmatrix} = \begin{vmatrix} x+y+z & y & z \\ x+y+z & x & y \\ x+y+z & z & x \end{vmatrix}$$

$\therefore x+y+z$ is a factor of Δ.

Let $\omega^2+\omega+1=0$, and $\therefore \omega^3=1$. Multiply the second column by ω and the third by ω^2, the second row by ω^2 and the third by ω, which is equivalent to multiplying Δ by ω^6, which $= 1$.

$$\therefore \Delta = \begin{vmatrix} x & \omega y & \omega^2 z \\ \omega^2 z & x & \omega y \\ \omega y & \omega^2 z & x \end{vmatrix} = \begin{vmatrix} x + \omega y + \omega^2 z & \omega y & \omega^2 z \\ x + \omega y + \omega^2 z & x & \omega y \\ x + \omega y + \omega^2 z & \omega^2 z & x \end{vmatrix}$$

$\therefore x + \omega y + \omega^2 z$ is a factor of Δ.

Operate with ω^2 instead of ω, and therefore with ω instead of ω^2, and $x + \omega^2 y + \omega z$ will be seen to be a factor of Δ.

$$\therefore \Delta = m (x + y + z)(x + \omega y + \omega^2 z)(x + \omega^2 y + \omega z)$$

in which m is numerical. The principal diagonal of Δ is x^3, and the factors give mx^3,

$$\therefore m = +1.$$

$$\therefore \Delta = (x + y + z)(x + \omega y + \omega^2 z)(x + \omega^2 y + \omega z)$$
$$= x^3 + y^3 + z^3 - 3xyz.$$

Ex. 75.

Evaluate:

1.
$$\begin{vmatrix} 1 & 3 & -5 & -8 \\ 4 & 7 & 2 & -6 \\ 3 & 10 & 12 & 6 \\ -9 & 1 & 13 & 19 \end{vmatrix}$$

2.
$$\begin{vmatrix} -1 & -1 & 1 & 1 \\ 1 & -1 & 1 & 1 \\ 1 & 1 & -1 & 1 \\ 1 & 1 & 1 & -1 \end{vmatrix}$$

3.
$$\begin{vmatrix} -5 & 9 & 5 & -5 \\ 9 & -15 & 19 & 23 \\ 5 & 19 & -10 & -15 \\ -5 & -23 & 15 & -25 \end{vmatrix}$$

4.
$$\begin{vmatrix} 1 & 14 & 15 & 4 \\ 8 & 11 & 10 & 5 \\ 12 & 7 & 6 & 9 \\ 13 & 2 & 3 & 16 \end{vmatrix}$$

5.
$$\begin{vmatrix} 17 & 24 & 1 & 8 & 15 \\ 23 & 5 & 7 & 14 & 16 \\ 4 & 6 & 13 & 20 & 22 \\ 10 & 12 & 19 & 21 & 3 \\ 11 & 18 & 25 & 2 & 9 \end{vmatrix}$$

6.
$$\begin{vmatrix} 1 & 1 & 1 & 1 \\ a & b & c & d \\ a^2 & b^2 & c^2 & d^2 \\ a^3 & b^3 & c^3 & d^3 \end{vmatrix}$$

7. $\begin{vmatrix} 1 & 1 & 1 & 1 \\ a & b & c & d \\ a^2 & b^2 & c^2 & d^2 \\ a^4 & b^4 & c^4 & d^4 \end{vmatrix}$

8. $\begin{vmatrix} 1 & 1 & 1 & 1 \\ a & b & c & d \\ a^3 & b^3 & c^3 & d^3 \\ a^4 & b^4 & c^4 & d^4 \end{vmatrix}$

9. $\begin{vmatrix} 1 & 1 & 1 & 1 \\ a^2 & b^2 & c^2 & d^2 \\ a^3 & b^3 & c^3 & d^3 \\ a^4 & b^4 & c^4 & d^4 \end{vmatrix}$

10. $\begin{vmatrix} 0 & 1 & 1 & 1 \\ 1 & 0 & a^2 & b^2 \\ 1 & a^2 & 0 & c^2 \\ 1 & b^2 & c^2 & 0 \end{vmatrix}$

11. $\begin{vmatrix} (a+b)^2 & c^2 & c^2 \\ a^2 & (b+c)^2 & a^2 \\ b^2 & b^2 & (c+a)^2 \end{vmatrix}$

12. $\begin{vmatrix} by+cz & bx & cx \\ ay & cz+ax & cy \\ az & bz & ax+by \end{vmatrix}$

13. $\begin{vmatrix} a & b & c & 0 \\ 0 & a & b & c \\ a' & b' & c' & 0 \\ 0 & a' & b' & c' \end{vmatrix}$

14. $\begin{vmatrix} x & a & b & c \\ c & x & a & b \\ b & c & x & a \\ a & b & c & x \end{vmatrix}$

15. $\begin{vmatrix} x & a & b & c & d \\ d & x & a & b & c \\ c & d & x & a & b \\ b & c & d & x & a \\ a & b & c & d & x \end{vmatrix}$

16. $\begin{vmatrix} x & y & y & y & y \\ y & x & y & y & y \\ y & y & x & y & y \\ y & y & y & x & y \\ y & y & y & y & x \end{vmatrix}$

17. $\begin{vmatrix} x & x & y & x \\ y & y & y & x \\ y & x & x & x \\ y & x & y & y \end{vmatrix}$

18. $\begin{vmatrix} (b+c)^2 & c^2 & b^2 \\ c^2 & (c+a)^2 & a^2 \\ b^2 & a^2 & (a+b)^2 \end{vmatrix}$

19. $\begin{vmatrix} a+b+c+d & a-b-c+d & a-b+c-d \\ a-b-c+d & a+b+c+d & a+b-c-d \\ a-b+c-d & a+b-c-d & a+b+c+d \end{vmatrix}$

20. $\begin{vmatrix} (b+c+d)^2 & b^2 & c^2 & d^2 \\ a^2 & (c+d+a)^2 & c^2 & d^2 \\ a^2 & b^2 & (d+a+b)^2 & d^2 \\ a^2 & b^2 & c^2 & (a+b+c)^2 \end{vmatrix}$

§ **61.** Any determinant of the third order may readily be evaluated by the following method, called **The Method of Sarrus.** Let the determinant be

$$\begin{vmatrix} a_1 & b_1 & c_1 \\ a_2 & b_2 & c_2 \\ a_3 & b_3 & c_3 \end{vmatrix}$$

Repeat in order the first and second rows below the determinant (or the first and second columns to the left of it); thus,

$$\begin{array}{ccc} a_1 & b_1 & c_1 \\ a_2 & b_2 & c_2 \\ a_3 & b_3 & c_3 \\ a_1 & b_1 & c_1 \\ a_2 & b_2 & c_2 \end{array} \quad \text{or,} \quad \begin{array}{ccccc} a_1 & b_1 & c_1 & a_1 & b_1 \\ a_2 & b_2 & c_2 & a_2 & b_2 \\ a_3 & b_3 & c_3 & a_3 & b_3 \end{array}$$

Form the product of the three elements in the principal diagonal, and also of the three in each of the two lines immediately following the principal diagonal, and parallel to it. In this case, these products are:

$$a_1 b_2 c_3, \ a_2 b_3 c_1, \ a_3 b_1 c_2 ; \ (\text{or } a_1 b_2 c_3, \ b_1 c_2 a_3, \ c_1 a_2 b_3).$$

Next, form the product of the three elements in the secondary diagonal, and also of the three in each of the two lines immediately following the secondary diagonal, and parallel to it. In this case, these products are:

$$a_3 b_2 c_1, \ a_2 b_1 c_3, \ a_1 b_3 c_2 ; \ (\text{or } c_1 b_2 a_3, \ a_1 c_2 b_3, \ b_1 a_2 c_3).$$

From the sum of the former three products subtract the sum of the latter three; giving in this case:

$$a_1 b_2 c_3 + a_2 b_3 c_1 + a_3 b_1 c_2 - (a_3 b_2 c_1 + a_2 b_1 c_3 + a_1 b_3 c_2),$$

or, taking the products derived from the right-hand arrangement as given above:

$$a_1 b_2 c_3 + b_1 c_2 a_3 + c_1 a_2 b_3 - (c_1 b_2 a_3 + a_1 c_2 b_3 + b_1 a_2 c_3).$$

The given determinant is equal to either of these expressions, which are of the same value, as may easily be seen, for they differ merely in the order of their terms, and in the order of the factors of those terms.

In practice, it will soon be found sufficient merely to imagine the rows (or the columns) repeated.

§ 62. The following theorem, which is an immediate consequence of Theor. VIII. and Cor. 2, Theor. III., often affords the quickest and readiest means of evaluating a determinant with numerical elements.

So arrange the given determinant that none of the elements $b_m, c_m, \ldots k_m$ *shall be zero, and that all the elements of the* m*th row after* l_m *shall be zero, then*

$$
\begin{vmatrix}
a_1 & b_1 & c_1 \ldots l_1 \ldots s_1 \\
a_2 & b_2 & c_2 \ldots l_2 \ldots s_2 \\
a_3 & b_3 & c_3 \ldots l_3 \ldots s_3 \\
\cdots \cdots \cdots \quad \cdots \quad \cdots \\
a_n & b_n & c_n \quad l_n \quad s_n
\end{vmatrix}
$$

$$
= \frac{1}{b_m c_m \ldots k_m}
\begin{vmatrix}
\begin{vmatrix} a_1 & b_1 \\ a_m & b_m \end{vmatrix} & \begin{vmatrix} b_1 & c_1 \\ b_m & c_m \end{vmatrix} & , \ldots & \begin{vmatrix} k_1 & l_1 \\ k_m & l_m \end{vmatrix} & \ldots s_1 \\[2ex]
\begin{vmatrix} a_2 & b_2 \\ a_m & b_m \end{vmatrix} & \begin{vmatrix} b_2 & c_2 \\ b_m & c_m \end{vmatrix} & , \ldots & \begin{vmatrix} k_2 & l_2 \\ k_m & l_m \end{vmatrix} & \ldots s_2 \\[2ex]
\cdots & \cdots & \cdots & \cdots & \cdots \\[1ex]
\begin{vmatrix} a_{m-1} & b_{m-1} \\ a_m & b_m \end{vmatrix} & \begin{vmatrix} b_{m-1} & c_{m-1} \\ b_m & c_m \end{vmatrix} & , \ldots & \begin{vmatrix} k_{m-1} & l_{m-1} \\ k_m & l_m \end{vmatrix} & \ldots s_{m-1} \\[2ex]
\begin{vmatrix} a_m & b_m \\ a_{m+1} & b_{m+1} \end{vmatrix} & \begin{vmatrix} b_m & c_m \\ b_{m+1} & c_{m+1} \end{vmatrix} & , \ldots & \begin{vmatrix} k_m & l_m \\ k_{m+1} & l_{m+1} \end{vmatrix} & \ldots s_{m+1} \\[2ex]
\cdots & \cdots & \cdots & \cdots & \cdots \\[1ex]
\begin{vmatrix} a_m & b_m \\ a_n & b_n \end{vmatrix} & \begin{vmatrix} b_m & c_m \\ b_n & c_n \end{vmatrix} & , \ldots & \begin{vmatrix} k_m & l_m \\ k_n & l_n \end{vmatrix} & \ldots s_n
\end{vmatrix}
$$

If $m = 1$, the elements of the first row will be:

$$
\begin{vmatrix} a_1 & b_1 \\ a_2 & b_2 \end{vmatrix}, \begin{vmatrix} b_1 & c_1 \\ b_2 & c_2 \end{vmatrix}, \ldots \begin{vmatrix} k_1 & l_1 \\ k_2 & l_2 \end{vmatrix}, p_2, \ldots s_2.
$$

This method of evaluation is known as **Condensation.**

Examples.

Evaluate :
$$\Delta = \begin{vmatrix} 1 & 2 & 3 & -6 \\ 4 & 1 & 2 & 0 \\ 3 & 0 & 1 & 4 \\ 0 & 2 & 1 & 1 \end{vmatrix}$$

Here, since none of the *inner* elements of the first row is zero, we may take $m = 1$; then, operating on the first two rows, we mentally evaluate

$$\begin{vmatrix} 1 & 2 \\ 4 & 1 \end{vmatrix} \quad \begin{vmatrix} 2 & 3 \\ 1 & 2 \end{vmatrix} \quad \begin{vmatrix} 3 & -6 \\ 2 & 0 \end{vmatrix}$$

and write the results, which are $-7, 1, 12$, for the first row of the new determinant; similarly, we proceed with the first and third rows, and then with the first and fourth rows. This gives a determinant of the third order, which we divide by 6, the product of 2 and 3, the *inner* elements in the first row of Δ, and we thus obtain

$$\Delta = \tfrac{1}{6} \begin{vmatrix} -7 & 1 & 12 \\ -6 & 2 & 18 \\ 2 & -4 & 9 \end{vmatrix} = \begin{vmatrix} -7 & 1 & 4 \\ -3 & 1 & 3 \\ 2 & -4 & 3 \end{vmatrix}$$

This determinant may be evaluated by the Method of Sarrus, or the condensation may be repeated. Condensation gives

$$\Delta = \begin{vmatrix} -4 & -1 \\ 26 & 19 \end{vmatrix} = -76 + 26 = -50.$$

Ex. 76.

Evaluate by the Method of Sarrus :

1.
$$\begin{vmatrix} 1 & 1 & 1 \\ 1 & 2 & 4 \\ 1 & 3 & 9 \end{vmatrix}$$

2.
$$\begin{vmatrix} 1 & 2 & 3 \\ 2 & 3 & 1 \\ 3 & 1 & 2 \end{vmatrix}$$

3.
$$\begin{vmatrix} 2 & -1 & 1 \\ 1 & 2 & -1 \\ -1 & 1 & 2 \end{vmatrix}$$

4. $\begin{vmatrix} 7 & 6 & 7 \\ 1 & -2 & 1 \\ 3 & 1 & -2 \end{vmatrix}$　　**5.** $\begin{vmatrix} 0 & 17 & 3 \\ 21 & 0 & -4 \\ 13 & -2 & 0 \end{vmatrix}$　　**6.** $\begin{vmatrix} \frac{1}{2} & \frac{1}{3} & \frac{1}{4} \\ \frac{1}{3} & \frac{1}{4} & \frac{1}{5} \\ \frac{1}{5} & \frac{1}{6} & \frac{1}{7} \end{vmatrix}$

Evaluate by condensation :

7. $\begin{vmatrix} 1 & -1 & 2 & 0 \\ 1 & 2 & 0 & 3 \\ 2 & 0 & 3 & 1 \\ 0 & 3 & -1 & 4 \end{vmatrix}$　　　　　　**8.** $\begin{vmatrix} 0 & 1 & 1 & 8 \\ 5 & 0 & 1 & 1 \\ 4 & 1 & 0 & 1 \\ 3 & 1 & 1 & 0 \end{vmatrix}$

9. $\begin{vmatrix} -2 & 0 & 0 & 1 \\ 1 & -3 & -1 & -2 \\ 3 & 1 & -7 & 0 \\ 0 & 3 & 1 & -5 \end{vmatrix}$　　**10.** $\begin{vmatrix} 1 & 1 & 0 & -1 \\ 2 & 2 & 1 & 0 \\ 3 & 2 & 3 & 1 \\ 5 & 4 & 2 & 3 \end{vmatrix}$

11. $\begin{vmatrix} 0 & 1 & 1 & 1 & -1 \\ -1 & 0 & 1 & 1 & 1 \\ 1 & -1 & 0 & 1 & 1 \\ 1 & 1 & -1 & 0 & 1 \\ 1 & 1 & 1 & -1 & 0 \end{vmatrix}$

IV. MULTIPLICATION.

§ **63.** THEOREM IX. *The product of two determinants Δ_1 and Δ_2 of the same order, is a determinant such that the element in its pth row and qth column is the sum of the products of the elements of the pth row of Δ_1 each multiplied into the corresponding element of the qth column of Δ_2.*

Writing a_{pq}, x_{pq}, and A_{pq}, for the (p, q)th element of Δ_1, Δ_2, and of their product respectively, then

$$A_{pq} = a_{p,1} x_{1,q} + a_{p,2} x_{2,q} + a_{p,3} x_{3,p} + \text{etc.}$$

Before forming the product as above, Δ_1 or Δ_2 may either or both of them be transformed by rearranging the rows or the columns, or by changing rows into columns. The product of the same two determinants will therefore appear under different forms depending on the arrangement of its factor-determinants, but these forms will all have the same value. If one of the determinants to be multiplied together be of a lower order than the other, its order must be raised to that of the other. (Cor. 4, Theor. III.)

EXAMPLES.

1.

$$\begin{vmatrix} a_1 & b_1 \\ a_2 & b_2 \end{vmatrix} \times \begin{vmatrix} x_1 & y_1 \\ x_2 & y_2 \end{vmatrix} = \begin{vmatrix} a_1x_1 + b_1x_2, & a_1y_1 + b_1y_2 \\ a_2x_1 + b_2x_2, & a_2y_1 + b_2y_2 \end{vmatrix}$$

$$= \begin{vmatrix} a_1 & b_1 \\ a_2 & b_2 \end{vmatrix} \times \begin{vmatrix} x_1 & x_2 \\ y_1 & y_2 \end{vmatrix} = \begin{vmatrix} a_1x_1 + b_1y_1, & a_1x_2 + b_1y_2 \\ a_2x_1 + b_2y_1, & a_2x_2 + b_2y_2 \end{vmatrix}$$

$$= \begin{vmatrix} a_1 & a_2 \\ b_1 & b_2 \end{vmatrix} \times \begin{vmatrix} x_1 & x_2 \\ y_1 & y_2 \end{vmatrix} = \begin{vmatrix} a_1x_1 + a_2y_1, & a_1x_2 + a_2y_2 \\ b_1x_1 + b_2y_1, & b_1x_2 + b_2y_2 \end{vmatrix}$$

$$= \begin{vmatrix} a_1 & a_2 \\ b_1 & b_2 \end{vmatrix} \times \begin{vmatrix} x_1 & y_1 \\ x_2 & y_2 \end{vmatrix} = \begin{vmatrix} a_1x_1 + a_2x_2, & a_1y_1 + a_2y_2 \\ b_1x_1 + b_2x_2, & b_1y_1 + b_2y_2 \end{vmatrix}$$

2.

$$\begin{vmatrix} a_1 & b_1 & c_1 \\ a_2 & b_2 & c_2 \\ a_3 & b_3 & c_3 \end{vmatrix} \times \begin{vmatrix} x_1 & x_2 & x_3 \\ y_1 & y_2 & y_3 \\ z_1 & z_2 & z_3 \end{vmatrix}$$

$$\begin{vmatrix} a_1x_1 + b_1y_1 + c_1z_1 & a_1x_2 + b_1y_2 + c_1z_2 & a_1x_3 + b_1y_3 + c_1z_3 \\ a_2x_1 + b_2y_1 + c_2z_1 & a_2x_2 + b_2y_2 + c_2z_2 & a_2x_3 + b_2y_3 + c_2z_3 \\ a_2x_1 + b_3y_1 + c_3z_1 & a_3x_2 + b_3y_2 + c_3z_2 & a_3x_3 + b_3y_3 + c_3z_3 \end{vmatrix}$$

3.

$$\begin{vmatrix} a_1 & b_1 & c_1 & d_1 \\ a_2 & b_2 & c_2 & d_2 \\ a_3 & b_3 & c_3 & d_3 \\ a_4 & b_4 & c_4 & d_4 \end{vmatrix} \times \begin{vmatrix} x_1 & x_2 \\ y_1 & y_2 \end{vmatrix} = \begin{vmatrix} a_1 & b_1 & c_1 & d_1 \\ a_2 & b_2 & c_2 & d_2 \\ a_3 & b_3 & c_3 & d_3 \\ a_4 & b_4 & c_4 & d_4 \end{vmatrix} \times \begin{vmatrix} x_1 & x_2 & x_3 & x_4 \\ y_1 & y_2 & y_3 & y_4 \\ 0 & 0 & 1 & z_4 \\ 0 & 0 & 0 & 1 \end{vmatrix}$$

$$= \begin{vmatrix} a_1x_1+b_1y_1 & a_1x_2+b_1y_2 & a_1x_3+b_1y_3+c_1 & a_1x_4+b_1y_4+c_1z_4+d_1 \\ a_2x_1+b_2y_1 & a_2x_2+b_2y_2 & a_2x_3+b_2y_3+c_2 & a_2x_4+b_2y_4+c_2z_4+d_2 \\ a_3x_1+b_3y_1 & a_3x_2+b_3y_2 & a_3x_3+b_3y_3+c_3 & a_3x_4+b_3y_4+c_3z_4+d_3 \\ a_4x_1+b_4y_1 & a_4x_2+b_4y_2 & a_4x_3+b_4y_3+c_4 & a_4x_4+b_4y_4+c_4z_4+d_4 \end{vmatrix}$$

Here x_3, y_3, x_4, y_4, z_4 are wholly arbitrary, and may be made all zero.

4. $| a_1 \ b_2 \ c_3 \ d_4 |^2 = \begin{vmatrix} a_1 & b_1 & c_1 & d_1 \\ a_2 & b_2 & c_2 & d_2 \\ a_3 & b_3 & c_3 & d_3 \\ a_4 & b_4 & c_4 & d_4 \end{vmatrix} \times \begin{vmatrix} b_1 & b_2 & b_3 & b_4 \\ -a_1 & -a_2 & -a_3 & -a_4 \\ d_1 & d_2 & d_3 & d_4 \\ -c_1 & -c_2 & -c_3 & -c_4 \end{vmatrix}$

$= \begin{vmatrix} 0 & |a_1 b_2|+|c_1 d_2| & |a_1 b_3|+|c_1 d_3| & |a_1 b_4|+|c_1 d_4| \\ |a_2 b_1|+|c_2 d_1| & 0 & |a_2 b_3|+|c_2 d_3| & |a_2 b_4|+|c_2 d_4| \\ |a_3 b_1|+|c_3 d_1| & |a_3 b_2|+|c_3 d_2| & 0 & |a_3 b_4|+|c_3 d_4| \\ |a_4 b_1|+|c_4 d_1| & |a_4 b_2|+|c_4 d_2| & |a_4 b_3|+|c_4 d_3| & 0 \end{vmatrix}$

This is a skew symmetrical determinant for

$$(| a_1 \ b_2 | + | c_1 \ d_2 |) + (| a_2 \ b_1 | + | c_2 \ d_1 |) = 0,$$

by Theor. II.; and the same holds for every other pair of conjugates.

§ 64. If from Δ, a determinant of order n, there be erased m rows and m columns, the determinant formed from the remaining rows and columns taken in order, is called a **Minor** of Δ of order $m - n$. The minors obtained by erasing one row and one column of any determinant are called the **Principal Minors** of that determinant.

Two minors which are so related that the rows and columns erased in forming one of them are exactly those not erased in forming the other, are called **Complementary Minors.**

Thus, $\quad | a_1 \ b_2 \ c_3 |, \quad | a_2 \ b_3 \ d_4 |, \quad | b_1 \ c_4 \ e_5 |$

are third-order minors of $| a_1 \ b_2 \ c_3 \ d_4 \ e_5 |$, and their complementaries are the second-order minors,

$$| d_4 \ c_5 |, \quad | c_1 \ e_5 |, \quad | a_2 \ d_3 |,$$

respectively.

$$| b_1 \ e_4 | \text{ and } | c_2 \ d_5 |$$

are second-order minors of $| a_1 \ b_2 \ c_3 \ d_4 \ e_5 f_6 |$, and their complementaries are the fourth-order minors,

$$| a_2 \ c_3 \ d_5 f_6 | \text{ and } | a_1 \ b_3 \ e_4 f_6 |$$

respectively,

The principal minors of $|\,a_1\,b_2\,c_3\,d_4\,e_5\,|$ are

$$|\,b_2\,c_3\,d_4\,e_5\,|,\quad |\,b_1\,c_3\,d_4\,e_5\,|,\quad |\,b_1\,c_2\,d_4\,e_5\,|,\ \ldots$$

$$|\,a_2\,c_3\,d_4\,e_5\,|,\quad |\,a_1\,c_3\,d_4\,e_5\,|,\ \ldots$$

complementary to $a_1, a_2, a_3, \ldots b_1, b_2, \ldots$, respectively.

Hence, if in any determinant, $A_{p,\,q}$ denote the complement of $a_{p,\,q}$ (page 000), $(-1)^{p+q}A_{p,\,q}$ will be the principal minor complementary to $a_{p,\,q}$.

THEOREM X. *If any* m *rows of a determinant be selected, and every possible minor of the* mth *order be formed from them, and if each be multiplied by its complementary and the product affected with* $+$ *or* $-$, *according as the sum of the numbers indicating the rows and the columns from which the minor is formed be even or odd, the sum of these products will be equal to the original determinant.*

Thus, the first two rows of $|\,a_1\,b_2\,c_3\,d_4\,|$ give the six minors,

$$|\,a_1\,b_2\,|,\quad |\,a_1\,c_2\,|,\quad |\,a_1\,d_2\,|,\quad |\,b_1\,c_2\,|,\quad |\,b_1\,d_2\,|,\quad |\,c_1\,d_2\,|,$$

whose complementaries are

$$|\,c_3\,d_4\,|,\quad |\,b_3\,d_4\,|,\quad |\,b_3\,c_4\,|,\quad |\,a_3\,d_4\,|,\quad |\,a_3\,c_4\,|,\quad |\,a_3\,b_4\,|,$$

and the sums of the numbers indicating the rows and the columns from which the first six minors are formed, are

$$(1+2+1+2),\quad (1+2+1+3),\quad (1+2+1+4),$$
$$(1+2+2+3),\quad (1+2+2+4),\quad (1+2+3+4),$$

$$\therefore\ |\,a_1\,b_2\,c_3\,d_4\,| = |\,a_1\,b_2\,|\ |\,c_3\,d_4\,| - |\,a_1\,c_2\,|\ |\,b_3\,d_4\,|$$
$$+ |\,a_1\,d_2\,|\ |\,b_3\,c_4\,| + |\,b_1\,c_2\,|\ |\,a_3\,d_4\,|$$
$$- |\,b_1\,d_2\,|\ |\,a_3\,c_4\,| + |\,c_1\,d_2\,|\ |\,a_3\,b_4\,|$$

Similarly, the second, third, and fifth columns being those selected,

$$
\begin{aligned}
\mid a_1\, b_2\, c_3\, d_4\, e_5 \mid = \; & \mid b_1\, c_2\, e_3 \mid \quad \mid a_4\, d_5 \mid - \mid b_1\, c_2\, e_4 \mid \quad \mid a_3\, d_5 \mid \\
+ & \mid b_1\, c_2\, e_5 \mid \quad \mid a_3\, d_4 \mid + \mid b_1\, c_3\, e_4 \mid \quad \mid a_2\, d_5 \mid \\
- & \mid b_1\, c_3\, e_5 \mid \quad \mid a_2\, d_4 \mid + \mid b_1\, c_4\, e_5 \mid \quad \mid a_2\, d_3 \mid \\
- & \mid b_2\, c_3\, e_4 \mid \quad \mid a_1\, d_5 \mid + \mid b_2\, c_3\, e_5 \mid \quad \mid a_1\, d_4 \mid \\
- & \mid b_2\, c_4\, e_5 \mid \quad \mid a_1\, d_3 \mid + \mid b_3\, c_4\, e_5 \mid \quad \mid a_1\, d_2 \mid
\end{aligned}
$$

Ex. 77.

Perform the following multiplications, expressing the results in determinant form :

1. $\begin{vmatrix} a_1 & b_1 \\ a_2 & b_2 \end{vmatrix} \begin{vmatrix} x_1 & x_2 \\ y_1 & y_2 \end{vmatrix}$ 2. $\begin{vmatrix} 3 & 5 \\ 4 & 6 \end{vmatrix} \begin{vmatrix} x & u \\ y & v \end{vmatrix}$ 3. $\begin{vmatrix} 2 & 5 \\ 3 & 4 \end{vmatrix} \begin{vmatrix} 3 & 5 \\ 4 & 2 \end{vmatrix}$

4. $\begin{vmatrix} 2 & 5 \\ 3 & 6 \end{vmatrix} \begin{vmatrix} -3 & -6 \\ 2 & 5 \end{vmatrix}$ 5. $\begin{vmatrix} 2 & -3 \\ -\frac{1}{2} & \frac{1}{3} \end{vmatrix} \begin{vmatrix} \frac{1}{2} & -2 \\ \frac{1}{3} & -3 \end{vmatrix}$

6. $\begin{vmatrix} 2 & 3 \\ 4 & -2 \end{vmatrix} \begin{vmatrix} 1 & 2 & 3 \\ 2 & 4 & 8 \\ 3 & 4 & -7 \end{vmatrix}$ 7. $\begin{vmatrix} \frac{2}{3} & \frac{2}{3} & \frac{1}{3} \\ \frac{6}{7} & \frac{3}{7} & \frac{2}{7} \\ \frac{8}{9} & \frac{4}{9} & \frac{1}{9} \end{vmatrix} \begin{vmatrix} \frac{2}{3} & \frac{6}{7} & \frac{8}{9} \\ \frac{2}{3} & \frac{3}{7} & \frac{4}{9} \\ \frac{1}{3} & \frac{2}{7} & \frac{1}{9} \end{vmatrix}$

8. $\begin{vmatrix} a-y & h \\ h & b-y \end{vmatrix} \begin{vmatrix} a+y & h \\ h & a+y \end{vmatrix}$

9. $\begin{vmatrix} a-y & h & g \\ h & b-y & f \\ g & f & c-y \end{vmatrix} \begin{vmatrix} a+y & h & g \\ h & b+y & f \\ g & f & c+y \end{vmatrix}$

10. $\begin{vmatrix} 1 & a & a^2 \\ 1 & b & b^2 \\ 1 & c & c^2 \end{vmatrix} \begin{vmatrix} a^2 & b^2 & c^2 \\ -a & -b & -c \\ 1 & 1 & 1 \end{vmatrix}$

11. $\begin{vmatrix} x & yi \\ yi & x \end{vmatrix} \begin{vmatrix} y & zi \\ zi & y \end{vmatrix} \begin{vmatrix} z & xi \\ xi & z \end{vmatrix}$

wherein $i^2 = -1$.

12. $\begin{vmatrix} a+bi & -c+di \\ c+di & a-bi \end{vmatrix} \begin{vmatrix} x+yi & u+vi \\ -u+vi & x-yi \end{vmatrix}$

13. $\begin{vmatrix} a+b & a & b \\ c & b+c & b \\ c & a & c+a \end{vmatrix} \begin{vmatrix} a+b+\frac{1}{2}c & -\frac{1}{2}a & -\frac{1}{2}b \\ -\frac{1}{2}c & b+c+\frac{1}{2}a & -\frac{1}{2}b \\ -\frac{1}{2}c & -\frac{1}{2}a & c+a+\frac{1}{2}b \end{vmatrix}$

14. $\begin{vmatrix} a & a & a & a \\ a & b & b & b \\ a & b & c & c \\ a & b & c & d \end{vmatrix} \begin{vmatrix} 1 & 0 & 0 & 0 \\ -1 & 1 & 0 & 0 \\ 0 & -1 & 1 & 0 \\ 0 & 0 & -1 & 1 \end{vmatrix}$ 15. $\begin{vmatrix} x & y & z \\ z & x & y \\ y & z & x \end{vmatrix} \begin{vmatrix} x' & z' & y' \\ y' & x' & z' \\ z' & y' & x' \end{vmatrix}$

16. $\begin{vmatrix} 1-x & x^2 & -x^3 \\ a & b & c & d \\ b & c & d & e \\ c & d & e & f \end{vmatrix} \begin{vmatrix} 1 & x & 0 & 0 \\ 0 & 1 & x & 0 \\ 0 & 0 & 1 & x \\ 0 & 0 & 0 & 1 \end{vmatrix}$

17. $\begin{vmatrix} a_1+a_2 & a_2+a_3 & a_3+a_1 \\ b_1+b_2 & b_2+b_3 & b_3+b_1 \\ c_1+c_2 & c_2+c_3 & c_3+c_1 \end{vmatrix} \begin{vmatrix} 1 & 1 & -1 \\ -1 & 1 & 1 \\ 1 & -1 & 1 \end{vmatrix}$

18. $\begin{vmatrix} 1 & -2 & 1 & 1 \\ 1 & 1 & -2 & 1 \\ 1 & 1 & 1 & -2 \\ -2 & 1 & 1 & 1 \end{vmatrix} \begin{vmatrix} a_1+a_2+a_3 & b_1+b_2+b_3 & c_1+c_2+c_3 & d_1+d_2+d_3 \\ a_2+a_3+a_4 & b_2+b_3+b_4 & c_2+c_3+c_4 & d_2+d_3+d_4 \\ a_3+a_4+a_1 & b_3+b_4+b_1 & c_3+c_4+c_1 & d_3+d_4+d_1 \\ a_4+a_1+a_2 & b_4+b_1+b_2 & c_4+c_1+c_2 & d_4+d_1+d_2 \end{vmatrix}$

19. $\begin{vmatrix} b_2 & c_2 & -b_1 & -c_1 \\ c_2 & 0 & -c_1 & 0 \\ 0 & 0 & 0 & -1 \\ 0 & 0 & -1 & 0 \end{vmatrix} \begin{vmatrix} a_1 & b_1 & c_1 & 0 \\ 0 & a_1 & b_1 & c_1 \\ a_2 & b_2 & c_2 & 0 \\ 0 & a_2 & b_2 & c_2 \end{vmatrix}$

20. $\begin{vmatrix} b_2 & c_2 & d_2 & -b_1 & -c_1 & -d_1 \\ c_2 & d_2 & 0 & -c_1 & -d_1 & 0 \\ d_2 & 0 & 0 & -d_1 & 0 & 0 \\ 0 & 0 & 0 & 0 & 0 & -1 \\ 0 & 0 & 0 & 0 & -1 & 0 \\ 0 & 0 & 0 & -1 & 0 & 0 \end{vmatrix} \begin{vmatrix} a_1 & b_1 & c_1 & d_1 & 0 & 0 \\ 0 & a_1 & b_1 & c_1 & d_1 & 0 \\ 0 & 0 & a_1 & b_1 & c_1 & d_1 \\ a_2 & b_2 & c_2 & d_2 & 0 & 0 \\ 0 & a_2 & b_2 & c_2 & d_2 & 0 \\ 0 & 0 & a_2 & b_2 & c_2 & d_2 \end{vmatrix}$

21. $\begin{vmatrix} 1 & -2x & x^2 \\ 1 & -2y & y^2 \\ 1 & -2z & z^2 \end{vmatrix} \begin{vmatrix} u^2 & v^2 & w^2 \\ u & v & w \\ 1 & 1 & 1 \end{vmatrix}$

22. $\begin{vmatrix} 1 & -3a & 3a^2 & -a^3 \\ 1 & -3b & 3b^2 & -b^3 \\ 1 & -3c & 3c^2 & -c^3 \\ 1 & -3d & 3d^2 & -d^3 \end{vmatrix} \begin{vmatrix} a^3 & b^3 & c^3 & d^3 \\ a^2 & b^2 & c^2 & d^2 \\ a & b & c & d \\ 1 & 1 & 1 & 1 \end{vmatrix}$

and deduce therefrom that

$$9\,(a-b)^2\,(a-c)^2\,(a-d)^2\,(b-c)^2\,(b-d)^2\,(c-d)^2$$
$$= [(a-b)^3\,(c-d)^3 + (b-d)^3(c-a)^3 + (a-d)^3(b-c)^3]^2.$$

23. If $s_m = a_1^m + a_2^m + a_3^m + \dots + a_n^m$, then will

$$\begin{vmatrix} 1 & a_1 & a_1^2 & \dots & a_1^n \\ 1 & a_2 & a_2^2 & \dots & a_2^n \\ 1 & a_3 & a_3^2 & \dots & a_3^n \\ \dots & \dots & \dots & \dots & \dots \\ 1 & a_n & a_n^2 & \dots & a_n^n \end{vmatrix}^2 = \begin{vmatrix} s_0 & s_1 & s_2 & \dots & s_n \\ s_1 & s_2 & s_3 & \dots & s_{n+1} \\ s_2 & s_3 & s_4 & \dots & s_{n+2} \\ \dots & \dots & \dots & \dots & \dots \\ s_n & s_{n+1} & s_{n+2} & \dots & s_{2n} \end{vmatrix}$$

24. If $\Delta = \mid a_1\,b_2\,c_3 \dots k_n \mid$ and $\Delta' = \mid A_1\,B_2\,C_3 \dots K_n \mid$ wherein $A_1,\ -A_2,\ A_3,\ \dots -B_1,\ B_2,\ -B_3$, etc., are the principal minors of Δ; and if $a_1,\ -a_2,\ a_3,\ \dots -\beta_1,$ $\beta_2,\ -\beta_3$, etc., are the principal minors of Δ', prove that

$$\Delta\Delta' = \Delta^n \text{ and } a_1\Delta = a_1\Delta^n,\ a_2\Delta = a_2\Delta^n, \text{ etc.}$$

NOTE. The determinant Δ' is called the *Reciprocal* of the determinant Δ; and the elements $A_1,\ A_2,\ \dots B_1,\ B_2$, etc., are called *Inverse Elements* with respect to $a_1,\ a_2,\ \dots b_1,\ b_2$, etc.

25. Prove that a minor of the order m, formed out of the inverse constituents, is equal to the complementary of the corresponding minor of the original determinant multiplied by the $(m-1)$th power of that determinant.

V. Applications.

§ **65.** To solve the simultaneous linear equations,

$$a_1x + b_1y + c_1z = d_1, \tag{1}$$
$$a_2x + b_2y + c_2z = d_2, \tag{2}$$
$$a_3x + b_3y + c_3z = d_3. \tag{3}$$

Let $\nabla = |a_1 b_2 c_3|$, $\nabla_a = |d_1 b_2 c_3|$, $\nabla_b = |a_1 d_2 c_3|$, $\nabla_c = |a_1 b_2 d_3|$, and let A_1, A_2, etc., denote the complements of a_1, a_2, etc., in ∇.

Multiply (1) by A_1, (2) by A_2, (3) by A_3, and add.

$$\therefore (a_1 A_1 + a_2 A_2 + a_3 A_3)x + (b_1 A_1 + b_2 A_2 + b_3 A_3)y$$
$$+ (c_1 A_1 + c_2 A_2 + c_3 A_3)z = (d_1 A_1 + d_2 A_2 + d_3 A_3).$$

Therefore, by Theor. III., Cor. 1, and Theor. VII.,

$$\nabla x = \nabla_a.$$

By using B_1, B_2, B_3 instead of A_1, A_2, A_3, we obtain

$$\nabla y = \nabla_b,$$

and using C_1, C_2, C_3 instead of A_1, A_2, A_3, gives

$$\nabla z = \nabla_c.$$

The method here exhibited is evidently applicable to the case of n linear equations containing n unknown quantities.

2. Given the above linear equations (1), (2), (3) to find the value of $ax + \beta y + \gamma z$.

By Theors. VIII. and VII.

$$\begin{vmatrix} ax + \beta y + \gamma z & a & \beta & \gamma \\ a_1 x + b_1 y + c_1 z & a_1 & b_1 & c_1 \\ a_2 x + b_2 y + c_2 z & a_2 & b_2 & c_2 \\ a_3 x + b_3 y + c_3 z & a_3 & b_3 & c_3 \end{vmatrix} = 0.$$

Therefore, substituting from the equations (1), (2), (3),

$$\begin{vmatrix} ax + \beta y + \gamma z & a & \beta & \gamma \\ d_1 & & a_1 & b_1 & c_1 \\ d_2 & & a_2 & b_2 & c_2 \\ d_3 & & a_3 & b_3 & c_3 \end{vmatrix} = 0;$$

$$\therefore (ax + \beta y + \gamma z)\nabla - a\nabla_a - \beta\nabla_b - \gamma\nabla_c = 0;$$
$$\therefore ax + \beta y + \gamma z = (a\nabla_a + \beta\nabla_b + \gamma\nabla_c) \div \nabla.$$

This result necessarily includes the *solution* of the equations found above.

3. To determine the condition that the homogeneous linear equations

$$a_1x + b_1y + c_1z = 0, \qquad (1)$$

$$a_2x + b_2y + c_2z = 0, \qquad (2)$$

$$a_3x + b_3y + c_3z = 0, \qquad (3)$$

may coexist for values of x, y, and z other than zero.

Multiply (1) by A_1, (2) by A_2, and (3) by A_3, and add.

$$\therefore \ \nabla x = 0,$$

and therefore if x be not zero,

$$\nabla = 0.$$

4. To find the condition that

$$a_1x^2 + b_1x + c_1 = 0 \qquad (1)$$

and

$$a_2x^2 + b_2x + c_2 = 0 \qquad (2)$$

may have a common root.

Multiply each of the given equations by x; the resulting equations together with the given equations constitute the four simultaneous equations,

$$a_1x^3 + b_1x^2 + c_1x \qquad = 0,$$

$$a_1x^2 + b_1x + c_1 = 0,$$

$$a_2x^3 + b_2x^2 + c_2x \qquad = 0,$$

$$a_2x^2 + b_2x + c_2 = 0,$$

which are to be satisfied by values of x^3, x^2, and x other than zero; hence, by No. 3 above,

$$\begin{vmatrix} a_1 & b_1 & c_1 & 0 \\ 0 & a_1 & b_1 & c_1 \\ a_2 & b_2 & c_2 & 0 \\ 0 & a_2 & b_2 & c_2 \end{vmatrix} = 0.$$

This is also the condition that

$$a_1x^2 + b_1x + c_1 \text{ and } a_2x^2 + b_2x + c_2$$

may have a common factor.

5. To find the condition that

$$ax^3 + 3bx^2 + 3cx + d$$

may have a square factor.

Let $(x - m)^2$ be the square factor. Divide the given expression by $x - m$, and the quotient by $x - m$; the two remainders thus obtained must both vanish. These remainders are

$$am^3 + 3bm^2 + 3cm + d$$

and $\qquad am^2 + 2bm + c.$

$$\therefore am^3 + 3bm^2 + 3cm + d = 0,$$

and $\qquad am^2 + 2bm + c = 0.$

Multiply the latter equation by m, and subtract the product from the former.

$$\therefore bm^2 + 2cm + d = 0.$$

Combining this equation with the second of preceding, the condition required is found to be

that $\qquad ax^2 + 2bx + c = 0$

and $\qquad bx^2 + 2cx + d = 0$

shall have a common root, and the condition for this has been found in No. 4 above; viz.,

$$\begin{vmatrix} a & 2b & c & 0 \\ 0 & a & 2b & c \\ b & 2c & d & 0 \\ 0 & b & 2c & d \end{vmatrix} = 0.$$

6. To find the condition that the expression

$$ax^2 + by^2 + cz^2 + 2fyz + 2gzx + 2hxy$$

may be the product of two linear factors.

Let the factors be $a_1x + \beta_1 y + \gamma_1 z$ and $a_2x + \beta_2 y + \gamma_2 z$.

Multiply these together, and equate the coefficients of the product with those of like powers of the variables in the given expression.

$$\therefore \ a = a_1a_2 \qquad\qquad b = \beta_1\beta_2 \qquad\qquad c = \gamma_1\gamma_2$$

$$2f = \beta_1\gamma_2 + \beta_2\gamma_1 \qquad 2g = a_1\gamma_2 + a_2\gamma_1 \qquad 2h = a_1\beta_2 + a_2\beta_1$$

$$\therefore \begin{vmatrix} a & h & g \\ h & b & f \\ g & f & c \end{vmatrix} = \frac{1}{8} \begin{vmatrix} a_1a_2 + a_2a_1 & a_1\beta_2 + a_2\beta_1 & a_1\gamma_2 + a_2\gamma_1 \\ \beta_1a_2 + \beta_2a_1 & \beta_1\beta_2 + \beta_2\beta_1 & \beta_1\gamma_2 + \beta_2\gamma_1 \\ \gamma_1a_2 + \gamma_2a_1 & \gamma_1\beta_2 + \gamma_2\beta_1 & \gamma_1\gamma_2 + \gamma_2\gamma_1 \end{vmatrix}$$

$$= \frac{1}{8} \begin{vmatrix} a_1 & a_2 & 0 \\ \beta_1 & \beta_2 & 0 \\ \gamma_1 & \gamma_2 & 0 \end{vmatrix} \times \begin{vmatrix} a_2 & \beta_2 & \gamma_2 \\ a_1 & \beta_1 & \gamma_1 \\ 0 & 0 & 0 \end{vmatrix} = 0.$$

Hence the required condition is that

$$\begin{vmatrix} a & h & g \\ h & b & f \\ g & f & c \end{vmatrix} = 0.$$

7. If
$$x = a_1X + \beta_1Y + \gamma_1Z$$
$$y = a_2X + \beta_2Y + \gamma_2Z \qquad\qquad (1)$$
$$z = a_3X + \beta_3Y + \gamma_3Z$$

and
$$ax^2 + by^2 + cz^2 + 2fyz + 2gzx + 2hxy$$
$$= AX^2 + BY^2 + CZ^2 + 2FYZ$$
$$+ 2GZX + 2HXY, \qquad\qquad (2)$$

then will

$$\begin{vmatrix} A & H & G \\ H & B & F \\ G & F & C \end{vmatrix} = \begin{vmatrix} a_1 & \beta_1 & \gamma_1 \\ a_2 & \beta_2 & \gamma_2 \\ a_3 & \beta_3 & \gamma_3 \end{vmatrix}^2 \begin{vmatrix} a & h & g \\ h & b & f \\ g & f & c \end{vmatrix}$$

Substitute for x, y, and z in (2) their values in (1), and equate coefficients of like powers of X, Y, and Z.

$$\therefore A = aa_1^2 + ba_2^2 + ca_3^2 + 2fa_2a_3 + 2ga_3a_1 + 2ha_1a_2$$

$$B = a\beta_1^2 + b\beta_2^2 + c\beta_3^2 + 2f\beta_2\beta_3 + 2g\beta_3\beta_1 + 2h\beta_1\beta_2$$

$$C = a\gamma_1^2 + b\gamma_2^2 + c\gamma_3^2 + 2f\gamma_2\gamma_3 + 2g\gamma_3\gamma_1 + 2h\gamma_1\gamma_2$$

$$F = a\beta_1\gamma_1 + b\beta_2\gamma_2 + c\beta_3\gamma_3 + f(\beta_2\gamma_3 + \beta_3\gamma_2)$$
$$+ g(\beta_3\gamma_1 + \beta_1\gamma_3) + h(\beta_1\gamma_2 + \beta_2\gamma_1)$$

$$G = a\gamma_1a_1 + b\gamma_2a_2 + c\gamma_3a_3 + f(\gamma_2a_3 + \gamma_3a_2)$$
$$+ g(\gamma_3a_1 + \gamma_1a_3) + h(\gamma_1a_2 + \gamma_2a_1)$$

$$H = aa_1\beta_1 + ba_2\beta_2 + ca_3\beta_3 + f(a_2\beta_3 + a_3\beta_2)$$
$$+ g(a_3\beta_1 + a_1\beta_3) + h(a_1\beta_2 + a_2\beta_1).$$

Now

$$\begin{vmatrix} a_1 & \beta_1 & \gamma_1 \\ a_2 & \beta_2 & \gamma_2 \\ a_3 & \beta_3 & \gamma_3 \end{vmatrix}^2 \begin{vmatrix} a & h & g \\ h & b & f \\ g & f & c \end{vmatrix}$$

$$= \begin{vmatrix} a_1 & a_2 & a_3 \\ \beta_1 & \beta_2 & \beta_3 \\ \gamma_1 & \gamma_2 & \gamma_3 \end{vmatrix} \begin{vmatrix} a & h & g \\ h & b & f \\ g & f & c \end{vmatrix} \begin{vmatrix} a_1 & \beta_1 & \gamma_1 \\ a_2 & \beta_2 & \gamma_2 \\ a_3 & \beta_3 & \gamma_3 \end{vmatrix}$$

$$= \begin{vmatrix} aa_1 + ha_2 + ga_3 & ha_1 + ba_2 + fa_3 & ga_1 + fa_2 + ca_3 \\ a\beta_1 + h\beta_2 + g\beta_3 & h\beta_1 + b\beta_2 + f\beta_3 & g\beta_1 + f\beta_2 + c\beta_3 \\ a\gamma_1 + h\gamma_2 + g\gamma_3 & h\gamma_1 + b\gamma_2 + f\gamma_3 & g\gamma_1 + f\gamma_2 + c\gamma_3 \end{vmatrix} \begin{vmatrix} a_1 & \beta_1 & \gamma_1 \\ a_2 & \beta_2 & \gamma_2 \\ a_3 & \beta_3 & \gamma_3 \end{vmatrix}$$

$$= \begin{vmatrix} A & H & G \\ H & B & F \\ G & F & C \end{vmatrix}$$

8. Eliminate x, y, and z from the equations

$$ax^2 + by^2 + cz^2 + 2fyz + 2gzx + 2hxy = 0, \qquad (1)$$

$$k_1x + l_1y + m_1z = 0, \qquad (2)$$

$$k_2x + l_2y + m_2z = 0. \qquad (3)$$

First Method. Let λ_1 and λ_2 be homogenous linear functions of x, y, and z, such that $\lambda_1(2) + \lambda_2(3) = (1)$.

$$\therefore ax^2 + by^2 + cz^2 + 2fyz + 2gzx + 2hxy$$
$$= (ax + hy + gz)x + (hx + by + fz)y + (gx + fy + cz)z$$
$$= (k_1x + l_1y + m_1z)\lambda_1 + (k_2x + l_2y + m_2z)\lambda_2.$$

$$\therefore ax + hy + gz = k_1\lambda_1 + k_2\lambda_2, \tag{4}$$

$$hx + by + fz = l_1\lambda_1 + l_2\lambda_2, \tag{5}$$

$$gx + fy + cz = m_1\lambda_1 + m_2\lambda_2. \tag{6}$$

Now eliminate x, y, z, λ_1, λ_2 from (2), (3), (4), (5), (6) by the method exhibited in No. 3, page 350.

$$\therefore \begin{vmatrix} a & h & g & k_1 & k_2 \\ h & b & f & l_1 & l_2 \\ g & f & c & m_1 & m_2 \\ k_1 & l_1 & m_1 & 0 & 0 \\ k_2 & l_2 & m_2 & 0 & 0 \end{vmatrix} = 0.$$

Second Method. Multiply (2) by x, y, and z, successively, and (3) by x and by y.

$$\therefore k_1x^2 \qquad\qquad\quad + m_1zx + l_1xy = 0,$$
$$l_1y^2 \quad + m_1yz \qquad\quad + k_1xy = 0,$$
$$m_1z^2 + l_1yz + k_1zx \qquad\quad = 0,$$
$$k_2x^2 \qquad\qquad\quad + m_2zx + l_2xy = 0,$$
$$l_2y^2 \quad + m_2yz \qquad\quad + k_2xy = 0.$$

Now eliminate x^2, y^2, z^2, yz, zx, xy from these five equations and (1), by the method of No. 3, page 350.

$$\therefore \begin{vmatrix} a & b & c & 2f & 2g & 2h \\ k_1 & 0 & 0 & 0 & m_1 & l_1 \\ 0 & l_1 & 0 & m_1 & 0 & k_1 \\ 0 & 0 & m_1 & l_1 & k_1 & 0 \\ k_2 & 0 & 0 & 0 & m_2 & l_2 \\ 0 & l_2 & 0 & m_2 & 0 & k_2 \end{vmatrix} = 0.$$

9. To solve the simultaneous equations

$$a_1 x^2 + 2 h_1 xy + b_1 y^2 = m_1, \qquad (1)$$

$$a_2 x^2 + 2 h_2 xy + b_2 y^2 = m_2. \qquad (2)$$

Write them in the form

$$(a_1 x + h_1 y) x + (h_1 x + b_1 y) y = m_1,$$

$$(a_2 x + h_2 y) x + (h_2 x + b_2 y) y = m_2.$$

Let $\quad \nabla = \begin{vmatrix} a_1 x + h_1 y & h_1 x + b_1 y \\ a_2 x + h_2 y & h_2 x + b_2 y \end{vmatrix}$

$$\therefore \ \nabla x = (h_2 x + b_2 y) m_1 - (h_1 x + b_1 y) m_2,$$

$$\nabla y = - (a_2 x + h_2 y) m_1 + (a_1 x + h_1 y) m_2.$$

$$\therefore \ (\nabla + |\ h_1\ m_2\ |) x + |\ b_1\ m_2\ | y = 0, \qquad (3)$$

$$- |\ a_1\ m_2\ | x + (\nabla - |\ h_1\ m_2\ |) y = 0. \qquad (4)$$

$$\therefore \ (\nabla^2 - |\ h_1\ m_2\ |^2) + |\ a_1\ m_2\ |\ |\ b_1\ m_2\ | = 0.$$

$$\therefore \ \nabla = \pm \surd(|\ h_1\ m_2\ |^2 - |\ a_1\ m_2\ |\ |\ b_1\ m_2\ |).$$

Hence ∇ may be treated as known, and then by (3),

$$y = - \frac{(\nabla + |\ h_1\ m_2\ |) x}{|\ b_1\ m_2\ |}.$$

Substitute this value of y in (1), and there will result a pure quadratic in x, from which the value of x may be immediately obtained.

10. To solve the simultaneous equations

$$a_1 x^2 + b_1 y^2 + 2 h_1 xy = m_1, \qquad (1)$$

$$a_2 x + b_2 y = m_2. \qquad (2)$$

This may be treated as a particular case of the preceding, or otherwise as follows:

Write the given equations in the form

$$(a_1x + h_1y)x + (h_1x + b_1y)y = m_1$$
$$a_2x + \qquad b_2y = m_2$$

Let $\quad \nabla = (a_1x + h_1y)b_2 - (h_1x + b_1y)a_2. \qquad (3)$

$$\therefore \nabla x = b_2m_1 - (h_1x + b_1y)m_2,$$
$$\nabla y = -a_2m_1 + (a_1x + h_1y)m_2.$$

$$\therefore (\nabla + h_1m_2)x + \qquad b_1m_2y - b_2m_1 = 0,$$
$$-a_1m_2x + (\nabla - h_1m_2)y + a_2m_1 = 0,$$
$$a_2x + b_2y - m_2 = 0.$$

$$\therefore \begin{vmatrix} \nabla + h_1m_2 & b_1m_2 & b_2m_1 \\ -a_1m_2 & \nabla - h_1m_2 & -a_2m_1 \\ a_2 & b_2 & m_2 \end{vmatrix} = 0.$$

$$\therefore \begin{vmatrix} m_2\nabla + h_1m_2{}^2 - a_2b_2m_1 & b_1m_2{}^2 - b_2{}^2m_1 & 0 \\ a_2{}^2m_1 - a_1m_2{}^2 & m_2\nabla - h_1m_2{}^2 + a_2b_2m_1 & 0 \\ a_2 & b_2 & m_2 \end{vmatrix} = 0.$$

$$\therefore m_2{}^2\nabla^2 - (a_2b_2m_1 - h_1m_2{}^2)^2$$
$$+ (a_2{}^2m_1 - a_1m_2{}^2)(b_2{}^2m_1 - b_1m_2{}^2) = 0,$$

which pure quadratic gives at once the two values of ∇, which may consequently be treated as known. Then from (2) and (3),

$$a_2x + b_2y = m_2,$$
$$(a_1b_2 - h_1a_2)x + (h_1b_2 - b_1a_2) = \nabla;$$

two linear equations from which to find x and y.

11. Eliminate x from the simultaneous equations

$$x^3 - px^2 + qx - r = 0, \qquad (1)$$
$$y = a_1 + b_1x + c_1x^2. \qquad (2)$$

Multiply (2) by x, and in the result substitute the value of x^3 given by (1).

$$\therefore \; xy = c_1 r + (a_1 - c_1 q) x + (b_1 + c_1 p) x^2$$
$$= a_2 + b_2 x + c_2 x^2, \text{ say.} \qquad (3)$$

Repeat with (3) and (1) instead of (2) and (1).

$$\therefore \; x^2 y = c_2 r + (a_2 - c_2 q) x + (b_2 + c_2 p) x^2$$
$$= a_3 + b_3 x + c_3 x^2, \text{ say.}$$

Eliminate x and x^2 from (2), (3), and (4).

$$\therefore \; \begin{vmatrix} a_1 - y & b_1 & c_1 \\ a_2 & b_2 - y & c_2 \\ a_3 & b_3 & c_3 - y \end{vmatrix} = 0,$$

which, on being expanded, gives a cubic in y.

12. To find the condition that

$$U \equiv ax^5 + bx^4 + cx^3 + dx^2 + ex + f$$

and $\qquad V \equiv \alpha x^3 + \beta x^2 + \gamma x + \delta$

may have a common factor, and to find that factor, apply the method of elimination exhibited in Example 4, page 350. The result is :

If
$$\begin{vmatrix} a & b & c & d & e & f & 0 & 0 \\ 0 & a & b & c & d & e & f & 0 \\ 0 & 0 & a & b & c & d & e & f \\ \alpha & \beta & \gamma & \delta & 0 & 0 & 0 & 0 \\ 0 & \alpha & \beta & \gamma & \delta & 0 & 0 & 0 \\ 0 & 0 & \alpha & \beta & \gamma & \delta & 0 & 0 \\ 0 & 0 & 0 & \alpha & \beta & \gamma & \delta & 0 \\ 0 & 0 & 0 & 0 & \alpha & \beta & \gamma & \delta \end{vmatrix} = 0,$$

U and V will have a common factor which, to a constant multiplier, will be

$$\begin{vmatrix} ax+b & c & d & e & f & 0 \\ a & b & c & d & e & f \\ ax+\beta & \gamma & \delta & 0 & 0 & 0 \\ a & \beta & \gamma & \delta & 0 & 0 \\ 0 & a & \beta & \gamma & \delta & 0 \\ 0 & 0 & a & \beta & \gamma & \delta \end{vmatrix}$$

If this determinant vanish *identically*, *i.e.*, if the constant multiplier be zero, U and V will have a common quadratic factor which, except as to a constant multiplier, will be

$$\begin{vmatrix} ax^2+bx+c & d & e & f \\ ax^2+\beta x+\gamma & \delta & 0 & 0 \\ ax+\beta & \gamma & \delta & 0 \\ a & \beta & \gamma & \delta \end{vmatrix}$$

If this determinant vanish identically, U and V will have a common cubic factor which will necessarily be V or V divided by a constant.

EXAMPLE.

Let it be required to find the common quadratic factor of

$$6x^5 - x^4 - x^3 + 10x^2 + 14x - 40$$

and $\qquad 2x^3 + x^2 - x + 10.$

Following the above-described method, it is found to be

$$\begin{vmatrix} 6x^2-x-1 & 10 & 14 & -40 \\ 2x^2+x-1 & 10 & 0 & 0 \\ 2x+1 & -1 & 10 & 0 \\ 2 & 1 & -1 & 10 \end{vmatrix}$$

$$=10\begin{vmatrix} 6x^2-x+7 & 14 & 10 \\ 2x^2+x-1 & 10 & 0 \\ 2x+1 & -1 & 10 \end{vmatrix}$$

$$=100\begin{vmatrix} 6x^2-3x+6 & 15 \\ 2x^2+x-1 & 10 \end{vmatrix}$$

$$= 1500 \begin{vmatrix} 2x^2 - x + 2 & 1 \\ 2x^2 + x - 1 & 2 \end{vmatrix} = 1500\,(2x^2 - 3x + 5).$$

Rejecting from this the constant multiplier, 1500, the common factor is $2x^2 - 3x + 5$, as may be proved either by actual division or by evaluation of the determinant for a linear factor.

Ex. 78.

Apply determinants to solve the following equations:

1. $3x + 7y = 8,$
   $4x + 9y = 11.$

2. $2x + 5y = 20,$
   $3x - 4y = 7.$

3. $3x - 5y + 4z = 5,$
   $7x + 2y - 3z = 2,$
   $4x + 3y - z = 7.$

4. $x - y + z = 6,$
   $7x - 9\frac{1}{2}y + 11z = 64,$
   $23x - 21y + 24z = 154.$

5. $\frac{1}{7}(x + 2y) = \frac{1}{8}(3y + 4z) = \frac{1}{9}(6z + 5x),$
   $x + y - z = 126.$

6. $1 + \frac{1}{2}u + \frac{1}{3}x + \frac{1}{4}y + \frac{1}{5}z = 0,$
   $\frac{1}{2} + \frac{1}{3}u + \frac{1}{4}x + \frac{1}{5}y + \frac{1}{6}z = 0,$
   $\frac{1}{3} + \frac{1}{4}u + \frac{1}{5}x + \frac{1}{6}y + \frac{1}{7}z = 0,$
   $\frac{1}{4} + \frac{1}{5}u + \frac{1}{6}x + \frac{1}{7}y + \frac{1}{8}z = 0.$

7. $\dfrac{24}{2x + 3y} - \dfrac{15}{3x + 4z} = 2,$

   $\dfrac{30}{3x + 4z} + \dfrac{37}{5y + 9z} = 3,$

   $\dfrac{222}{5y + 9z} - \dfrac{8}{2x + 3y} = 5.$

8. $115\,(13 - x) + 719\,(y - 19) - 590\,(37 - z) = 27,$

   $\dfrac{5\,(13 - x) + 2}{y - 19} = \dfrac{37 - z}{y - 21} = 4.$

9. $u + b = a(x + y),$
$x + b = (a + 1)(y + z),$
$y + b = (a + 2)(z + u),$
$z + b = (a + 3)(u + x).$

10. $(a + b + c)x = ay + bz + c,$
$(a + d + e)y = ax + dz + e,$
$(b + d + f)z = bx + dy + f.$
Generalize.

11. Given $x_1 = b_1 y_1, \quad y_2 = a_1 x_1 + y_1, \quad x_2 = b_2 y_2 + x_1,$
$y_3 = a_2 x_2 + y_2, \quad x_3 = b_3 y_3 + x_2, \quad y_4 = a_3 x_3 + y_3,$
$x_4 = b_4 y_4 + x_3 \, ;$

prove that

$$x_4 = -y_1 \begin{vmatrix} -b_4 & -1 & 0 & 0 & 0 & 0 & 0 \\ 1 & -a_3 & -1 & 0 & 0 & 0 & 0 \\ 0 & 1 & -b_3 & -1 & 0 & 0 & 0 \\ 0 & 0 & 1 & -a_2 & -1 & 0 & 0 \\ 0 & 0 & 0 & 1 & -b_2 & -1 & 0 \\ 0 & 0 & 0 & 0 & 1 & -a_1 & -1 \\ 0 & 0 & 0 & 0 & 0 & 1 & -b_1 \end{vmatrix}$$

12. Given $x = b_1/(a_1 + y), \quad y = b_2/(a_2 + z), \quad z = b_3/(a_3 + u),$
and $u = b_4/a_4 \, ;$ prove that

$$x = b_1 \begin{vmatrix} a_2 & -1 & 0 \\ b_3 & a_3 & -1 \\ 0 & b_4 & a^4 \end{vmatrix} \div \begin{vmatrix} a_1 & -1 & 0 & 0 \\ b_2 & a_2 & -1 & 0 \\ 0 & b_3 & a_3 & -1 \\ 0 & 0 & b_4 & a_4 \end{vmatrix}$$

(Take for variables x, xy, xyz, $xyzu$, and eliminate the last three.)

Solve

13. $ax + by - cz = 2ab,$
$by + cz - ax = 2bc,$
$cz + ax - by = 2ac.$

14. $(c + a)x - (c - a)y = 2bc,$
$(a + b)y - (a - b)z = 2ac,$
$(b + c)z - (b - c)x = 2ab.$

15. $(z+x)\,a - (z-x)\,b = 2yz,$
$(x+y)\,b - (x-y)\,c = 2xz,$
$(y+z)\,c - (y-z)\,a = 2xy.$

16. $x + ay + a^2z + a^3u + a^4 = 0,$
$x + by + b^2z + b^3u + b^4 = 0,$
$x + cy + c^2z + c^3u + c^4 = 0,$
$x + dy + d^2z + d^3u + d^4 = 0.$

17. $x + ay + a^2z + a^3u = d,$
$x + by + b^2z + b^3u = a,$
$x + cy + c^2z + c^3u = b,$
$x + dy + d^2z + d^3u = c;$

and if a, b, c, d, are the roots of the quartic

$$v^4 - p_1v^3 + p_2v^2 - p_3v + p_4 = 0,$$

determine x, y, z, u, in terms of p_1, p_2, p_3, p_4.

18. $\quad u + \quad x + \quad y + \quad z = k,$
$au + bx + cy + dz = l,$
$a^2u + b^2x + c^2y + d^2z = m,$
$a^3u + b^3x + c^3y + d^3z = n.$

19. Show that either of the following systems of equations can be reduced to the other:

(1) $x_1 + x_2 + x_3 = u_1,$ (2) $y_1 + ay_2 + a^2y_3 = v_1,$
$ax_1 + bx_2 + cx_3 = u_2,$ $y_1 + by_2 + b^2y_3 = v_2,$
$a^2x_1 + b^2x_2 + c^2x_3 = u_3;$ $y_1 + cy_2 + c^2y_3 = v_3.$
Generalize.

20. There is a certain rational integral expression whose value depends on that of x, and into which x enters in no degree higher than the third. Its value is 4 when $x = 0$, is 9 when $x = 1$, is 20 when $x = 2$, and is 49 when $x = 3$. Find the expression.

Solve

21. $\dfrac{x}{a+k}+\dfrac{y}{b+k}+\dfrac{z}{c+k}+\dfrac{u}{d+k}=1,$

$\dfrac{x}{a+l}+\dfrac{y}{b+l}+\dfrac{z}{c+l}+\dfrac{u}{d+l}=1,$

$\dfrac{x}{a+m}+\dfrac{y}{b+m}+\dfrac{z}{c+m}+\dfrac{u}{d+m}=1,$

$\dfrac{x}{a+n}+\dfrac{y}{b+n}+\dfrac{z}{c+n}+\dfrac{u}{d+n}=1.$

22. $\dfrac{a}{bz+cy}+\dfrac{(a-b)(c-a)}{b-c}=0,$

$\dfrac{b}{cx+az}+\dfrac{(b-c)(a-b)}{c-a}=0,$

$\dfrac{c}{ay+bx}+\dfrac{(c-a)(b-c)}{a-b}=0.$

23. $\begin{vmatrix} 1 & x & x & x \\ x & 1 & c & b \\ x & c & 1 & a \\ x & b & a & 1 \end{vmatrix}=0.$

24. $\begin{vmatrix} 0 & 1 & 1 & 1 \\ 1 & (a^2+b^2)^2 & a^4 & b^4 \\ 1 & x^4 & (x^2+b^2)^2 & b^4 \\ 1 & x^4 & a^4 & (x^2+a^2)^2 \end{vmatrix}=0.$

25. $\begin{vmatrix} a-b & x & x & x \\ x & b-c & x & x \\ x & x & c-d & x \\ x & x & x & d-a \end{vmatrix}=0.$

26. $\begin{vmatrix} a^3 & b^3 & c^3 \\ (a+x)^3 & (b+x)^3 & (c+x)^3 \\ (2a+x)^3 & (2b+x)^3 & (2c+x)^3 \end{vmatrix}=0.$

27. Determine a, b, and c so that the two systems of equations

$$ax + by - cz = l, \qquad a_1x + \beta_1y + \gamma_1z = l_1,$$
$$ax - by + cz = m, \qquad a_2x + \beta_2y + \gamma_2z = m_1,$$
$$-ax + by + cz = n; \qquad a_3x + \beta_3y + \gamma_3z = n_1;$$

may be satisfied by the same values of x, y, z.

Apply to the case

$$ax + by - cz = 4, \qquad 2x - y + 3z = 9,$$
$$ax - by + cz = 8, \qquad 3x + 2y - 2z = 1,$$
$$-ax + by + cz = 16, \qquad -x + y + z = 4.$$

28. Solve $\dfrac{2x + 3y - 4z}{x + 5} = \dfrac{3x + 4y - 2z}{5x} = \dfrac{4x + 2y - 3z}{4x - 1}$

$$= \dfrac{x + y - z}{6}.$$

29. Eliminate x, y, and z from

$$(a_1x + b_3y + b_2z + c_1)/u = (b_3x + a_2y + b_1z + c_2)/v$$
$$= (b_2x + b_1y + a_3z + c_3)/w$$
$$= 1 - c_1x - c_2y - c_3z;$$
$$ux + vy + wz = 1.$$

30. Determine a, given

$$x + y + z + w = 0, \qquad ax + by + cz + dw = 0,$$
$$\frac{x}{a} + \frac{y}{b} + \frac{z}{c} + \frac{w}{e} = 0, \qquad \frac{x}{a^2} + \frac{y}{b^2} + \frac{z}{c^2} + \frac{aw}{e^2} = 0.$$

31. Solve $l_1(l_1x + m_1y + n_1z) = am_1^2 + bn_1^2,$
$$l_2(l_2x + m_2y + n_2z) = am_2^2 + bn_2^2,$$
$$l_3(l_3x + m_3y + n_3z) = am_3^2 + bn_3^2.$$
$$l_2l_3 + m_2m_3 + n_2n_3 = l_3l_1 + m_3m_1 + n_3n_1$$
$$= l_1l_2 + m_1m_2 + n_1n_2 = 0.$$

32. If x_1, y_1; x_2, y_2; x_3, y_3 are the values of x and y that satisfy each possible pair of the equations

$$a_1x + b_1y + c_1 = 0,$$
$$a_2x + b_2y + c_2 = 0,$$
$$a_3x + b_3y + c_3 = 0,$$

prove that

$$|\,1\;x_2y_3\,| = |\,a_1b_2c_3\,|^2 \div \{|\,a_1b_2\,|\,|\,a_2b_3\,|\,|\,a_3b_1\,|\}.$$

33. The equations

$$x + 3y + 5z + 3u = 34, \qquad x + 2y + 5z + 4u = 36,$$
$$x + \ y + 2z + \ u = 13, \qquad x + 3y + 8z + 5u = 51,$$

have for sole solution $x = 1$, $y = 2$, $z = 3$, $u = 4$, but on attempting to find the value of u by indeterminate multipliers, on adding together the equations multiplied respectively by 1, a, β, γ, and equating to zero the coefficients of x, y, and z in the resulting equation, we obtain the incompatible equations,

$$1 + \ a + \ \beta + \ \gamma = 0,$$
$$3 + \ a + 2\beta + 3\gamma = 0,$$
$$5 + 2a + 5\beta + 8\gamma = 0.$$

Explain the paradox.

34. Eliminate x, y, and z from

$$ax + by + cz - 1 = b_1x + a_1y - z + c$$
$$= c_1x - y + a_1z + b = -x + c_1y + b_1z + a = 0.$$

35. Eliminate u, v, w, x, y, z from

$$a_1u + b_1v + c_1w = 0, \qquad a_1x + \beta_1y + \gamma_1z = u,$$
$$a_2u + b_2v + c_2w = 0, \qquad a_2x + \beta_2y + \gamma_2z = v,$$
$$a_3u + b_3v + c_3w = 0; \qquad a_3x + \beta_3y + \gamma_3z = w;$$

and prove thereby that

$$|\,a_1b_2c_3\,| \times |\,a_1\beta_2\gamma_3\,|$$
$$\equiv |\,a_1a_1 + b_1a_2 + c_1a_3,\ a_2\beta_1 + b_2\beta_2 + c_2\beta_3,\ a_3\gamma_1 + b_3\gamma_2 + c_3\gamma_3\,|.$$

36. Eliminate u, v, and w from

$$au + hv + gw = \lambda u,$$
$$hu + bv + fw = \lambda v,$$
$$gu + fv + cw = \lambda w;$$

and u, v, w, x, y, z from the three preceding equations combined with the three following:

$$ax + hy + gz = u - \lambda x,$$
$$hx + by + fz = v - \lambda y,$$
$$gx + fy + cz = w - \lambda z;$$

and reduce the two resultants to the same *form*.

37. Eliminate first, x, y, z, second, f, g, h, from

$$aw + hy - gz = 0,$$
$$bw + fz - hx = 0,$$
$$cw + gx - fy = 0.$$

38. Show that

$$\begin{vmatrix} a_1^2 + b_1^2 + c_1^2 & a_1a_2 + b_1b_2 + c_1c_2 \\ a_1a_2 + b_1b_2 + c_1c_2 & a_2^2 + b_2^2 + c_2^2 \end{vmatrix}$$
$$\equiv |\, a_1b_2\,|^2 + |\, b_1c_2\,|^2 + |\, c_1a_2\,|^2.$$

39. Prove that

$$|\, a_1b_2c_3\,|^2 + |\, b_1c_2d_3\,|^2 + |\, c_1d_2a_3\,|^2 + |\, d_1a_2b_3\,|^2$$
$$\equiv \begin{vmatrix} a_1^2 + b_1^2 + c_1^2 + d_1^2 & a_1a_2 + b_1b_2 + c_1c_2 + d_1d_2 & a_1a_3 + b_1b_3 + c_1c_3 + d_1d_3 \\ a_2a_1 + b_2b_1 + c_2c_1 + d_2d_1 & a_2^2 + b_2^2 + c_2^2 + d_2^2 & a_2a_3 + b_2b_3 + c_2c_3 + d_2d_3 \\ a_3a_1 + b_3b_1 + c_3c_1 + d_3d_1 & a_3a_2 + b_3b_2 + c_3c_2 + d_3d_2 & a_3^2 + b_3^2 + c_3^2 + d_3^2 \end{vmatrix}.$$

Generalize.

40. Let $\Delta_1 \equiv |\, a_1b_2c_3\,|$, $\Delta_2 \equiv |\, a_1\beta_2\gamma_3\,|$, and $\Delta_1\Delta_2 \equiv |\, A_1B_2C_3\,|$,

then will
$$\begin{vmatrix} 0 & a_0 & \beta_0 & \gamma_0 \\ a_0 & A_1 & A_2 & A_3 \\ b_0 & B_1 & B_2 & B_3 \\ c_0 & C_1 & C_2 & C_3 \end{vmatrix}$$

$$\equiv |\, a_0b_1c_2\,|\,|\, a_0\beta_1\gamma_2\,| + |\, a_0b_1c_3\,|\,|\, a_0\beta_1\gamma_3\,| + |\, a_0b_2c_3\,|\,|\, a_0\beta_2\gamma_3\,|.$$

State this theorem in the cases

1°. $\Delta_1 \equiv \Delta_2$, $a_0 = a_0$, $\beta_0 = b_0$, $\gamma_0 = c_0$;

2°. $\Delta_1 \equiv \Delta_2$, $a_0 = a_0$, $\beta_0 = b_0 = \gamma_0 = c_0 = 0$.

41. Given
$$\left. \begin{array}{l} u_1 \equiv a_1 x + b_1 y + c_1 z + d_1 = 0, \\ u_2 \equiv a_2 x + b_2 y + c_2 z + d_2 = 0, \\ u_3 \equiv a_3 x + b_3 y + c_3 z + d_3 = 0, \\ u_4 \equiv a_4 x + b_4 y + c_4 z + d_4 = 0, \\ u_5 \equiv a_5 x + b_5 y + c_5 z + d_5 = 0. \end{array} \right\} \quad (1)$$

1°. Determine the value of x that will satisfy

$$\left. \begin{array}{l} a_1 u_1 + a_2 u_2 + a_3 u_3 + a_4 u_4 + a_5 u_5 = 0, \\ b_1 u_1 + b_2 u_2 + b_3 u_3 + b_4 u_4 + b_5 u_5 = 0, \\ c_1 u_1 + c_2 u_2 + c_3 u_3 + c_4 u_4 + c_5 u_5 = 0. \end{array} \right\} \quad (2)$$

2°. Eliminate z from the group of equations (1), taken two by two in every possible way ; from the resulting ten equations form two equations by a method similar to that by which the set (2) was formed from the group (1); and determine the value of x that will satisfy these two equations. Show that this value of x is the same as the value obtained by the solution of the set (2).

Apply the preceding to the equations
$$\begin{array}{r} x - y + 2z - 3 = 0, \\ 3x + 2y - 5z - 5 = 0, \\ 4x + y + 4z - 21 = 0, \\ x + 3y + 3z - 14 = 0. \end{array}$$

Generalize

42. Eliminate x, y, z, and k from

$$a_1 x + \beta_1 y + \gamma_1 z = 0, \quad a_2 x + \beta_2 y + \gamma_2 z = 0.$$
$$\frac{a_1 x + b_3 y + b_2 z}{a_1 + a_2 k} = \frac{b_3 x + a_2 y + b_1 z}{\beta_1 + \beta_2 k} = \frac{b_2 x + b_1 y + a_3 z}{\gamma_1 + \gamma_2 k}.$$

43. Eliminate x, y, z, u, k_1, and k_2 from

$$a_1x + \beta_1y + \gamma_1z + \delta_1u = 0,$$
$$a_2x + \beta_2y + \gamma_2z + \delta_2u = 0,$$
$$a_3x + \beta_3y + \gamma_3z + \delta_3u = 0,$$

$$\frac{ax + hy + gz + lu}{a_1 + a_2k_1 + a_3k_2} = \frac{hx + by + fz + mu}{\beta_1 + \beta_2k_1 + \beta_3k_2}$$

$$= \frac{gx + fy + cz + nu}{\gamma_1 + \gamma_2k_1 + \gamma_3k_2} = \frac{lx + my + nz + du}{\delta_1 + \delta_2k_1 + \delta_3k_2}.$$

44. If $\dfrac{a_1x + b_1y + c_1z}{u} = \dfrac{a_2x + b_2y + c_2z}{v} = \dfrac{a_3x + b_3y + c_3z}{w}$,

then $\dfrac{x}{A_1u + A_2v + A_3w} = \dfrac{y}{B_1u + B_2v + B_3w}$

$$= \frac{z}{C_1u + C_2v + C_3w},$$

in which A_1, A_2, etc., are the inverse elements of a_1, a_2, etc., with respect to $|\, a_1b_2c_3 \,|$.

45. Eliminate x, y, and z from

$$
\begin{array}{llll}
x^2 & + y^2 & + z^2 & = k^2, \\
(x - a)^2 & + y^2 & + z^2 & = (k - d)^2, \\
(x - a_1)^2 & + (y - b_1)^2 + z^2 & & = (k - d_1)^2, \\
(x - a_2)^2 & + (y - b_2)^2 + (z - c_2)^2 & & = (k - d_2)^2.
\end{array}
$$

46. Apply the results of Examples 1 and 2, pages 348 and 349, to prove that if the value of a determinant be zero, the determinant may be transformed into another of the same order in which all the elements of a row or of a column shall be zero.

$$
\text{Transform} \quad
\begin{vmatrix}
2 & -3 & -4 & 3 \\
3 & 4 & 1 & -2 \\
6 & 5 & -2 & -5 \\
10 & -12 & 3 & 4
\end{vmatrix}
$$

into a determinant of the fourth order with a column of zeros.

Similarly transform $\begin{vmatrix} 1 & 3 & 2 & -4 \\ 4 & -2 & -13 & 5 \\ 2 & 8 & 7 & -11 \\ 3 & -1 & -9 & 3 \end{vmatrix}$.

Apply the above to prove the rule for the multiplication of determinants.

47. Eliminate a, b, c, d, e, f from

$$ax^2 + 2bxy \qquad\qquad + cy^2 \qquad\qquad + 2ex + 2fy + d = 0,$$
$$ax + b(xp + y) + cyp \qquad\qquad + e + fp \quad = 0,$$
$$a \; + \; b(2p + xq) + c(p^2 + yq) \qquad\qquad + fq \quad = 0,$$
$$b(3q + xr) + c(3pq + yr) \qquad\qquad + fr \quad = 0,$$
$$b(4r + xs) + c(4pr + 3q^2 + ys) + fs \quad = 0,$$
$$b(5s + xt) + c(5ps + 10qr + yt) + ft \quad = 0,$$

and evaluate the resulting determinant.

[Omit the first three equations, — this eliminates a, e, and d; in the remaining three, take for variables $b + cp$, cq, and $bx + cy + f$.]

48. If $-A \; + Bz + Cy + Du = 0,$
$\qquad Az - B \; + Cx + Dv = 0,$
$\qquad Ay + Bx - C \; + Dw = 0,$
$\qquad Au + Bv + Cw - D \; = 0,$

then will

$$\frac{A^2}{1 - x^2 - v^2 - w^2 + 2xvw} = \frac{B^2}{1 - u^2 - y^2 - w^2 + 2uyw}$$
$$= \frac{C^2}{1 - u^2 - v^2 - z^2 + 2uvz} = \frac{D^2}{1 - x^2 - y^2 - z^2 + 2xyz}.$$

49. Eliminate $u, v,$ and w from

$$ux_1 + vy_1 + wz_1 = 0,$$
$$ux_2 + vy_2 + wz_2 = 0,$$

$$\begin{vmatrix} a & h & g & u \\ h & b & f & v \\ g & f & c & w \\ u & v & w & 0 \end{vmatrix} = 0.$$

50. Show that the system of equations

$$\frac{a(b-c)}{a-a}+\frac{b(c-a)}{b-\beta}+\frac{c(a-b)}{c-\gamma}=0,$$

$$\frac{a(\beta-\gamma)}{a-a}+\frac{\beta(\gamma-a)}{\beta-b}+\frac{\gamma(a-\beta)}{\gamma-c}=0,$$

is satisfied by either

$$a-b+\beta-a=b-c+\gamma-\beta=c-a+a-\gamma,$$

or $a\beta\gamma=ab\gamma=a\beta c.$

51. If $\dfrac{bx+ay-cz}{a^2+b^2}=\dfrac{cx-by+az}{a^2+c^2}=\dfrac{-ax+cy+bz}{b^2+c^2},$

then will $\begin{vmatrix} x & y & z \\ c & a & b \\ b & c & a \end{vmatrix}=0.$

52. Eliminate x, y, and z from

$$a(x-1)+b(y-1)+c(z-1)=0,$$
$$x+y+z=1,$$

$$\begin{vmatrix} x & y & z \\ c & a & b \\ b & c & a \end{vmatrix}=0.$$

[Reduce the determinant to the form

$$\begin{vmatrix} ax+by+cz & x+y+z \\ ab+bc+ca & a+b+c \end{vmatrix}=0.]$$

53. If $\begin{aligned} lx+my+nz &= 0, \\ ax-by-cz &= u, \\ -ax+by-cz &= v, \\ -ax-by+cz &= w, \\ aux+bvy+cwz &= 1, \end{aligned}$ $\begin{vmatrix} l & m & n \\ x & y & z \\ u & v & w \end{vmatrix}=0,$

then will

$$\frac{m^2+n^2}{(bn+cm)^2}=\frac{n^2+l^2}{(cl+an)^2}=\frac{l^2+m^2}{(am+bl)^2}$$
$$=x^2+y^2+z^2,$$

and $4a^2b^2c^2(x^2+y^2+z^2)^3-(a^2+b^2+c^2)(x^2+y^2+z^2)+1=0.$

54. Exhibit in a single equation the result of eliminating u, x, y, z from

$$ax + hy + gz = a_1 u + \lambda_1 x,$$
$$hx + by + fz = b_1 u + \lambda_1 y,$$
$$gx + fy + cz = c_1 u + \lambda_1 z,$$
$$a_1 x + b_1 y + c_1 z = 0 ;$$

and v, x_1, y_1, z_1 from

$$ax_1 + hy_1 + gz_1 = a_1 v + \lambda_2 x,$$
$$hx_1 + by_1 + fz_1 = b_1 v + \lambda_2 y,$$
$$gx_1 + fy_1 + cz_1 = c_1 v + \lambda_2 z,$$
$$a_1 x_1 + b_1 y_1 + c_1 z_1 = 0.$$

55. If $\quad a_1 + b_1 y + c_1 u = a_2 v + b_2 + c_2 z = a_3 x + b_3 w + c_3 = s,$
and $ux = vy = wz = 1,$
then will

$$\begin{vmatrix} a_1 - s & b_1 & c_1 \\ a_2 & b_2 - s & c_2 \\ a_3 & b_3 & c_3 - s \end{vmatrix} = 0.$$

56. Eliminate λ from

$$\frac{x^2}{a - \lambda} + \frac{y^2}{b - \lambda} + \frac{z^2}{c - \lambda} = 0,$$

$$\frac{xx'}{a - \lambda} + \frac{yy'}{b - \lambda} + \frac{zz'}{c - \lambda} = 0.$$

57. Eliminate λ from

$$\frac{x^2}{a - \lambda} + \frac{y^2}{b - \lambda} + \frac{z^2}{c - \lambda} + \frac{u^2}{g - \lambda} = 0,$$

$$\frac{xx'}{a - \lambda} + \frac{yy'}{b - \lambda} + \frac{zz'}{c - \lambda} + \frac{uu'}{g - \lambda} = 0.$$

58. Eliminate $u, v,$ and w from

$$p^2 / u + q^2 / v + r^2 / w = 0,$$
$$l^2 / u + m^2 / v + n^2 / w = 0,$$
$$P(q - r)u + Q(r - p)v + R(p - q)w = 0.$$

59. If
$$(a_1 - a_2)^2 + (\beta_1 - \beta_2)^2 = r_1^2 + r_2^2,$$
$$(a_1 - a_3)^2 + (\beta_1 - \beta_3)^2 = r_1^2 + r_3^2,$$
$$(a_1 - a_4)^2 + (\beta_1 - \beta_4)^2 = r_1^2 + r_4^2,$$
$$(a_2 - a_3)^2 + (\beta_2 - \beta_3)^2 = r_2^2 + r_3^2,$$
$$(a_2 - a_4)^2 + (\beta_2 - \beta_4)^2 = r_2^2 + r_4^2,$$
$$(a_3 - a_4)^2 + (\beta_3 - \beta_4)^2 = r_3^2 + r_4^2,$$
then will $r_1^{-2} + r_2^{-2} + r_3^{-2} + r_4^{-2} = 0$,

and $\qquad A_1 r_1^{-2} + A_2 r_2^{-2} + A_3 r_3^{-2} + A_4 r_4^{-2} = 0$,

in which A is either a or β.

60. Eliminate u, x, y, z from
$$u + x + y + z = 0,$$
$$au + bx + cy + dz = 0,$$
$$/u + /x + /y + /z = 0,$$
$$a/u + b/x + c/y + d/z = 0.$$

61. Eliminate u, v, w, x, y, z from
$$lx + my + nz = 0,$$
$$lu + mv + nw = 0,$$
$$fyz + gxz + hxy = 0,$$
$$\frac{ux}{b^2 - c^2} = \frac{vy}{c^2 - a^2} = \frac{wz}{a^2 - b^2}.$$

62. Eliminate x, y, z from
$$x + y + z = /u + /v + /w,$$
$$ax + by + cz = /(ax) + /(by) + /(cz),$$
$$\sqrt{(x/u)} + \sqrt{(y/v)} + \sqrt{(z/w)} = 0,$$
$$x/(1 - bc) + y/(1 - ca) + z/(1 - ab) = 0,$$
and reduce the resultant to the form
$$Au + Bv + Cw = 0.$$

63. Eliminate x and y from
$$(1 + x)/(a - y) = /a + x/a,$$
$$(1 + x)/(b - y) = /b + x/\beta,$$
$$(1 + x)/(c - y) = /c + x/\gamma.$$

64. Eliminate x, y, z, s from

$$(s-y)(s-z) = ayz,$$
$$(s-z)(s-x) = bzx,$$
$$(s-x)(s-y) = cxy,$$
$$x + y + z = 2s.$$

65. Given $xyz = a + yz(y+z) = b + zx(z+x)$
$$= c + xy(x+y),$$

show that $4(xyz)^3 - (ab + bc + ca)(xyz) + abc = 0.$

66. Determine λ, μ, ν, given

$$\begin{vmatrix} a-a & b-\beta & c-\gamma \\ l & m & n \\ \lambda & \mu & \nu \end{vmatrix} = 0.$$

$$l\lambda + m\mu + n\nu = 0,$$
$$\lambda^2 + \mu^2 + \nu^2 = 1.$$

67. Find u, v, w, given

$$lu + mv + nw = 0,$$
$$l_1 u + m_1 v + n_1 w = 0,$$
$$u^2 + v^2 + w^2 = 1.$$

68. If $\lambda(ax + hy + gz + lw) + \mu(hx + by + fz + mw)$
$$+ \nu(gx + fy + cz + nw) = 0,$$
$$a\lambda + h\mu + g\nu = u\lambda,$$
$$h\lambda + b\mu + f\nu = u\mu,$$
$$g\lambda + f\mu + c\nu = u\nu,$$

show that $\dfrac{ux + lw}{(a-u)f - gh} + \dfrac{uy + mw}{(b-u)g - fh} + \dfrac{uz + nw}{(c-u)h - fg} = 0.$

69. Resolve a system of three equations in three unknowns of which one equation is quadratic and two are linear.

70. Apply the method of Example 9, p. 355, to resolve

$$x^2 + y^2 + z^2 = 1,$$
$$a_1 x + b_1 y + c_1 z = ux,$$

$$a_2x + b_2y + c_2z = uy,$$
$$a_3x + b_3y + c_3z = uz;$$

and compare the values of u given by the resolution and that obtained by eliminating x, y, and z from the last three equations.

Generalize.

71. Eliminate u_1, v_1, u_2, v_2 from

$$x_1u_2 = x_2u_1, \qquad ax_1 + du_2 = bx_2 + cu_1,$$
$$y_1v_2 = y_2v_1, \qquad av_2 + dy_1 = bv_1 + cy_2,$$
$$(ad - bc)\, x_1u_2 = x_2\,y_2 - x_1y_1 + u_1v_1 - u_2v_2.$$

72. Eliminate x, y, and z from

$$x + y + z = 0,$$
$$ax^2 + by^2 + cz^2 + 2fyz + 2gxz + 2hxy = 0,$$
$$a_1x^3 + b_2y^3 + c_3z^3 + 3a_2x^2y + 3a_3x^2z + 3b_1xy^2$$
$$+ 3b_3y^2z + 3c_1xz^2 + 3c_2yz^2 + 6dxyz = 0.$$

73. Eliminate x, y, and z from

$$x^2 + y^2 - 2gxy = 0,$$
$$y^2 + z^2 - 2hyz = 0,$$
$$z^2 + x^2 - 2kzx = 0;$$

and assuming the resulting relation to hold among g, h, and k, find the H.C.F. of the functions

$$u^2 + v^2 - 2guv - (1 - g^2),$$
$$v^2 + w^2 - 2hvw - (1 - h^2),$$
$$w^2 + u^2 - 2kwu - (1 - k^2).$$

74. Eliminate x, y, and z from

$$(f_1 - u)\, yz + g_1zx + h_1xy = 0,$$
$$f_2yz + (g_2 - u)\, zx + h_2xy = 0,$$
$$f_3yz + g_3zx + (h_3 - u)\, xy = 0.$$

75. Eliminate x, y, and z from

$$a_1x^2+b_1y^2+c_1z^2+2(f_1-u)yz+2g_1zx+2h_1xy=0,$$
$$a_2x^2+b_2y^2+c_2z^2+2f_2yz+2(g_2-u)zx+2h_2xy=0,$$
$$a_3x^2+b_3y^2+c_3z^2+2f_3yz+2g_3zx+2(h_3-u)xy=0.$$

76. Expand

$$\begin{vmatrix} x^2-a^2 & y^2 & z^2 \\ x^2 & y^2-b^2 & z^2 \\ x^2 & y^2 & z^2-c^2 \end{vmatrix}=0.$$

77. Expand

$$\begin{vmatrix} a^2-y^2-z^2 & y^2 & z^2 \\ x^2 & b^2-x^2-z^2 & z^2 \\ x^2 & y^2 & c^2-x^2-y^2 \end{vmatrix}=0,$$

and reduce the expanded equation to the form

$$\frac{a^2x^2}{a^2-s^2}+\frac{b^2y^2}{b^2-s^2}+\frac{c^2z^2}{c^2-s^2}=0,$$
$$s^2\equiv x^2+y^2+z^2.$$

78. If a, β, γ be the values of u satisfying

$$\begin{vmatrix} x+a-u & y & z \\ x & y+b-u & z \\ x & y & z+c-u \end{vmatrix}=0,$$

and if

$$\begin{vmatrix} a-y-z & y & z \\ x & b-z-x & z \\ x & y & c-x-y \end{vmatrix}=0,$$

then will

$$(\sigma-a)(\sigma-\beta)(\sigma-\gamma)=0,$$

in which

$$\sigma\equiv a+\beta+\gamma-a-b-c.$$

79. Solve $x^3+2x^2y+2xy(y-2)+y^2-4=0,$
$$x^2+2xy+2y^2-5y+2=0.$$

80. Solve
$$x^3 + 3xy^2 - 6xy - x + y^2 = 0,$$
$$3x^2y - 3x^2 + y^3 - 3y^2 - y + 3 = 0.$$

81. Solve
$$(x^2 - x + 1)(y^2 - y - 1) = 3,$$
$$(x + 1)(y + 1) = 6.$$
[Transform by $u = x + y$, $v = xy$, and eliminate u.]

82. Solve
$$x^3 + x^3y^3 + y^3 = 17,$$
$$x + xy + y = 5.$$
[Transform by $u = x + y + xy$, $v = (x + y)xy$].

83. Solve
$$x^2 + y^2 + z^2 = (a + u)^2,$$
$$(a - x)^2 + y^2 + z^2 = (\beta + u)^2,$$
$$(a_1 - x)^2 + (b_1 - y)^2 + z^2 = (\gamma + u)^2,$$
$$(a_2 - x)^2 + (b_2 - y)^2 + (c_2 - z)^2 = (\delta + u)^2.$$

84. If $a \equiv a_1\lambda + a_2\mu + a_3\nu$, $\beta \equiv b_1\lambda + b_2\mu + b_3\nu$,
$\gamma \equiv c_1\lambda + c_2\mu + c_3\nu$, $\delta \equiv d_1\lambda + d_2\mu + d_3\nu$.

$$ax_1^3 + \beta x_1^2 + \gamma x_1 + \delta = 0,$$
$$ax_2^3 + \beta x_2^2 + \gamma x_2 + \delta = 0,$$
$$ax_3^3 + \beta x_3^2 + \gamma x_3 + \delta = 0,$$

in which x_1, x_2, and x_3 are the roots of
$$x^3 + px^2 + qx + r = 0,$$
show that
$$\begin{vmatrix} 1 & p & q & r \\ a_1 & b_1 & c_1 & d_1 \\ a_2 & b_2 & c_2 & d_2 \\ a_3 & b_3 & c_3 & d_3 \end{vmatrix} = 0.$$

What does this equation become if $x_1 = x_2$?
What does it become if $x_1 = x_2 = x_3$?

85. If
$$\frac{a_1x_1^3 + b_1x_1^2 + c_1x_1 + d_1}{a_2x_2^3 + b_1x_2^2 + c_1x_2 + d_2} = \frac{a_2x_1^3 + b_2x_1^2 + c_2x_1 + d_2}{a_2x_2^3 + b_2x_2^2 + c_2x_2 + d_2}$$
$$= \frac{a_3x_1^3 + b_3x_1^2 + c_3x_1 + d_3}{a_3x_2^3 + b_3x_2^2 + c_3x_2 + d_3},$$

show that x_1 and x_2 are the roots of the quadratic

$$\begin{vmatrix} x^2 & -x & 1 \\ A & B & C \\ B & C & D \end{vmatrix} = 0,$$

in which A, B, C, and D are the four determinants of the third order that can be made from the array

$$\left\{ \begin{array}{llll} a_1 & b_1 & c_1 & d_1 \\ a_2 & b_2 & c_2 & d_2 \\ a_3 & b_3 & c_3 & d_3. \end{array} \right.$$

86. If $u_1^2 + v_1^2 = u_2^2 + v_2^2 = 1$, then both

$$\{a^2(x - u_1)^2 + b^2(y - v_1)^2\}(xu_2 + yv_2 - 1)^2$$
$$= \{a^2(x - u_2)^2 + b^2(y - v_2)^2\}(xu_1 + yv_1 - 1)^2$$

and $\{a^2(x - u_1)^2 + b^2(y - v_1)^2\} \{(xv_2 - yu_2)^2 - (x - u_2)^2$
$$- (y - v_2)^2\}$$
$$= \{a^2(x - u_2)^2 + b^2(y - v_2)^2\} \{(xv_1 - yu_1)^2$$
$$- (x - u_1)^2 - (y - v_1)^2\}$$

are satisfied by

$$\begin{vmatrix} x & y & 1 \\ u_1 & v_1 & 1 \\ u_2 & v_2 & 1 \end{vmatrix} = 0.$$

87. If $r_1 h_2 h_3 (h_1^2 + l^2) = r_2 h_3 h_1 (h_2^2 + l^2) = r_3 h_1 h_2 (h_3^2 + l^2)$, then will

$$\begin{vmatrix} h_1 & h_2 h_3 & r_2 r_3 \\ h_2 & h_3 h_1 & r_3 r_1 \\ h_3 & h_1 h_2 & r_1 r_2 \end{vmatrix} = 0.$$

88. If $\dfrac{x^2}{a^2} + \dfrac{y^2}{b^2} = \dfrac{x_1^2}{a^2} + \dfrac{y_1^2}{b^2} = \dfrac{x_2^2}{a^2} + \dfrac{y_2^2}{b^2} = 1$,

$$\frac{xx_1}{a^2} + \frac{yy_1}{b^2} + \frac{a_2^2}{2\delta_2^2} = \frac{xx_2}{a^2} + \frac{yy_2}{b^2} + \frac{a_1^2}{2\delta_1^2}$$
$$= \frac{x_1 x_2}{a^2} + \frac{y_1 y_2}{b^2} + \frac{a^2}{2\delta^2} = 1,$$

show that $\begin{vmatrix} 1 & x & y \\ 1 & x_1 & y_1 \\ 1 & x_2 & y_2 \end{vmatrix} = \dfrac{abaa_1a_2}{2\delta\delta_1\delta_2}.$

89. Given that $a_1x^2 + 2b_1x + c_1 = 0$ and $a_2x^2 + 2b_2x + c_2 = 0$ have a common root, determine it.

Apply to case of

$$\begin{cases} 990\,x^2 - 441\,x - 5390 = 0, \\ 825\,x^2 - 428\,x - 4620 = 0. \end{cases}$$

90. Given that $ax^3 + 3bx^2 + 3cx + d$ has a square factor, find it.

Apply to $\quad 2940\,x^3 + 812\,x^2 - 8385\,x - 6300.$

91. Determine the condition that $ax^3 + 3bx^2 + 3cx + d = 0$ and $ax^2 + 2\beta x + \gamma = 0$ shall have a common root, and find it.

Apply to $\quad \begin{cases} 30\,x^3 + x^2 + 35\,x + 204 = 0, \\ 110\,x^2 - 23\,x - 357 = 0. \end{cases}$

92. Determine the condition that

$$ax^4 + 4bx^3 + 6cx^2 + 4dx + e$$

and $\quad ax^3 + 3\beta x^2 + 3\gamma x + \delta$

shall have a common linear factor, and find the common factor.

93. Determine the condition that

$$ax^4 + 4bx^3 + 6cx^2 + 4dx + e$$

and $\quad ax^4 + 4\beta x^3 + 6\gamma x^2 + 4\delta x + \epsilon$

may have a common quadratic factor, and find the common factor.

Apply to $\quad \begin{cases} 60\,x^4 - 4\,x^3 + 37\,x^2 - \ x + 28, \\ 90\,x^4 + 3\,x^3 + 84\,x^2 + 22\,x + 56. \end{cases}$

94. Determine the conditions that
$$a_1 x^2 + 2 b_1 x + c_1 = 0,$$
$$a_2 x^2 + 2 b_2 x + c_2 = 0,$$
$$a_3 x^2 + 2 b_3 x + c^3 = 0,$$
shall have a common root.

95. Determine the condition that
$$a x^4 + 4 b x^3 + 6 c x^2 + 4 d x + e = 0$$
shall have two equal roots.

96. Determine the conditions that
$$a x^4 + 4 b x^3 + 6 c x^2 + 4 d x + e = 0$$
may have three equal roots.

97. Determine the conditions that
$$a x^5 + 5 b x^4 + 10 c x^3 + 10 d x^2 + 5 e x + g = 0$$
may have three equal roots.

98. Determine the conditions that $a x + b y$ may be a common factor of
$$a_1 x^2 + 2 b_1 x y + c_1 y^2$$
and $a_2 x^3 + 3 b_2 x^2 y + 3 c_2 x y^2 + d_2 y^3.$

99. Determine the remainders in the process for finding the H.C.F. of $f(x)^m$ and $F(x)^{m+n}$.

100. If $x^m + b_1 x^{m-1} + b_2 x^{m-2} + \cdots\cdots$ be divided by
$$x^n + a_1 x^{n-1} + a_2 x^{n-2} + \cdots\cdots,$$
then will the coefficient of the rth term of the quotient be

$$(-1)^{r-1} \begin{vmatrix} 1 & 1 & 0 & 0 & \cdots\cdots \\ b_1 & a_1 & 1 & 0 & \cdots\cdots \\ b_2 & a_2 & a_1 & 1 & \cdots\cdots \\ b_3 & a_3 & a_2 & a_1 & \cdots\cdots \\ \cdots & \cdots & \cdots & \cdots & \cdots\cdots \\ \cdots & \cdots & \cdots & \cdots & \cdots\cdots \\ b_{r-1} & a_{r-1} & a_{r-2} & a_{r-3} & \cdots\cdots \end{vmatrix}.$$

101. For what value of a will

$$(a - 2) x^2 - 2x + 5a - 2$$
and $ax^2 - 5x + 4a$

have a common factor?

102. For what values of y will

$$x^3 - 4x^2y + xy^2 - (y - 1)^2$$
and $(x - 1)^2 y + (y - 1)^2$

have a common factor?

103. Find the relations that must hold among a, b, and c, that

$$ax^2 + bx + c$$
and $a(1 - c) x^2 + b\{(1 - c) + ac\} x + c$

may have a common factor.

104. If $2x^4 - x^2 + a = 0$ and $x^2 - x + b = 0$ have a common root, then must

$$(4b - 1)(2b^2 - a) + (a + 2b^2 - b)^2 = 0;$$

and if $b + 2 = 0$, find the values of a and the resulting equations.

105. If $a_1 x^2 + 2b_1 x + c_1 = 0$ and $a_2 x^2 + 2b_2 x + c_2 = 0$ have a common root,

$$(a_1 c_1 - b_1^2) x^2 + (a_1 c_2 + c_1 a_2 - 2b_1 b_2) x + (a_2 c_2 - b_2^2) = 0$$

will have equal roots.

106. If $a_1 x^2 + 2b_1 x + c_1 = 0$ and $a_2 x^2 + 2b_2 x + c_2 = 0$ have a common root, their other roots are given by

$$a_1 a_2 \,|\, b_1 c_2 \,|\, x^2 + |\, a_1 c_2 \,|^2 x + c_1 c_2 \,|\, a_1 b_2 \,| = 0.$$

107. Eliminate x_1, x_2, x_3 from

$$a_1 (x_1 + x_3) = - 2b_1, \quad a_1 x_1 x_3 = c_1,$$
$$a_2 (x_2 + x_3) = - 2b_2, \quad a_2 x_2 x_3 = c_2.$$

Form the quadratic whose roots are x_1, x_2; and the cubic whose roots are x_1, x_2, x_3.

108. Find the condition that must be fulfilled in order that

$$\frac{ux^2}{u+a^2} + \frac{uy^2}{u+b^2} = u+r$$

shall have equal roots in u.

109. Find the condition that

$$\frac{x^2}{u+a^2} + \frac{y^2}{u+b^2} + \frac{z^2}{u+c^2} = 1$$

may have equal roots in u.

110. If $ax^4 + 4bx^3 + 6cx^2 + 4dx + e = 0$ have a double root, it will be given by

$$3(ax^2 + 2bx + c)^2 = I.$$

[For the values of I and J, see (28) and (29), p. 305.]

111. Show that if the quartic

$$ax^4 + 4bx^3 + 6cx^2 + 4dx + e = 0$$

have three equal roots, then will

$$\frac{ad - bc}{2(ac - b^2)} = \frac{ae + 2bd - 3c^2}{3(ad - bc)} = \frac{3(be - cd)}{ae + 2bd - 3c^2}$$

$$= \frac{2(ce - d^2)}{be - cd},$$

and prove that these equations are equivalent to $I = J = 0$.

112. Show that the conditions that the quartic in 111 shall have three equal roots are the same as the conditions that

$$ax^2 + 2bx + c = 0,$$
$$bx^2 + 2cx + d = 0,$$
$$cx^2 + 2dx + e = 0,$$

shall have a common root, and express them as determinants. (See problems 94 and 96 above.)

113. Find the relation that must exist between g and h in order that

$$z^6 - 10g^2z^3 + 12h^4z + 5g^4 = 0$$

may have a pair of equal roots.

For what values of z will each of the following equations have a pair of equal roots?

114. $x^3 + (z - 1)x^2 + (z - 8)x - 6(z - 2) = 0.$

115. $x^3 + 2zx^2 + (z^2 - 5z - 75)x - 5(z^2 + 5z - 50) = 0.$

116. $x^3 + (3z - 2)x^2 - (6z + 15)x - 45z = 0.$

117. $x^3 - 2(z + 2)x^2 + (8z + 3)x - 6z = 0.$

118. $x^3 + (z + 6)x^2 + (4z + 11)x + 3(z + 2) = 0.$

119. $x^3 - 3x + 2z = 0.$ **122.** $x^3 + 3zx - (1 - z)^2 - 4z^3 = 0.$

120. $x^3 - 3x^2 + z^2 = 0.$ **123.** $2x^3 - 3zx + z^3 + 1 = 0.$

121. $x^3 - 3zx + z^3 = 0.$ **124.** $x^4 - 9x^2 + 4x + z = 0.$

125. $x^4 + 4x^3 + 44x^2 - 96x + z = 0.$

126. $\{x^5 - (1 - z)x^4\}5^5 - 4^4z(1 - z)^4 = 0.$

127. $x^4 + x^3 + x^2 + z = 0.$

128. Find the relation that must hold among the coefficients that

$$x^6 + 6bx^5 + 15cx^4 + 20dx^3 + 15ex^2 + 6fx + g^2$$

may break up into the cubic factors

$$x^3 + 3a_1x^2 + 3b_1x + g$$

and $x^3 + 3a_2x^2 + 3b_2x + g.$

129. Show that the discriminant of

$$a(x^5 + y^5) + b(vx^3 + v^{-1}y^3)xy$$
$$+ c(v^2x + v^{-2}y)x^2y^2$$

is a rational integral function of a, b, c, and $(v^5 + v^{-5})$ and of the second degree in the last of these.

130. Prove that
$$(h^2 - ab)(ax^2 + by^2 + 2fy + 2gx + 2hxy)$$
$$- af^2 - bg^2 + 2fgh$$
can be resolved into linear factors, and find them.

131. Show that if
$$a + b + c = 0$$
and $a^2b + b^2c + c^2a + 2mabc = 0,$
then will
$$ax^2 + by^2 + cz^2 + 2(mc + a)yz + 2(ma + b)xz$$
$$+ 2(mb + c)xy$$
be the product of linear factors.　Find them.

132. If $x^2 + 2hxy + y^2 - 5x - 7y + 6$ has linear factors, find them.

133. For what values of λ will
$$2(x^2 + 5y^2 + z^2 - yz - 7xz + 2xy)$$
$$- \lambda(x^2 + 3y^2 + 2z^2)$$
be resolvable into linear factors?

Find the factors in each case.

134. For what values of λ will
$$80x^2 + 8y^2 - 4z^2 - 3yz + 5zx + 33xy$$
$$- \lambda(x^2 + y^2 + z^2)$$
be resolvable into linear factors?

Find the factors in each case.

135. If
$$(l_1x + m_1y + n_1z)^2 + (l_2x + m_2y + n_2z)^2$$
$$+ (l_3x + m_3y + n_3z)^2 \equiv x^2 + y^2 + z^2,$$
show that
$$80x^2 + 8y^2 - 4z^2 - 3yz + 5zx + 33xy$$
$$\equiv a(l_1x + m_1y + n_1z)^2 + \beta(l_2x + m_2y + n_2z)^2$$
$$+ \gamma(l_3x + m_3y + n_3z)^2,$$
in which a, β, and γ are the values of λ found in the preceding problem.

136. Find the condition that
$$(x^2+yz)(b-c)(1+kbc)+(y^2+zx)(c-a)(1+kca)$$
$$+(z^2+xy)(a-b)(1+kab)$$
may break up into linear factors.

137. Find the condition that
$$k(a_1x^2 + b_1y^2 + c_1z^2 + 2f_1yz + 2g_1xz + 2h_1xy)$$
$$+ (a_2x^2 + b_2y^2 + c_2z^2 + 2f_2yz + 2g_2xz + 2h_2xy)$$
shall be resolvable into linear factors.

138. Determine k so that
$$4x^2 - 9y^2 - 2z^2 - 3yz + 2xz + 3xy$$
$$+ k(x - 3y + z)(x + y - 5z)$$
may be resolvable into a pair of linear factors.

139. Find the condition that
$$k(a_1x^3 + 3b_1x^2 + 3c_1x + d_1)$$
$$+ l(a_2x^3 + 3b_2x^2 + 3c_2x + d_2)$$
shall have a square factor, and this condition being fulfilled, find the square factor.

140. Given
$$x^2/a^2 + y^2/b^2 - z^2/c^2 = 0,$$
$$lx + my + nz = 0,$$
find the condition that the ratios $x : y : z$ shall be each single-valued.

141. Given
$$f^2x^2 + g^2y^2 + h^2z^2 - 2ghyz - 2fhxz - 2fgxy = 0$$
and $lx + my + nz = 0,$
find the condition that the ratios $x : y : z$ shall be each single-valued.

142. If $\quad \alpha x + \beta y + \gamma z = 0,$

and $f/x + g/y + h/z = 0,$

and if the ratios $x : y : z$ are each single-valued, then will

$$(f\alpha)^{\frac{1}{2}} + (g\beta)^{\frac{1}{2}} + (h\gamma)^{\frac{1}{2}} = 0.$$

143. Find the condition that if

$$ax^2 + by^2 + cz^2 + 2fyz + 2gxz + 2hxy = 0,$$

and
$$\begin{vmatrix} a & h & g & 0 & z & -y & 0 \\ h & b & f & -z & 0 & x & 0 \\ g & f & c & y & -x & 0 & 0 \\ 0 & -z & y & 1 & 0 & 0 & l \\ z & 0 & -x & 0 & 1 & 0 & m \\ -y & x & 0 & 0 & 0 & 1 & n \\ 0 & 0 & 0 & l & m & n & l^2 + m^2 + n^2 \end{vmatrix} = 0,$$

the ratios $x : y : z$ shall be each single-valued.

144. Given $ax^2 + by^2 + cz^2 + 2fyz + 2gxz + 2hxy = 0$

and $lx + my + nz = 0,$

find the condition that the ratios $x : y : z$ shall be each single-valued.

145. Eliminate x, y, and z from

$$f/x + g/y + h/z = 0,$$
$$\alpha x + \beta y + \gamma z = 0,$$
$$\{\alpha x (s - x)\}^{\frac{1}{2}} + \{\beta y (s - y)\}^{\frac{1}{2}} + \{\gamma z (s - z)\}^{\frac{1}{2}} = 0,$$
$$2s = x + y + z.$$

146. If $\quad u \equiv ax^2 + by^2 + cz^2 + 2fyz + 2gxz + 2hxy$

be resolvable into linear factors, and if $u = 0$, then will

$$\begin{vmatrix} a & h & ax+hy+gz \\ h & b & hx+by+fz \\ ax+hy+gz & hx+by+fz & 0 \end{vmatrix} = 0.$$

147. Prove that the roots of

$$\begin{vmatrix} a-x & h & g \\ h & b-x & f \\ g & f & c-x \end{vmatrix} = 0$$

are all three real.

[See last part of problem 36 of this Exercise.]

148. Show that the roots of

$$\begin{vmatrix} a-x & h & g & k \\ h & b-x & f & l \\ g & f & c-x & m \\ k & l & m & 0 \end{vmatrix} = 0$$

lie between the roots of the equation of problem 147.

149. Find the condition that the equation of problem 148 shall have equal roots.

150. Show that the roots of

$$\begin{vmatrix} a-a_1x & h-h_1x & g-g_1x \\ h-h_1x & b-b_1x & f-f_1x \\ g-g_1x & f-f_1x & c-c_1x \end{vmatrix} = 0$$

are real.

[Reduce to a determinant of the form of that in problem 147.]

151. Reduce

$$\begin{vmatrix} ax+a_1 & mx+m_1 & lx+l_1 \\ hx+h_1 & bx+b_1 & kx+k_1 \\ gx+g_1 & fx+f_1 & cx+c_1 \end{vmatrix}$$

to a determinant of the third order with x in the principal diagonal elements only.

152. If $A, B, C, -F, G, -H$ denote the minors of

$$\begin{vmatrix} a & h & g \\ h & b & f \\ g & f & c \end{vmatrix},$$

show that

$$\begin{vmatrix} ax+ hy+ gz & hx+ by+ fz & gx+ fy+ cz \\ Ax+Hy+Gz & Hx+By+Fz & Gx+Fy+Cz \\ x & y & z \end{vmatrix}$$

can be resolved into linear factors.

Find the factors in the case

$$a = 3, \; b = 4, \; c = 5, \; f = 1, \; g = 2, \; h = 3.$$

153. Show that $13\,x^2 + 5y^2 - 16\,xy - 2$ is a factor of the resultant of the elimination of z from

$$10\,y^2 + 13\,z^2 - 6\,yz = 242$$
and $5\,z^2 + 10\,x^2 - 2\,xz = 98.$

154. Eliminate y from

$$a^2(x^2 + xy + y^2) - axy(x + y) + x^2y^2 = 0,$$
$$a^2(y^2 + yz + z^2) - ayz(y + z) + y^2z^2 = 0,$$

and show that

$$a^2(x^2 + xz + z^2) - axz(x + z) + x^2z^2$$

is a factor of the resultant.

155. If $a_1, \; a_2$ are the roots of $a_1x^2 + 2b_1x + c_1 = 0,$

       $\beta_1, \; \beta_2$ "   "   "   " $a_2x^2 + 2b_2x + c_2 = 0,$

       $\gamma_1, \; \gamma_2$ "   "   "   " $a_3x^2 + 2b_3x + c_3 = 0,$

form the equations whose roots are

     (i.) $2\,u_1 \equiv \beta_1\gamma_2 + \beta_2\gamma_1, \quad 2\,u_2 \equiv \beta_1\gamma_1 + \beta_2\gamma_2,$

     (ii.) $2\,v_1 \equiv a_1\gamma_2 + a_2\gamma_1, \quad 2\,v_2 \equiv a_1\gamma_1 + a_2\gamma_2,$

     (iii.) $2\,w_1 \equiv a_1\beta_2 + a_2\beta_1, \quad 2\,w_2 \equiv a_1\beta_1 + a_2\beta_2,$

and show that the elements inverse to u, v, and w, respectively in

$$\begin{vmatrix} 1 & b_1 & b_2 & b_3 \\ b_1 & a_1c_1 & a_1a_2w & a_1a_3v \\ b_2 & a_1a_2w & a_2c_2 & a_2a_3u \\ b_3 & a_1a_3v & a_2a_3u & a_3c_3 \end{vmatrix}$$

all vanish.

[The required equations are the elements inverse to c_1, c_2, and c_3 in this determinant.]

156. If $ax^2 + by^2 + cz^2 + 2fyz + 2gxz + 2hxy$ be resolvable into linear factors, the coefficients of y in these factors will be the roots of

$$au^2 - 2hu + b = 0,$$

and those of z will be the roots of

$$au^2 - 2gu + c = 0,$$

and f must be a root of

$$au^2 - 2hgu + h^2c + g^2b - abc = 0.$$

Example. For what values of u will

$$x^2 + 12y^2 - 21z^2 - uyz - 4xz + 7xy$$

be the product of linear factors?

157. Apply problem 155 to determine the conditions that

$$ax^2 + by^2 + cz^2 + dw^2 + 2fyz + 2gxz + 2hxy$$
$$+ 2lxw + 2myw + 2nzw$$

shall be the product of linear factors.

For what value of n will

$$x^2 + 10y^2 + 9z^2 + 5w^2 + 18yz + 6xz + 6xy$$
$$+ 2xw - 2yw + 2nzw$$

be the product of linear factors? Find the factors.

158. If $\quad x^3 + 3a_1x^2y + 3b_1xy^2 + c_1y^3 + 3a_2x^2z + 3b_2xz^2$
$$+ c_2z^3 + 3a_3y^2z + 3b_3yz^2 + 6exyz$$

be resolvable into linear factors, determine the condition that these factors should vanish simultaneously for values of x, y, z, other than zero.

159. If $\quad u \equiv ax^2 + by^2 + cz^2 + 2fyz + 2gxz + 2hxy,$
and $v \equiv a_1x^2 + b_1y^2 + c_1z^2 + 2f_1yz + 2g_1xz + 2h_1xy,$
find the equation expressing the condition that

$u + kv$ shall be the product of linear factors. If this equation have equal roots in k, show that the resultant of the elimination of y between $u = 0$ and $v = 0$ has equal roots in x/z.

160. Find the square root of the resultant of the elimination of u from

$$au^4 - 2xu^2 + a = 0,$$
$$au^2 - 2yu - a = 0.$$

161. Eliminate x from $ax^2 + bx + c = 0$ and $x^3 = 1$.

162. Eliminate x from $ax^4 + bx^3 + cx^2 + dx + e = 0$ and $x^5 = 1$.

163. Eliminate x from

$$x^4 + 6Ax^2 - 4Bx + C = 0$$
$$\text{and } 2y^3 + 2xy^2 + x^2y + 3Ay + B = 0,$$

and show how to apply the resultant to obtain a solution of the quartic in x.

164. Given $\beta = b - a$, $\beta_1 = c - b$, $\beta_2 = d - c$, $\beta_3 = e - d$,

$\gamma = \beta_1 - \beta$, $\gamma_1 = \beta_2 - \beta_1$, $\gamma_2 = \beta_3 - \beta_2$,

$\delta = \gamma_1 - \gamma$, $\delta_1 = \gamma_2 - \gamma_1$,

$\eta = \delta_1 - \delta$,

show that

$$\begin{vmatrix} a & \beta & \gamma \\ \beta & \gamma & \delta \\ \gamma & \delta & \eta \end{vmatrix} = \begin{vmatrix} a & b & c \\ b & c & d \\ c & d & e \end{vmatrix}.$$

165. Show that if $\begin{vmatrix} a & b & c \\ b & c & d \\ c & d & e \end{vmatrix} = 0$, then will

$$(ac - b^2) \begin{vmatrix} ax^2 + 2bxy + cy^2 & bx^2 + 2cxy + dy^2 \\ bx^2 + 2cxy + dy^2 & cx^2 + 2dxy + ey^2 \end{vmatrix}$$
$$= \begin{vmatrix} ax + by & bx + cy \\ bx + cy & cx + dy \end{vmatrix}^2.$$

166. Given $a_0x^4 + 4a_1x^3 + 6a_2x^2 + 4a_3x + a_4 = 0$,

$$a_0u + a_1v + a_2w = 0,$$
$$\beta_0u + \beta_1v + \beta_2w = 0,$$
$$\gamma_0u + \gamma_1v + \gamma_2w = 0,$$
$$a_ma_n = \beta_m\beta_n = \gamma_m\gamma_n = a_{m+n},$$

show that

$$\begin{vmatrix} a_0 & a_1 & a_2 \\ \beta_0 & \beta_1 & \beta_2 \\ \gamma_0 & \gamma_1 & \gamma_2 \end{vmatrix} = 27 \begin{vmatrix} a_0 & a_1 & a_2 \\ a_1 & a_2 & a_3 \\ a_2 & a_3 & a_4 \end{vmatrix}.$$

167. If $S_m \equiv a^m + \beta^m + \gamma^m + \delta^m + $ etc., in which a, β, γ, δ, etc. are the roots of

$$a_0x^n + a_1x^{n-1} + a_2x^{n-2} + \cdots + a_n = 0,$$

show that

$$S_m = (-a_0^{-1})_m \begin{vmatrix} a_1 & a_0 & 0 & 0 & \ldots & 0 \\ 2a_2 & a_1 & a_0 & 0 & \ldots & 0 \\ 3a_3 & a_2 & a_1 & a_0 & \ldots & 0 \\ \vdots & & & & & \vdots \\ ma_m & a_{m-1} & a_{m-2} & a_{m-3} & \ldots & a_1 \end{vmatrix}$$

and that

$$a_m = \frac{(-a_0)^m}{1 \times 2 \times 3 \cdots m} \begin{vmatrix} S_1 & 1 & 0 & 0 & \ldots & 0 \\ S_2 & S_1 & 2 & 0 & \ldots & 0 \\ S_3 & S_2 & S_1 & 3 & \ldots & 0 \\ \vdots & & & & \ldots & \vdots \\ S_m & S_{m-1} & S_{m-2} & S_{m-3} & \ldots & S_1 \end{vmatrix}.$$

168. If $S_n \equiv a^n + b^n + c^n + $ etc., then will

$$\begin{vmatrix} S_0 & S_1 \\ S_1 & S_2 \end{vmatrix} \equiv \Sigma(a-b)^2,$$

$$\begin{vmatrix} S_0 & S_1 & S_2 \\ S_1 & S_2 & S_3 \\ S_2 & S_3 & S_4 \end{vmatrix} \equiv \Sigma(a-b)^2(a-c)^2(b-c)^2,$$

$$\begin{vmatrix} S_0 & S_1 & S_2 & S_3 \\ S_1 & S_2 & S_3 & S_4 \\ S_2 & S_3 & S_4 & S_5 \\ S_3 & S_4 & S_5 & S_6 \end{vmatrix} \equiv \Sigma(a-b)^2(a-c)^2(a-d)^2 \\ (b-c)^2(b-d)^2(c-d)^2.$$

Generalize.

169. If $a_1, a_2, a_3, \ldots a_n$ are the roots of $f(x)^n = 0$, show that

$$f(x)^n = \begin{vmatrix} x & a_1 & a_1 & a_1 & \ldots & a_1 & a_1 \\ a_1 & x & a_2 & a_2 & & a_2 & a_2 \\ a_2 & a_2 & x & a_3 & & a_3 & a_3 \\ a_3 & a_3 & a_3 & x & & a_4 & a_4 \\ & & & & & & \\ a_{n-1} & a_{n-1} & a_{n-1} & a_{n-1} & & x & a_n \\ c & c & c & c & & c & c \end{vmatrix}.$$

170. If $S_m = a_1^m + a_2^m + a_3^m + \cdots + a_n^m$, show that

$$\begin{vmatrix} 1 & x & \ldots & x^n \\ S_0 & S_1 & \ldots & S_n \\ S_1 & S_2 & \ldots & S_{n+1} \\ \vdots & \vdots & & \\ S_{n-1} & S_n & \ldots & S_{2n-1} \end{vmatrix} = \begin{vmatrix} S_0 & S_1 & \ldots & S_{n-1} \\ S_1 & S_2 & & S_n \\ S_2 & S_3 & & S_{n+1} \\ & & & \\ S_{n-1} & S_n & & S_{2n-2} \end{vmatrix}$$

$$\times \begin{vmatrix} x & a_1 & a_1 & \ldots & a_1 & a_1 \\ a_1 & x & a_2 & \ldots & a_2 & a_2 \\ a_2 & a_2 & x & \ldots & a_3 & a_3 \\ & & & & & \\ a_{n-1} & a_{n-1} & a_{n-1} & & x & a_n \\ 1 & 1 & 1 & & 1 & 1 \end{vmatrix}.$$

171. Resolve into factors,

$$\begin{vmatrix} 0 & 1 & x & x^2 & x^3 \\ 1 & S_0 & S_1 & S_2 & S_3 \\ y & S_1 & S_2 & S_3 & S_4 \\ y^2 & S_2 & S_3 & S_4 & S_5 \\ y^3 & S_3 & S_4 & S_5 & S_6 \end{vmatrix}$$

in which $S_n \equiv a^n + b^n + c^n$.

172. Show that

$$\begin{vmatrix} S_1 & 1 & 0 & 0 & \ldots & 0 \\ S_2 & S_1 & 2 & 0 & \ldots & 0 \\ S_3 & S_2 & S_1 & 3 & \ldots & 0 \\ \ldots & \ldots & \ldots & \ldots & \ldots & \ldots \\ \ldots & \ldots & \ldots & \ldots & \ldots & m-1 \\ S_m & S_{m-1} & S_{m-2} & S_{m-3} & \ldots & S_1 \end{vmatrix}$$

$$= (-1)^n 1 \times 2 \times 3 \cdots (n-1) n a_1 a_2 \cdots a_n, \text{ if } m = n,$$

but $= 0$, if $m > n$,

in which $S_r \equiv a_1{}^r + a_2{}^r + a_3{}^r + \cdots + a_n{}^r$,

r being any positive integer.

173. If $u_m = a_1 a_1{}^m + a_2 a_2{}^m + a_3 a_3{}^m + a_4 a_4{}^m + a_5 a_5{}^m$, prove that

$$\begin{vmatrix} u_0 & u_1 & u_2 & u_3 & u_4 & u_5 \\ u_1 & u_2 & u_3 & u_4 & u_5 & u_6 \\ u_2 & u_3 & u_4 & u_5 & u_6 & u_7 \\ u_3 & u_4 & u_5 & u_6 & u_7 & u_8 \\ u_4 & u_5 & u_6 & u_7 & u_8 & u_9 \\ u_5 & u_6 & u_7 & u_8 & u_9 & u_{10} \end{vmatrix} = 0.$$

174. Obtain G in determinant form by eliminating x between

$$ax^3 + 3bx^2 + 3cx + d = 0$$

and $ax + b - y = 0$.

(See § 52, p. 297.)

175. Obtain I in determinant form by eliminating x between

$$ax^4 + 4bx^3 + 6cx^2 + 4dx + e = 0$$

and $ax + b - y = 0$.

(See Ex. 70, prob. 1, p. 312.)

176. Express H, Δ, I, J, and $I^3 - 27 J^2$ in determinant form, given as data the propositions stated in prob. 26, Ex. 68, and probs. 14 and 15, Ex. 70.

177. Express H_s, I_s, J_s, G_s of the equation

$$S_0 x^4 + 4 S_1 x^3 + 6 S_2 x^2 + 4 S_3 x + S_4 \equiv \Sigma_1{}^4 (x + r_1)^4 = 0$$

in terms of H, I, J, G of the equation

$$a_0 x^4 + 4 a_1 x^3 + 6 a_2 x^2 + 4 a_3 x + a_4 = 0,$$

of which r_1, r_2, r_3, r_4 are the roots.

178. Express H_S, I_S, J_S, G_S, and $I_S^3 - 27 J_S^2$, as functions of the differences of the roots of the quartic.

179. Express Δ_S as a function of the differences of the roots of the cubic.

180. If $x = \lambda_1 u + \mu_1 v,$
$$y = \lambda_2 u + \mu_2 v,$$

transforms
$$ax^2 + 2hxy + by^2$$

into $Au^2 + 2Huv + Bv^2$,

find the value of
$$\begin{vmatrix} A, & H \\ H, & B \end{vmatrix} \div \begin{vmatrix} a, & h \\ h, & b \end{vmatrix}.$$

181. If $x = \lambda_1 u + \mu_1 v + \nu_1 w,$
$$y = \lambda_2 u + \mu_2 v + \nu_2 w,$$
$$z = \lambda_3 u + \mu_3 v + \nu_3 w,$$

transforms
$$ax^2 + by^2 + cz^2 + 2fyz + 2gxz + 2hxy$$

into $Au^2 + Bv^2 + Cw^2 + 2Fvw + 2Guw + 2Huv$,

find the value of
$$\begin{vmatrix} A & H & G \\ H & B & F \\ G & F & C \end{vmatrix} \div \begin{vmatrix} a & h & g \\ h & b & f \\ g & f & c \end{vmatrix}.$$

182. If $x = (\lambda_1 y + \mu_1 z)/(\lambda_2 y + \mu_2 z),$
$$a_1 x^2 + 2b_1 x + c_1$$
$$\equiv (A_1 y^2 + 2B_1 yz + C_1 z^2)/(\lambda_2 y + \mu_2 z)^2,$$
and $a_2 x^2 + 2b_2 x + c_2$
$$\equiv (A_2 y^2 + 2B_2 yz + C_2 z^2)/(\lambda_2 y + \mu_2 z)^2,$$

then will
$$\begin{vmatrix} A_1 & 2B_1 & C_1 & 0 \\ 0 & A_1 & 2B_1 & C_1 \\ A_2 & 2B_2 & C_2 & 0 \\ 0 & A_2 & 2B_2 & C_2 \end{vmatrix} = \begin{vmatrix} a_1 & 2b_1 & c_1 & 0 \\ 0 & a_1 & 2b_1 & c_1 \\ a_2 & 2b_2 & c_2 & 0 \\ 0 & a_2 & 2b_2 & c_2 \end{vmatrix} \times \begin{vmatrix} \lambda_1 & \mu_1 \\ \lambda_2 & \mu_2 \end{vmatrix}^4.$$

183. If the quartic $(a, b, c, d, e)(x, 1)^4 = 0$ be transformed by the homographic transformation

$$x = (\lambda_1 y + \mu_1)/(\lambda_2 y + \mu_2),$$

then will

$$H_y = M^2 H_x, \qquad G_y = M^3 G_x,$$
$$I_y = M^4 I_x, \qquad J_y = M^6 J_x,$$

in which $\quad M \equiv \begin{vmatrix} \lambda_1 & \mu_1 \\ \lambda_2 & \mu_2 \end{vmatrix}.$

[M is called the modulus of the transformation.]

184. Find Δ_y/Δ_x for the homographic transformation of the cubic $(a, b, c, d)(x, 1)^3 = 0$.

185. If $\quad \lambda_1 = (p^2 - q^2)/(p^2 + q^2), \quad \mu_1 = 2pq/(p^2 + q^2),$
$\lambda_2 = -2pq/(p^2 + q^2), \quad \mu_2 = (p^2 - q^2)/(p^2 + q^2),$
$x = \lambda_1 u + \mu_1 v, \qquad y = \lambda_2 u + \mu_2 v,$

then will $\quad x^2 + y^2 \equiv u^2 + v^2.$

[A transformation that changes $x_1^2 + x_2^2 + x_3^2 + \cdots + x_n^2$ into $u_1^2 + u_2^2 + u_3^2 + \cdots + u_n^2$ is termed an orthogonal transformation of the nth order.]

186. Form an orthogonal transformation of the third order and determine the value of its modulus.
(See prob. 181 above.)

187. Form an orthogonal tranformation of the fourth order.

188. Show that $H, I, J, G, I^3 - 27 J^2$ are the same for both the quartics

$$\begin{vmatrix} a_0 x^2 + 2 b_0 xy + c_0 y^2 & a_1 x^2 + 2 b_1 xy + c_1 y^2 \\ a_1 x^2 + 2 b_1 xy + c_1 y^2 & a_2 x^2 + 2 b_2 xy + c_2 y^2 \end{vmatrix} = 0,$$

$$\begin{vmatrix} a_0 x^2 + 2 a_1 xy + a_2 y^2 & b_0 x^2 + 2 b_1 xy + b_2 y^2 \\ b_0 x^2 + 2 b_1 xy + b_2 y^2 & c_0 x^2 + 2 c_1 xy + c_2 y^2 \end{vmatrix} = 0.$$

189. Apply Example 7, p. 336, to solve the cubic
$$x^3 + 3\,Hx + G = 0.$$

190. Form the equation whose roots are the products in pairs of those of $x^3 + px^2 + qx + r = 0$.

[$a_1 = \beta\gamma, \therefore aa_1 = a\beta\gamma = -r. \therefore xy + r = 0$; eliminate x.]

191. Form the equation whose roots are the products in pairs of those of $x^4 + px^3 + qx^2 + rx + s = 0$.

192. a, β, γ being the roots of the cubic $(a, b, c, d)(x, 1)^3 = 0$, form the equation whose roots are $a\beta + \gamma$, $\beta\gamma + a$, $\gamma a + \beta$.

193. If a, β, γ be the roots of
$$\begin{vmatrix} x & 0 & 0 & -a \\ 1 & x & 0 & b \\ 0 & 1 & x & -c \\ 0 & 0 & 1 & d \end{vmatrix} = 0,$$
then will $\beta\gamma, a\gamma, a\beta$ be the roots of
$$\begin{vmatrix} y^2 & 0 & 0 & 0 & 1 \\ 0 & y & 0 & 1 & 0 \\ 0 & 0 & 1 & 0 & 0 \\ 0 & a & b & c & d \\ a & b & c & d & 0 \end{vmatrix} = 0.$$

194. If $(a, b, c, d)(x, y)^3 \equiv A(x + \theta_1 y)^3 + B(x + \theta_2 y)^3$, show that θ_1, θ_2 are the roots of
$$\begin{vmatrix} a & b & c \\ b & c & d \\ 1 & \theta & \theta^2 \end{vmatrix} = 0.$$

195. If the cubic $(a, b, c, d)(x, 1)^3 = 0$ be transformed into a cubic in y by means of the equation
$$y = a(ax + b) + \beta(ax^2 + 3bx + 2c),$$
show that this cubic is

$$\begin{vmatrix} y-ab-2\beta c & aa-3\beta b & -\beta a \\ \beta d & y-ab+\beta c & -ab \\ ad & 3ac+\beta d & y+2ab+\beta c \end{vmatrix}=0.$$

196. Determine the condition that the roots of

$$(a, b, c, d)(x, 1)^3 = 0$$

may be formed from those of

$$(m, n, p, q)(x, 1)^3 = 0$$

by adding the same quantity to each.

197. Given $ax^3 + 3bx^2 + 3cx + d = 0$
and $gy^2 + 2hy + k = 0$,

express in the form of a determinant equated to zero, the equation whose roots are $lx + my$.

198. Being given the cubic $(a, b, c, d)(x, 1)^3 = 0$, express m, p, and q in terms of a, b, c, and d, so that the values which $my^2 + 2py + q$ takes when y is replaced successively by a, β, γ, the three roots of the cubic, are the three roots in the order β, γ, a.

199. Determine the relation that must exist among the coefficients of the cubic $(a, b, c, d)(x, 1)^3 = 0$, in order that

$$Aa + B\beta + C\gamma = 0,$$

a, β, γ being the roots of the cubic.

200. If a, β, γ, a_1, β_1, γ_1, are the roots of the cubics

$$(a, b, c, d)(x, 1)^3 = 0, \quad (a_1, b_1, c_1, d)(x, 1)^3 = 0,$$

form the equation whose roots are $aa_1 + \beta\beta_1 + \gamma\gamma_1$, etc.

201. a, β, γ, δ being the roots of a quartic, form the equation whose roots are

$$\beta\gamma\delta + \gamma\delta + \beta\delta + \beta\gamma + \beta + \gamma + \delta, \text{ etc.}$$

202. Also the equation whose roots are

$$(a-\beta)(a-\gamma)(a-\delta),\ (\beta-\gamma)(\beta-\delta)(\beta-a),\ \ldots\ldots$$

203. And the equation whose roots are

$$(a-\beta)(\gamma-\delta),\ (a-\gamma)(\delta-\beta),\ (a-\delta)(\beta-\gamma).$$

From this result prove that if $I^3 - 27J^2 = 0$, the quartic will have equal roots.

(See prob. 14, Ex. 70, p. 312.)

204. If a, β, γ, δ be the roots of $(a, b, c, d, e)(x, 1)^4 = 0$, then will the equation whose roots are $(a-\beta)^2$, $(a-\gamma)^2$, etc., be

$$\begin{vmatrix} -3a & a^2z & \frac{1}{4}a^2z^2+Hz+4I \\ a^2z & a(\frac{1}{4}a^2z^2+4Hz+I) & 6J \\ \frac{1}{4}a^2z^2+4Hz+I & aIz+6J & -2Jz \end{vmatrix} = 0.$$

205. a, β, γ, δ being the roots of the quartic

$$(a, b, c, d, e)(x, 1)^4 = 0,$$

express the product

$$\{x-(\beta-\gamma)^2\}\{x-(a-\delta)^2\}$$

in terms of a, b, c, d, and a single root of the reducing cubic $t^3 - It + 2J = 0$, and hence form the equation of the squares of the differences of the quartic.

206. Solve $\{x-a(a\beta+\gamma\delta)\}\{x-a(a\gamma+\beta\delta)\}\{x-a(a\delta+\beta\gamma)\}$

$$= 16 \begin{vmatrix} a & b & c \\ b & c & d \\ c & d & e \end{vmatrix}$$

in which a, β, γ, δ are the roots of

$$(a, b, c, d, e)(x, 1)^4 = 0.$$

207. Find the relation which connects the coefficients of the cubics

$$U \equiv (a,\ b,\ c,\ d\ )(x,\ 1)^3,$$
$$V \equiv (a',\ b',\ c',\ d')(x,\ 1)^3,$$

when it is possible to determine the ratio λ/μ so that $\lambda U + \mu V$ may be a perfect cube.

208. If the roots a, β, γ, δ of $(a, b, c, d, e)(x, 1)^4 = 0$ are all unequal, and if there exist unequal magnitudes θ and ϕ, such that

$$(a + \theta)^4 : (\beta + \theta)^4 : (\gamma + \theta)^4 : (\delta + \theta)^4$$
$$:: (a + \phi)^4 : (\beta + \phi)^4 : (\gamma + \phi)^4 : (\delta + \phi)^4,$$

show that $\begin{vmatrix} a & b & c \\ b & c & d \\ c & d & e \end{vmatrix} = 0,$

and form the quadratic determining θ and ϕ.

209. If $\quad \dfrac{1}{a - \beta} + \dfrac{2}{\gamma - a} + \dfrac{1}{a - \delta} = 0,$

in which a, β, γ, δ are the roots of

$$(a, b, c, d, e)(x, 1)^4 = 0,$$

show that $\begin{vmatrix} a & b & c \\ b & c & d \\ c & d & e \end{vmatrix} = 0.$

210. If $a - \sqrt{\beta}$, a, and $a + \sqrt{\beta}$ be three of the roots of the quartic $(a, b, c, d, e)(x, 1)^4 = 0$, show that

$$\begin{vmatrix} 5, & 0, & 3H, & -G, & 0 \\ 0, & 5, & 0, & 3H, & -G \\ 27H, & 21G, & 5(a^2I - 3H^2), & 0, & 0 \\ 0, & 27H, & 21G, & 5(a^2I - 3H^2), & 0 \\ 0, & 0, & 27H, & 21G, & 5(a^2I - 3H^2) \end{vmatrix} = 0.$$

211. Show how to solve the quartic $(a, b, c, d, e)(x, 1)^4 = 0$ by assuming

$$ax^4 + 4bx^3 + 6cx^2 + 4dx + e$$
$$\equiv a(x^2 + 2u_1x + v_1)(x^2 + 2u_2x + v_2)$$
$$u_1u_2 = c - t,$$

and eliminating u_1, u_2, v_1, v_2.

$$\left[\begin{vmatrix} 1 & 1 & 0 \\ u_1 & u_2 & 0 \\ v_1 & v_2 & 0 \end{vmatrix}\begin{vmatrix} 1 & u_2 & v_2 \\ 1 & u_1 & v_1 \\ 0 & 0 & 0 \end{vmatrix}\equiv\begin{vmatrix} 2 & u_1+u_2 & v_1+v_2 \\ u_1+u_2 & 2u_1u_2 & u_1v_2+u_2v_1 \\ v_1+v_2 & u_1v_2+u_2v_1 & 2v_1v_2 \end{vmatrix}=0.\right.$$

$$\therefore\ \begin{vmatrix} a & b & c+2t \\ b & c-t & d \\ c+2t & d & e \end{vmatrix}=0;\ \text{i.e., } 4t^3 - It + J = 0.\,]$$

212. Show that the reduction of the quartic

$$(a, b, c, d, e)(x, 1)^4 = 0$$

to the biquadratic form

$$(x^2 + 1)^2 + Ax(x^2 + 1) + Bx^2 = 0$$

depends upon the solution of the cubic

$$(ax^4 + 4bx^3 + 6cx^2 + 4dx + e)(ax + b)^2$$
$$= a(ax^3 + 3bx^2 + 3cx + d)^2.$$

213. If a, β, and a_1, β_1 are the roots of

$$ax^2 + 2bx + c = 0 \text{ and } a_1x^2 + 2b_1x + c_1 = 0,$$

respectively, show that

$$\begin{vmatrix} a & 2b & c & 0 \\ 0 & a & 2b & c \\ a_1 & 2b_1 & c_1 & 0 \\ 0 & a_1 & 2b_1 & c_1 \end{vmatrix}$$
$$\equiv a^2a_1^2(a - a_1)(a - \beta_1)(\beta - a_1)(\beta - \beta_1).$$

214. Similarly, resolve the resultant of

$$ax^3 + 3bx^2 + 3cx + d = 0 \text{ and } a_1x^2 + 2b_1x + c_1 = 0$$

into a product of the differences of the roots of the two equations.

215. By eliminating g, h, k, and l from

$$a = (ga + h)(ka + l), \quad \beta = (gb + h)(kb + l),$$
$$\gamma = (gc + h)(kc + l), \quad \delta = (gd + h)(kd + l),$$

prove that

$$\begin{vmatrix} 1 & a & a & aa \\ 1 & b & \beta & b\beta \\ 1 & c & \gamma & c\gamma \\ 1 & d & \delta & d\delta \end{vmatrix} \equiv \begin{vmatrix} 1 & ab+cd & a\beta+\gamma\delta \\ 1 & bc+da & \beta\gamma+\delta a \\ 1 & ac+bd & a\gamma+\beta\delta \end{vmatrix}$$

$$\equiv (a-\beta)(b-c)(\gamma-\delta)(d-a)$$
$$\quad -(a-b)(\beta-\gamma)(c-d)(\delta-a)$$

$$\equiv (a-\gamma)(c-d)(\delta-\beta)(b-a)$$
$$\quad -(a-c)(\gamma-\delta)(d-b)(\beta-a)$$

$$\equiv (a-\delta)(d-b)(\beta-\gamma)(c-a)$$
$$\quad -(a-d)(\delta-\beta)(b-c)(\gamma-a).$$

216. Similarly, prove that

$$\begin{vmatrix} 1 & a & aa \\ 1 & b & b\beta \\ 1 & c & c\gamma \end{vmatrix} \equiv (a-b)(\beta-\gamma)c - (a-\beta)(b-c)a,$$

and that

$$\begin{vmatrix} 1 & a+a & aa \\ 1 & b+\beta & b\beta \\ 1 & c+\gamma & c\gamma \end{vmatrix}$$
$$\equiv (a-b)(\beta-\gamma)(c-a)+(a-\beta)(b-c)(\gamma-a).$$

217. a, β, γ, δ and a_1, β_1, γ_1, δ_1 being the roots of two quartics, prove that if $I^3 J_1^2 = I_1^3 J^2$, then also will

$$\begin{vmatrix} 1 & a & a_1 & aa_1 \\ 1 & \beta & \beta_1 & \beta\beta_1 \\ 1 & \gamma & \gamma_1 & \gamma\gamma_1 \\ 1 & \delta & \delta_1 & \delta\delta_1 \end{vmatrix} = 0.$$

218. If u, v, w denote the roots of $a^3x^3 - aIx + 2J = 0$, and u_1, v_1, w_1 the roots of $a_1^3x^3 - a_1 I_1 x + 2J_1 = 0$, a, β, γ, δ the roots, and I and J the invariants, of $(a,b,c,d,e)(x,1)^4 = 0$, and a_1, β_1, γ_1, δ_1 the roots, and I_1 and J_1 the invariants, of $(a_1,b_1,c_1,d_1,e_1)(x,1)^4 = 0$, then will

$$\begin{vmatrix} 1 & a & a_1 & aa_1 \\ 1 & \beta & \beta_1 & \beta\beta_1 \\ 1 & \gamma & \gamma_1 & \gamma\gamma_1 \\ 1 & \delta & \delta_1 & \delta\delta_1 \end{vmatrix} = 4 \begin{vmatrix} 1 & u & u_1 \\ 1 & v & v_1 \\ 1 & w & w_1 \end{vmatrix}.$$

219. If a, β, γ, δ be the roots of $(a, b, c, d, e)(x, 1)^4 = 0$, and a_1, β_1, γ_1, δ_1 those of $(a_1, b_1, c_1, d_1, e_1)(x, 1)^4 = 0$, form the equation whose roots are the twelve different values of

$$\begin{vmatrix} 1 & a & a_1 & aa_1 \\ 1 & \beta & \beta_1 & \beta\beta_1 \\ 1 & \gamma & \gamma_1 & \gamma\gamma_1 \\ 1 & \delta & \delta_1 & \delta\delta_1 \end{vmatrix}.$$

220. If $(a - a_1)(\beta - \beta_1) + (a_1 - \beta)(\beta_1 - a) = 0$,

in which a and β are the roots of $ax^2 + 2bx + c = 0$, and a_1 and β_1 are the roots of $a_1 x^2 + 2b_1 x + c_1 = 0$, show that

$$ac_1 - 2bb_1 + ca_1 = 0,$$

and that $ac_1 - 2bb_1 + ca_1$ is a factor of the invariant J of the quartic

$$(ax^2 + 2bx + c)(a_1 x^2 + 2b_1 x + c_1) = 0.$$

221. Reduce

$$\{x^2 + y^2 - (1-k)gx + m^2\}^2$$
$$= (1+k)^2 \{h^2(x^2 + y^2) - g^2 y^2\}$$

to the form

$$\{x^2 + y^2 + Ax + B\}^2 = ax^2 + bx + c,$$

and show that $b^2 - 4ABb + 4A^2c \equiv 0$.

222. So determine k and l in terms of a, b, and c, that

$$(x^2 + y^2 + z^2)^2 + 2ax^2 + 2by^2 + 2cz^2 + 2kx + l = 0$$

may for $y = 0$ assume the form

$$(x^2 + z^2 + m)(x^2 + z^2 + n) = 0;$$

and for $z = 0$, the form

$$(x^2 + y^2 + m_1)(x^2 + y^2 + n_1) = 0.$$

223. If a, β, γ, δ, four of the roots of the quintic

$$(a, b, c, d, e, f)(x, 1)^5 = 0,$$

be connected by the relation $a + \beta = \gamma + \delta$, show that ϵ, the fifth root, will be given by the equations

$$z = a\epsilon + b, \quad z^3 - 8Hz + 16G = 0.$$

224. If a, β, γ, δ, four of the roots of the quintic

$$(a, b, c, d, e, f)(x, 1)^5 = 0,$$

be connected by the relation

$$\Sigma(2a - \beta - \gamma)(2\beta - \gamma - a)(2\gamma - a - \beta) = 0,$$

then will ϵ, the fifth root, be determined by

$$z = a\epsilon + b, \quad z^3 - 8Hz + 16G = 0.$$

225. If $(a - \beta)(\gamma - \delta) + (\beta - \gamma)(\delta - a) = 0$, in which a, β, γ, δ are four of the roots of the quintic

$$ax^5 + b = c(x + 1)^5,$$

then will

$$J^3 - 2^7 3^2 JK + 2^{13} 3^3 L = 0,$$

in which

$$J \equiv b^2 c^2 + c^2 a^2 + a^2 b^2 - 2abc(a + b + c),$$
$$K \equiv a^2 b^2 c^2 (bc + ca + ab),$$
$$L \equiv a^4 b^4 c^4.$$

226. Reduce $(a, b, c, d, e, f)(x, y)^5$ to the form

$$k_1(x + h_1 y)^5 + k_2(x + h_2 y)^5 + k_3(x + h_3 y)^5,$$

and hence prove that $(\cdots)(x, 1)^5 = 0$ can be reduced to the form

$$l(x + 1)^5 - mx^5 - n = 0.$$

227. Resolve $\begin{vmatrix} 0 & 1 & 1 & 1 \\ 1 & 0 & c^2 & b^2 \\ 1 & c^2 & 0 & a^2 \\ 1 & b^2 & a^2 & 0 \end{vmatrix}$ into linear factors.

228. a, β, γ being the roots of $x^3 - px^2 + qx - r = 0$, express

$$\begin{vmatrix} 0 & a & \beta & \gamma \\ \gamma & 0 & \gamma & \beta \\ \beta & \gamma & 0 & a \\ a & \beta & a & 0 \end{vmatrix} \div \begin{vmatrix} 0 & 1 & 1 & 1 \\ 1 & 0 & \gamma & \beta \\ 1 & \gamma & 0 & a \\ 1 & \beta & a & 0 \end{vmatrix}$$

in terms of p, q, and r.

229. Find the value of

$$\begin{vmatrix} 0 & a & b & c \\ a & 0 & c & b \\ b & c & 0 & a \\ c & b & a & 0 \end{vmatrix} \div \begin{vmatrix} 0 & 1 & 1 & 1 & 1 \\ 1 & 0 & a & b & c \\ 1 & a & 0 & b & c \\ 1 & b & b & 0 & a \\ 1 & c & c & a & 0 \end{vmatrix}.$$

230. Show that

$$\begin{vmatrix} 0 & a^2 & b^2 & c^2 \\ a^2 & 0 & \gamma^2 & \beta^2 \\ b^2 & \gamma^2 & 0 & a^2 \\ c^2 & \beta^2 & a^2 & 0 \end{vmatrix}$$

$$= 16 p^2 (p^2 - a a)(p^2 - b\beta)(p^2 - c\gamma)$$
$$2p^2 \equiv aa + b\beta + c\gamma.$$

231. If $\begin{vmatrix} 0 & 1 & 1 & 1 & 1 \\ 1 & 0 & a^2 & \beta^2 & \gamma^2 \\ 1 & a^2 & 0 & c^2 & b^2 \\ 1 & \beta^2 & c^2 & 0^2 & a^2 \\ 1 & \gamma^2 & b^2 & a^2 & 0 \end{vmatrix} = 0$,

then will

$$(a^2 + b^2 + c^2 + a^2 + \beta^2 + \gamma^2)(a^2 a^2 + b^2 \beta^2 + c^2 \gamma^2)$$
$$= 2 a^2 a^2 (a^2 + a^2) + 2 b^2 \beta^2 (b^2 + \beta^2) + 2 c^2 \gamma^2 (c^2 + \gamma^2)$$
$$+ a^2 \beta^2 \gamma^2 + b^2 \gamma^2 a^2 + c^2 a^2 \beta^2 + a^2 b^2 c^2.$$

232. If $a^2 + \beta^2 + a\beta = c^2,$
$\beta^2 + \gamma^2 + \beta\gamma = a^2,$
$\gamma^2 + a^2 + \gamma a = b^2,$

then will

$$\begin{vmatrix} 0 & 1 & 1 & 1 & 1 \\ 1 & 0 & a^2 & \beta^2 & \gamma^2 \\ 1 & a^2 & 0 & c^2 & b^2 \\ 1 & \beta^2 & c^2 & 0 & a^2 \\ 1 & \gamma^2 & b^2 & a^2 & 0 \end{vmatrix} = 0.$$

233. If
$$x^2 + y^2 - 2axy = c^2, \quad \text{and} \quad \begin{vmatrix} 1 & \gamma & \beta \\ \gamma & 1 & a \\ \beta & a & 1 \end{vmatrix} = 0,$$
$$y^2 + z^2 - 2\beta yz = a^2,$$
$$z^2 + x^2 - 2\gamma zx = b^2,$$
then will
$$\begin{vmatrix} 0 & 1 & 1 & 1 & 1 \\ 1 & 0 & x^2 & y^2 & z^2 \\ 1 & x^2 & 0 & c^2 & b^2 \\ 1 & y^2 & c^2 & 0 & a^2 \\ 1 & z^2 & b^2 & a^2 & 0 \end{vmatrix} = 0.$$

234. Prove that
$$\begin{vmatrix} 0 & x & y & z \\ -x & 0 & c & b \\ -y & -c & 0 & a \\ -z & -b & -a & 0 \end{vmatrix} \equiv (ax - by + cz)^2,$$

and generalize the theorem.

235. Evaluate
$$\begin{vmatrix} x & y & z \\ y & x & y \\ z & z & x \end{vmatrix} \quad \text{and use the result to prove that}$$
$$u^3 + v^3 + w^3 - 3uvw$$
$$\equiv (a^3 + b^3 + c^3 - 3abc)(x^3 + y^3 + z^3 - 3xyz),$$
wherein
$$u = ax + by + cz, \quad v = cx + ay + bz, \quad w = bx + cy + az.$$

236. Evaluate
$$\begin{vmatrix} 1 & x & y \\ x & 1 & z \\ y & z & 1 \end{vmatrix} \quad \text{and use the result to prove that}$$
$$1 - (2x^2 - 1)^2 - (2y^2 - 1)^2 - (2z - 1)^2$$
$$+ 2(2x^2 - 1)(2y^2 - 1)(2z^2 - 1)$$
$$= -4(1 - x^2 - y^2 - z^2 + 2xyz)(1 - x^2 - y^2 - z^2 - 2xyz).$$

237. $\begin{vmatrix} a & b & c & d \\ -b & a & -d & c \\ -c & d & a & -b \\ -d & -c & b & a \end{vmatrix} \equiv \begin{vmatrix} a+ib & c+id \\ -c+id & a-id \end{vmatrix}^2$

$$\equiv (a^2 + b^2 + c^2 + d^2)^2, \; i^2 + 1 = 0.$$

Apply this identity to prove that the product of the sum of four squares by the sum of four squares can be reduced to the sum of four squares.

238. Prove that $\begin{vmatrix} a & pb & qc & pqd \\ -b & a & -qd & qc \\ -c & pd & a & -pb \\ -d & -c & b & a \end{vmatrix}$

$$\equiv (a^2 + pb^2 + qc^2 + pqd^2)^2.$$

Hence prove that

$$(a^2 + pb^2 + qc^2 + pqd^2)(A^2 + pB^2 + qC^2 + pqD^2)$$
$$\equiv (aA + pbB + qcC + pqdD)^2$$
$$+ p(-aB + bA - qcD + qdC)^2$$
$$+ q(-aC + pbD + cA - pdB)^2$$
$$+ pq(-aD - bC + cB + dA)^2.$$

239. Show that $\begin{vmatrix} a^2 & 2ab & b^2 \\ a\alpha & a\beta+ab & b\beta \\ a^2 & 2a\beta & \beta^2 \end{vmatrix} \equiv (a\beta + ab)^3,$

and generalize.

If $u \equiv (x - a_1)(x - a_2)(x - a_3) \cdots (x - a_n)$

$$\equiv x^n - p_1 x^{n-1} + p_2 x^{n-2} \cdots (-)^n p_n, \text{ show that:}$$

240. $\begin{vmatrix} 0 & 1 & 1 & 1 & 1 & \cdots \\ 1 & x & a_2 & a_3 & a_4 & \cdots \\ 1 & a_1 & x & a_3 & a_4 & \cdots \\ 1 & a_1 & a_2 & x & a_4 & \cdots \\ 1 & a_1 & a_2 & a_3 & x & \cdots \\ \cdots & \cdots & \cdots & \cdots & \cdots & \cdots \end{vmatrix}$

$$= -nx^{n-1} + (n-1)p_1 x^{n-2} - (n-2)p_2 x^{n-3} + \cdots$$

241.
$$\begin{vmatrix} 0 & 1 & 1 & 1 & \ldots \\ 1 & a_1 & x & x & \ldots \\ 1 & x & a_2 & x & \ldots \\ 1 & x & x & a_3 & \ldots \\ \ldots & \ldots & \ldots & \ldots & \ldots \end{vmatrix}$$

$$= (-1)^n \{ nx^{n-1} - (n-1)p_1 x^{n-2} + \ldots \}.$$

242.
$$\begin{vmatrix} x & a_2 & a_3 & \ldots, \\ a_1 & x & a_3 & \ldots \\ a_1 & a_2 & x & \ldots \\ \ldots & \ldots & \ldots & \ldots \end{vmatrix} = u + \Sigma \frac{a_r u}{x - a_r}.$$

243. Show that $\begin{vmatrix} a+x & x & x & x \\ x & \beta+x & x & x \\ x & x & \gamma+x & x \\ x & x & x & \delta+x \end{vmatrix}$

$$\equiv \alpha\beta\gamma\delta x (/a + /\beta + /\gamma + /\delta + /x).$$

244. Prove that $\begin{vmatrix} 0 & 1 & 1 & 1 & 1 \\ 1 & c & a & a & a \\ 1 & b & c & a & a \\ 1 & b & b & c & a \\ 1 & b & b & b & c \end{vmatrix} \equiv \dfrac{(c-a)^4 - (c-b)^4}{a-b}.$

and generalize.

245. Writing $f(x)$ for $(c_1 - x)(c_2 - x) \ldots (c_n - x)$, show that

$$\begin{vmatrix} c_1 & a & a & a & \ldots \\ b & c_2 & a & a & \ldots \\ b & b & c_3 & a & \ldots \\ b & b & b & c_4 & \ldots \\ \ldots & \ldots & \ldots & \ldots & \ldots \end{vmatrix} \equiv \frac{af(b) - bf(a)}{a - b}.$$

246. Show that $\begin{vmatrix} 0 & a_2 & a_3 & a_4 & a_5 \\ b_1 & 0 & a_3 & a_4 & a_5 \\ b_1 & b_2 & 0 & a_4 & a_5 \\ b_1 & b_2 & b_3 & 0 & a_5 \\ b_1 & b_2 & b_3 & b_4 & 0 \end{vmatrix}$

$$\equiv b_1 a_2 a_3 a_4 a_5 + b_1 b_2 a_3 a_4 a_5 + b_1 b_2 b_3 a_4 a_5 + b_1 b_2 b_3 b_4 a_5.$$

247.
$$\begin{vmatrix} a & b & b & b & b & b & b & \cdots \\ c & b & c & c & c & c & c & \cdots \\ a & a & c & a & a & a & a & \cdots \\ b & b & b & a & b & b & b & \cdots \\ c & c & c & c & b & c & c & \cdots \\ a & a & a & a & a & c & a & \cdots \\ b & b & b & b & b & b & a & \cdots \\ \cdots & \cdots & \cdots & \cdots & \cdots & \cdots & \cdots & \cdots \end{vmatrix}$$

$$\equiv \{(a-b)(b-c)(c-a)\}^n$$
$$\times \left\{ 1 + n\left(\frac{a}{c-a} + \frac{b}{a-b} + \frac{c}{b-c} \right) \right\}.$$

248.
$$\begin{vmatrix} 0 & 1 & 1 & 1 & 1 & 1 & 1 & \cdots \\ 1 & a & b & b & b & b & b & \cdots \\ 1 & c & b & c & c & c & c & \cdots \\ 1 & a & a & c & a & a & a & \cdots \\ 1 & b & b & b & a & b & b & \cdots \\ 1 & c & c & c & c & b & c & \cdots \\ 1 & a & a & a & a & a & c & \cdots \\ \cdots & \cdots & \cdots & \cdots & \cdots & \cdots & \cdots & \cdots \end{vmatrix}$$

$$\equiv n\{(a-b)(b-c)(c-a)\}^{n-1}$$
$$\times (a^2 + b^2 + c^2 - bc - ca - ab).$$

Prove the two following identities, and generalize them :

249.
$$\begin{vmatrix} a-x & b & c \\ b & c-x & a \\ c & a & b-x \end{vmatrix} \equiv (x - S_0)(x^2 - S_1 S_2),$$

$$S_n \equiv a + \omega^n b + \omega^{2n} c, \quad \omega^2 + \omega + 1 = 0.$$

250.
$$\begin{vmatrix} a-x & b & c & d \\ b & c-x & d & a \\ c & d & a-x & b \\ d & a & b & c-x \end{vmatrix} \equiv (x - S_0)(x - S_2)(x^2 - S_1 S_3),$$

$$S_n \equiv a + i^n b + i^{2n} c + i^{3n} d, \quad i^2 + 1 = 0.$$

251. $\begin{vmatrix} (S-u)^2 & x^2 & y^2 & z^2 \\ u^2 & (S-x)^2 & y^2 & z^2 \\ u^2 & x^2 & (S-y)^2 & z^2 \\ u^2 & x^2 & y^2 & (S-z)^2 \end{vmatrix}$

$$\equiv 2S^5 uxyz (/u + /x + /y + /z - 4/S)$$

wherein

$$S \equiv u + x + y + z.$$

252. $\begin{vmatrix} 0 & 1 & 1 & 1 \\ 1 & (y+z)^2 & y^2 & z^2 \\ 1 & x^2 & (x+z)^2 & z^2 \\ 1 & x^2 & y^2 & (x+y)^2 \end{vmatrix}$

$$\equiv (x+y+z)^2 (x^2+y^2+z^2-2yz-2xz-2xy).$$

253. $\begin{vmatrix} 0 & 1 & 1 & 1 & 1 \\ 1 & (S-u)^2 & x^2 & y^2 & z^2 \\ 1 & u^2 & (S-x)^2 & y^2 & z^2 \\ 1 & u^2 & x^2 & (S-y)^2 & z^2 \\ 1 & u^2 & x^2 & y^2 & (S-z)^2 \end{vmatrix}$

$$\equiv S^3 \{ u^2(S-2u) + x^2(S-2x) + y^2(S-2y)$$
$$+ z^2(S-2z) + 2uxyz(/u + /x + /y + /z) \},$$
$$S \equiv u + x + y + z.$$

Prove the four identities next following, in which

$$U \equiv (x - 2a_1)(x - 2a_2) \cdots\cdots (x - 2a_n).$$

254. $\begin{vmatrix} (x-a_1)^2 & a_2^2 & a_3^2 & \cdots \\ a_1^2 & (x-a_2)^2 & a_3^2 & \cdots \\ a_1^2 & a_2^2 & (x-a_3)^2 & \cdots \\ \cdots & \cdots & \cdots & \cdots \end{vmatrix}$

$$\equiv U x^{n-1} \left\{ x + \Sigma \frac{a_m^2}{x - 2a_m} \right\}.$$

255. $\begin{vmatrix} 0 & 1 & 1 & 1 & \cdots \\ 1 & (x-a_1)^2 & a_2^2 & a_3^2 & \cdots \\ 1 & a_1^2 & (x-a_2)^2 & a_3^2 & \cdots \\ 1 & a_1^2 & a_2^2 & (x-a_3)^2 & \cdots \\ \cdots & \cdots & \cdots & \cdots & \cdots \end{vmatrix}$

$$\equiv -nU x^{n-1} \Sigma \frac{1}{x - 2a_m}.$$

256.
$$\begin{vmatrix} (x-a_1)^2 & a_1a_2 & a_1a_3 & \ldots \\ a_1a_2 & (x-a_2)^2 & a_2a_3 & \ldots \\ a_1a_3 & a_2a_3 & (x-a_3)^2 & \ldots \\ \ldots & \ldots & \ldots & \ldots \end{vmatrix}$$

$$\equiv Ux^{n-1}\left\{ x + \Sigma\, \frac{a_m}{x-2a_m} \right\}.$$

257.
$$\begin{vmatrix} 0 & a_1 & a_2 & a_3 & \ldots \\ a_1 & (x-a_1)^2 & a_1a_2 & a_1a_3 & \ldots \\ a_2 & a_1a_2 & (x-a_2)^2 & a_2a_3 & \ldots \\ a_3 & a_1a_3 & a_2a_3 & (x-a_3)^2 & \ldots \\ \ldots & \ldots & \ldots & \ldots & \ldots \end{vmatrix}$$

$$\equiv -\, Ux^{n-1}\, \Sigma\, \frac{a_m^{\,2}}{x-2a_m}.$$

258.
$$\begin{vmatrix} x-A_1 & a_2 & a_3 & \ldots & a_n \\ a_1 & x-A_2 & a_3 & \ldots & a_n \\ a_1 & a_2 & x-A_3 & \ldots & a_n \\ \ldots & \ldots & \ldots & \ldots & \ldots \\ a_1 & a_2 & a_3 & \ldots & x-A_n \end{vmatrix} \equiv x(x-S)^{n-1},$$

$$S \equiv a_1 + a_2 + \cdots + a_n, \quad A_m = S - a_m.$$

259. If $f(a,b) \equiv \dfrac{a-k}{b-k} + \dfrac{a-l}{b-l} + \dfrac{a-m}{b-m} + \dfrac{a-n}{b-n}$,

show that

$$\begin{vmatrix} f(a,a) & f(b,a) & f(c,a) & f(d,a) \\ f(a,b) & f(b,b) & f(c,b) & f(d,b) \\ f(a,c) & f(b,c) & f(c,c) & f(d,c) \\ f(a,d) & f(b,d) & f(c,d) & f(d,d) \end{vmatrix} \equiv 0.$$

260. Expand $\begin{vmatrix} \lambda-c & 2a & 3a & 2a \\ 2b & \lambda-c & 2a & 3a \\ 3b & 2b & \lambda-c & 2a \\ 2b & 3b & 2b & \lambda-c \end{vmatrix}$.

261. Expand $\begin{vmatrix} a & b & c & d-\lambda \\ b & c & d+\frac{1}{3}\lambda & e \\ c & d-\frac{1}{3}\lambda & e & f \\ d+\lambda & e & f & g \end{vmatrix}$.

262. Show that if

$$(x-a)(x-b)(x-c) \equiv x^3 - px^2 + qx - r,$$

$$\begin{vmatrix} a+b-c & 4a & 6a & 4a \\ 4b & a+b-c & 4a & 6a \\ 6b & 4b & a+b-c & 4a \\ 4b & 6b & 4b & a+b-c \end{vmatrix}$$

$$= \begin{vmatrix} p & q & r & 0 \\ -4 & p & q & r \\ 0 & -16 & p & q \\ 0 & 0 & -4 & p \end{vmatrix}.$$

263. If $a+b+c=0$, $ab+bc+ca=q$, $abc=r$, then will

$$\begin{vmatrix} \lambda - b^2c^2 & ab(a^2+c^2) & ac(a^2+b^2) \\ ab(b^2+c^2) & \lambda - a^2c^2 & bc(a^2+b^2) \\ ac(b^2+c^2) & bc(a^2+c^2) & \lambda - a^2b^2 \end{vmatrix}$$

$$= \lambda^3 - \lambda q(q^3 - 2r^2) + q^6.$$

264. Show that

$$\begin{vmatrix} \lambda - 2b\beta - 2c\gamma & a\beta + ab & a\gamma + ac \\ a\beta + ab & \lambda - 2a\alpha - 2c\gamma & b\gamma + \beta c \\ a\gamma + ac & b\gamma + \beta c & \lambda - 2a\alpha - 2b\beta \end{vmatrix}$$

$$= \{(\lambda + a\alpha + b\beta + c\gamma)^2$$
$$- (a^2+b^2+c^2)(a^2+\beta^2+\gamma^2)\}(\lambda - 2a\alpha - 2b\beta - 2c\gamma).$$

265. Show that

$$\begin{vmatrix} \lambda + a\alpha + a'\alpha' & ba + b'\alpha' & ca + c'\alpha' & \cdots \\ a\beta + a'\beta' & \lambda + b\beta + b'\beta' & c\beta + c'\beta' & \cdots \\ a\gamma + a'\gamma' & b\gamma + b'\gamma' & \lambda + c\gamma + c'\gamma' & \cdots \\ \cdots & \cdots & \cdots & \cdots \end{vmatrix}$$

$$= \lambda^{n-2}\{\lambda^2 + \lambda(A+B) + AB - CD\};$$

wherein n is the order of the determinant, and

$$A = a\alpha' + bb' + cc' + \cdots$$
$$B = a\alpha' + \beta\beta' + \gamma\gamma' + \cdots$$
$$C = a\alpha' + b\beta' + c\gamma' + \cdots$$
$$D = a\alpha' + \beta b' + \gamma c' + \cdots$$

266. Show that

$$\begin{vmatrix} 1 & a & a & a^2 \\ 1 & \beta & \beta & \beta^2 \\ 1 & \gamma & \gamma' & \gamma\gamma' \\ 1 & \delta & \delta' & \delta\delta' \end{vmatrix} \equiv \begin{vmatrix} 1 & a \\ 1 & \beta \end{vmatrix} \times \begin{vmatrix} 1 & 1 & 1 \\ a+\beta & \gamma+\delta' & \delta+\gamma' \\ a\beta & \gamma\delta' & \delta\gamma' \end{vmatrix}$$

and generalize the proposition.

267. Show that

$$\begin{vmatrix} a_1 & b_1 & c_1 & d_1 \\ a_2 & b_2 & c_2 & d_2 \\ a_3 & b_3 & 0 & 0 \\ a_4 & b_4 & 0 & 0 \end{vmatrix} \equiv \begin{vmatrix} a_3 & b_4 \end{vmatrix} \times \begin{vmatrix} c_1 & d_2 \end{vmatrix}$$

and generalize the proposition.

268. Given $(x_1 - x_2)^2 + (y_1 - y_2)^2 = a^2$,

$$(x_2 - x_3)^2 + (y_2 - y_3)^2 = b^2,$$
$$(x_3 - x_4)^2 + (y_3 - y_4)^2 = c^2,$$
$$(x_4 - x_1)^2 + (y_4 - y_1)^2 = d^2,$$
$$(x_1 - x_3)^2 + (y_1 - y_3)^2 = h^2,$$
$$(x_2 - x_4)^2 + (y_2 - y_4)^2 = k^2,$$

$$S \equiv \begin{vmatrix} 1 & 1 & x_1 & y_1 \\ 0 & 1 & x_2 & y_2 \\ 1 & 1 & x_3 & y_3 \\ 0 & 1 & x_4 & y_4 \end{vmatrix},$$

then will $16S^2 = 4h^2k^2 - (a^2 - b^2 + c^2 - d^2)^2$.

If $hk = ac + bd$ and $2s = a + b + c + d$,

then will $S^2 = (s-a)(s-b)(s-c)(s-d)$.

269. $\begin{vmatrix} a_1 & b_1 & c_1 \\ a_2 & b_2 & c_2 \\ a_3 & b_3 & c_3 \end{vmatrix} \begin{vmatrix} x_1 & y_1 & z_1 \\ x_2 & y_2 & z_2 \\ x_3 & y_3 & z_3 \end{vmatrix} = 0$

is satisfied by any one of twelve systems of three equations each; find them.

270. Given $x_1^2 + y_1^2 + z_1^2 = 1$, $\quad x_1x_2 + y_1y_2 + z_1z_2 = 0$,

$$x_2^2 + y_2^2 + z_2^2 = 1, \quad x_2x_3 + y_2y_3 + z_2z_3 = 0,$$
$$x_3^2 + y_3^2 + z_3^2 = 1, \quad x_3x_1 + y_3y_1 + z_3z_1 = 0,$$

prove that

$$x_1^2 + x_2^2 + x_3^2 = 1, \qquad x_1y_1 + x_2y_2 + x_3y_3 = 0,$$
$$y_1^2 + y_2^2 + y_3^2 = 1, \qquad y_1z_1 + y_2z_2 + y_3z_3 = 0,$$
$$z_1^2 + z_2^2 + z_3^2 = 1, \qquad z_1x_1 + z_2x_2 + z_3x_3 = 0,$$

and if

$$\lambda \equiv \begin{vmatrix} x_1 & y_1 & z_1 \\ x_2 & y_2 & z_2 \\ x_3 & y_3 & z_3 \end{vmatrix}, \text{ then will } \lambda = \pm 1,$$

and

$$\begin{vmatrix} x_1 - \lambda & y_1 & z_1 \\ x_2 & y_2 - \lambda & z_2 \\ x_3 & y_3 & z_3 - \lambda \end{vmatrix} = 0.$$

271.
$$\begin{vmatrix} x & a & a & b & b & y \\ c & x & a & b & y & d \\ c & c & x & y & d & d \\ d & d & y & x & c & c \\ d & y & b & a & x & c \\ y & b & b & a & a & x \end{vmatrix} \equiv \begin{vmatrix} x-y & a-b & a-b \\ c-d & x-y & a-b \\ c-d & c-d & x-y \end{vmatrix} \times \begin{vmatrix} x+y & a+b & a+b \\ c+d & x+y & a+b \\ c+d & c+d & c+d \end{vmatrix}.$$

272.
$$\begin{vmatrix} x & a & a & p & b & b & y \\ c & x & a & p & b & y & d \\ c & c & x & p & y & d & d \\ q & q & q & 2r & q & q & q \\ d & d & y & p & x & c & c \\ d & y & b & p & a & x & c \\ y & b & b & p & a & a & x \end{vmatrix} \equiv 2 \begin{vmatrix} x-y & a-b & a-b \\ c-d & x-y & a-b \\ c-d & c-d & x-y \end{vmatrix} \times \begin{vmatrix} r & p & p & p \\ q & x+y & a+b & a+b \\ q & c+d & x+y & a+b \\ q & c+d & c+d & x+y \end{vmatrix}.$$

273.
$$\begin{vmatrix} x_1^2 & x_1y_1 & y_1^2 & y_1z_1 & z_1^2 & z_1x_1 \\ x_2^2 & x_2y_2 & y_2^2 & y_2z_2 & z_2^2 & z_2x_2 \\ x_3^2 & x_3y_3 & y_3^2 & y_3z_3 & z_3^2 & z_3x_3 \\ x_4^2 & x_4y_4 & y_4^2 & y_4z_4 & z_4^2 & z_4x_4 \\ x_5^2 & x_5y_5 & y_5^2 & y_5z_5 & z_5^2 & z_5x_5 \\ x_6^2 & x_6y_6 & y_6^2 & y_6z_6 & z_6^2 & z_6x_6 \end{vmatrix}$$
$$\equiv \begin{vmatrix} Y_{12}Z_{45}-Z_{12}Y_{45} & Z_{12}X_{45}-X_{12}Z_{45} & X_{12}Y_{45}-Y_{12}X_{45} \\ Y_{23}Z_{56}-Z_{23}Y_{56} & Z_{23}X_{56}-X_{23}Z_{56} & X_{23}Y_{56}-Y_{23}X_{56} \\ Y_{34}Z_{61}-Z_{34}Y_{61} & Z_{34}X_{61}-X_{34}Z_{61} & X_{34}Y_{61}-Y_{34}X_{61} \end{vmatrix}.$$

$$X_{m,\,n} \equiv y_m z_n - y_n z_m,$$
$$Y_{m,\,n} \equiv z_m x_n - z_n x_m,$$
$$Z_{m,\,n} \equiv x_m y_n - x_n y_m.$$

274. If a, b, c, d, e, f denote the six determinants that can be formed from the array

$$\begin{cases} x, y, z, w, \\ a, \beta, \gamma, \delta, \end{cases}$$

then will $ad + be + cf \equiv 0$.

275. Prove that

$$| a_1 c_2 e_3 f_4 \dots t_n | \, | b_1 d_2 e_3 f_4 \dots t_n |$$
$$- | a_1 d_2 e_3 f_4 \dots t_n | \, | b_1 c_2 e_3 f_4 \dots t_n |$$
$$\equiv | a_1 b_2 e_3 f_4 \dots t_n | \, | c_1 d_2 e_3 f_4 \dots t_n |.$$

276. If A_n, B_n, C_n are the inverse elements of $| a_1 b_2 c_3 |$ with respect to a_n, b_n, c_n, show that

$$\begin{vmatrix} 1 & 1 & 1 \\ a_1 & a_2 & a_3 \\ A_1 & A_2 & A_3 \end{vmatrix} + \begin{vmatrix} 1 & 1 & 1 \\ b_1 & b_2 & b_3 \\ B_1 & B_2 & B_3 \end{vmatrix} + \begin{vmatrix} 1 & 1 & 1 \\ c_1 & c_2 & c_3 \\ C_1 & C_2 & C_3 \end{vmatrix} \equiv 0.$$

277. If $| A_1 B_2 C_3 |$ be the reciprocal of $| a_1 b_2 c_3 |$, and

$$\begin{vmatrix} a_1 & a_2 & a_3 \\ b_1 & b_2 & b_3 \\ c_1 & c_2 & c_3 \end{vmatrix} \begin{vmatrix} a_1 & b_1 & c_1 \\ a_2 & b_2 & c_2 \\ a_3 & b_3 & c_3 \end{vmatrix} \equiv \begin{vmatrix} u & w_1 & v_1 \\ w_1 & v & u_1 \\ v_1 & u_1 & w \end{vmatrix}$$

element for element, and U, V, W be the principal diagonal elements of $| A_1 B_2 C_3 |^2$, then will

$$uU + vV + wW$$
$$= 3 | a_1 b_2 c_3 |^2 + 2 u u_1{}^2 + 2 v v_1{}^2 + 2 w w_1{}^2 - 6 u_1 v_1 w_1.$$

278. If

$$\begin{vmatrix} a & b & c & d \\ b & a & d & c \\ c & d & a & b \\ d & c & b & a \end{vmatrix} \equiv aA + bB + cC + dD,$$

prove that

$$(a + b + c + d)^3 (A + B + C + D)^3$$
$$\equiv \begin{vmatrix} A & B & C & D \\ B & A & D & C \\ C & D & A & B \\ D & C & B & A \end{vmatrix}.$$

279. If
$$a \equiv \begin{vmatrix} x & y & z \\ u & x & y \\ z & u & x \end{vmatrix}, \qquad b \equiv \begin{vmatrix} y & z & u \\ x & y & z \\ u & x & y \end{vmatrix},$$
$$c \equiv \begin{vmatrix} z & u & x \\ y & z & u \\ x & y & z \end{vmatrix}, \qquad d \equiv \begin{vmatrix} u & x & y \\ z & u & x \\ y & z & u \end{vmatrix},$$

prove that

$$x^3 = a \begin{vmatrix} a & b & c \\ d & a & b \\ c & d & a \end{vmatrix}^3 \div \begin{vmatrix} \alpha & \beta & \gamma \\ \delta & \alpha & \beta \\ \gamma & \delta & \alpha \end{vmatrix},$$

in which

$$\alpha \equiv \begin{vmatrix} a & b & c \\ d & a & b \\ c & d & a \end{vmatrix},$$

with similar expressions for β, γ, δ.

280. Given $x^2 - yz = a$, $y^2 - zx = b$, $z^2 - xy = c$, show that

$$(ax + by + cz)^2 = \begin{vmatrix} a & b & c \\ c & a & b \\ b & c & a \end{vmatrix}.$$

281. If $yz - u^2 = a^2$, $zx - v^2 = b^2$, $xy - w^2 = c^2$,
$$vw - xu = d^2, \quad wu - yv = e^2, \quad uv - zw = f^2,$$
prove that

$$\begin{vmatrix} x & w & v \\ w & y & u \\ v & u & z \end{vmatrix} = \begin{vmatrix} a^2 & f^2 & e^2 \\ f^2 & b^2 & d^2 \\ e^2 & d^2 & c^2 \end{vmatrix},$$

and solve the equations.

282. If $|A_0 B_1 C_2 \ldots K_n|$ be the reciprocal of $|a^0 b c^2 \ldots k^n|$, prove that $(-1)^{n-r} A_r / A_n = $ sum of all the products of $b, c, d, \ldots k$, taken $n-r$ at a time.

283. Evaluate

$$\begin{vmatrix} L & M & N \\ l & m & n \\ l^2 & m^2 & n^2 \end{vmatrix} \div \begin{vmatrix} 1 & 1 & 1 \\ l & m & n \\ l^2 & m^2 & n^2 \end{vmatrix},$$

in which

$$L \equiv (a, b, c, d)(l, 1)^3,$$
$$M \equiv (a, b, c, d)(m, 1)^3,$$
$$N \equiv (a, b, c, d)(n, 1)^3.$$

284. Prove that

$$\begin{vmatrix} x & 0 & 0 & x_1^2 & x_1 x_2 & x_2^2 \\ 0 & y & 0 & y_1^2 & y_1 y_2 & y_2^2 \\ 0 & 0 & z & z_1^2 & z_1 z_2 & z_2^2 \\ 0 & z & y & 2 y_1 z_1 & y_1 z_2 + y_2 z_1 & 2 y_2 z_2 \\ z & 0 & x & 2 z_1 x_1 & z_1 x_2 + z_2 x_1 & 2 z_2 x_2 \\ y & x & 0 & 2 x_1 y_1 & x_1 y_2 + x_2 y_1 & 2 x_2 y_2 \end{vmatrix} \equiv |x\, y_1\, z_2|^3.$$

285. Prove that

$$\begin{vmatrix} x & 0 & 0 & x_1^2 & x_2^2 & l \\ 0 & y & 0 & y_1^2 & y_2^2 & m \\ 0 & 0 & z & z_1^2 & z_2^2 & n \\ 0 & z & y & 2 y_1 z_1 & 2 y_2 z_2 & 2p \\ z & 0 & x & 2 z_1 x_1 & 2 z_2 x_2 & 2q \\ y & x & 0 & 2 x_1 y_1 & 2 x_2 y_2 & 2r \end{vmatrix}$$

$$\equiv -2\Delta\{l X_1 X_2 + m Y_1 Y_2 + n Z_1 Z_2$$
$$+ p(Y_1 Z_2 + Y_2 Z_1)$$
$$+ q(Z_1 X_2 + Z_2 X_1)$$
$$+ r(X_1 Y_2 + X_2 Y_1)\},$$

in which

$$\Delta \equiv \begin{vmatrix} x & x_1 & x_2 \\ y & y_1 & y_2 \\ z & z_1 & z_2 \end{vmatrix}. \quad X_1 \text{ is the minor of } x_1 \text{ in } \Delta, \text{ etc.}$$

286. Evaluate

$$\begin{vmatrix} A & H & G & N & x \\ H & B & F & M & y \\ G & F & C & L & z \\ N & M & L & D & w \\ x & y & z & w & 0 \end{vmatrix},$$

in which A, B, C, D, etc., are the complements of a, b, c, d, etc., in

$$\begin{vmatrix} a & h & g & n \\ h & b & f & m \\ g & f & c & l \\ n & m & l & d \end{vmatrix}.$$

287. If $|\, a_1\, b_2\, c_3\,| = 0$, then will

$$\begin{vmatrix} x & a_1 & b_1 \\ y & a_2 & b_2 \\ z & a_3 & b_3 \end{vmatrix} \begin{vmatrix} x & c_1 & a_1 \\ y & c_2 & a_2 \\ z & c_3 & a_3 \end{vmatrix} - \begin{vmatrix} x & b_1 & c_1 \\ y & b_2 & c_2 \\ z & b_3 & c_3 \end{vmatrix}^2 \text{ be a square.}$$

288. If

$$\begin{vmatrix} a & b & c & d & 0 & 0 \\ a_1 & b_1 & c_1 & d_1 & 0 & 0 \\ a_2 & b_2 & c_2 & d_2 & 0 & 0 \\ 0 & a & b & 0 & c & d \\ 0 & a_1 & b_1 & 0 & c_1 & d_1 \\ 0 & a_2 & b_2 & 0 & c_2 & d_2 \end{vmatrix} = 0, \text{ then will}$$

$$\begin{vmatrix} a & c & d \\ a_1 & c_1 & d_1 \\ a_2 & c_2 & d_2 \end{vmatrix} - \begin{vmatrix} a & b & d \\ a_1 & b_1 & d_1 \\ a_2 & b_2 & d_2 \end{vmatrix} \begin{vmatrix} b & c & d \\ b_1 & c_1 & d_1 \\ b_2 & c_2 & d_2 \end{vmatrix} = 0.$$

289. Show that

$$\begin{vmatrix} 0 & b & c & d \\ a & c & d & e \\ 2b & d & e & f \\ 3c & e & f & g \end{vmatrix} + \begin{vmatrix} a & 0 & c & d \\ b & b & d & e \\ c & 2c & e & f \\ d & 3d & f & g \end{vmatrix} + \begin{vmatrix} a & b & 0 & d \\ b & c & c & e \\ c & d & 2d & f \\ d & e & 3e & g \end{vmatrix}$$

$$+ \begin{vmatrix} a & b & c & 0 \\ b & c & d & d \\ c & d & e & 2e \\ d & e & f & 3f \end{vmatrix} \equiv 0.$$

290. Given $A_1 \equiv |l_2 m_3 p_4|^2 \div \{|l_2 m_3| \, |l_3 m_4| \, |l_4 m_2|\},$

$\qquad A_2 \equiv |l_1 m_3 p_4|^2 \div \{|l_1 m_3| \, |l_3 m_4| \, |l_4 m_1|\},$

$\qquad A_3 \equiv |l_1 m_2 p_4|^2 \div \{|l_1 m_2| \, |l_2 m_4| \, |l_4 m_1|\},$

$\qquad A_4 \equiv |l_1 m_2 p_3|^2 \div \{|l_1 m_2| \, |l_2 m_3| \, |l_3 m_1|\},$

find the value of $A_1 l_1 + A_2 l_2 + A_3 l_3 + A_4 l_4.$

291. Given $u_{11} \equiv a_1 x + b_1 y + c_1 z, \qquad u_{22} \equiv b_2 x + a_2 y + d_1 z,$

$\qquad u_{12} \equiv b_1 x + b_2 y + e_2 z, \qquad u_{23} \equiv e_2 x + d_1 y + d_2 z,$

$\qquad u_{13} \equiv c_1 x + e_2 y + c_2 z, \qquad u_{33} \equiv c_2 x + d_2 y + a_3 z,$

$\qquad u_{21} = u_{12}, \quad u_{31} = u_{13}, \quad u_{32} = u_{23},$

$\qquad u_1 \equiv x u_{11} + y u_{12} + z u_{13},$

$\qquad u_2 \equiv x u_{21} + y u_{22} + z u_{23},$

$\qquad u_3 \equiv x u_{31} + y u_{32} + z u_{33},$

$\qquad U \equiv x u_1 + y u_2 + z u_3,$

then will

$$|u_{11}, u_{22}, u_{33}| \, z^2 = 4 \begin{vmatrix} u_{11} & u_{12} & u_1 \\ u_{21} & u_{22} & u_2 \\ u_1 & u_2 & 0 \end{vmatrix} + 6 U |u_{11} u_{22}|.$$

Generalize.

292.
$$\begin{vmatrix} a_0 & a_1 & a_2 & a_3 & a_4 & a_5 & a_6 & a_7 \\ a_1 & a_2 + a_0 & a_3 & a_4 & a_5 & a_6 & a_7 & 0 \\ a_2 & a_3 + a_1 & a_4 + a_0 & a_5 & a_6 & a_7 & 0 & 0 \\ a_3 & a_4 + a_2 & a_5 + a_1 & a_6 + a_0 & a_7 & 0 & 0 & 0 \\ a_4 & a_5 + a_3 & a_6 + a_2 & a_7 + a_1 & a_0 & 0 & 0 & 0 \\ a_5 & a_6 + a_4 & a_7 + a_3 & a_2 & a_1 & a_0 & 0 & 0 \\ a_6 & a_7 + a_5 & a_4 & a_3 & a_2 & a_1 & a_0 & 0 \\ a_7 & a_6 & a_5 & a_4 & a_3 & a_2 & a_1 & a_0 \end{vmatrix}$$

is divisible by

$$\begin{vmatrix} a_2 - a_0 & a_3 & a_4 & a_5 & a_6 & -a_7 \\ a_3 - a_1 & a_4 - a_0 & a_5 & a_6 & a_7 & 0 \\ a_4 - a_2 & a_5 - a_1 & a_6 - a_0 & a_7 & 0 & 0 \\ a_5 - a_3 & a_6 - a_2 & a_7 - a_1 & -a_0 & 0 & 0 \\ a_6 - a_4 & a_7 - a_3 & -a_2 & -a_1 & -a_0 & 0 \\ a_7 - a_5 & -a_4 & -a_3 & -a_2 & -a_1 & -a_0 \end{vmatrix}.$$

Generalize the result obtained.

293. Simplify

$$\begin{vmatrix} a_{11}a_{22} - a_{12}{}^2 & a_{11}a_{32} - a_{13}a_{12} \ldots\ldots a_{11}a_{n2} - a_{1n}a_{12} \\ a_{11}a_{23} - a_{12}a_{13} & a_{11}a_{33} - a_{13}{}^2 \quad \ldots\ldots a_{11}a_{n3} - a_{1n}a_{13} \\ a_{11}a_{24} - a_{12}a_{14} & a_{11}a_{34} - a_{13}a_{14} \ldots\ldots a_{11}a_{n4} - a_{1n}a_{14} \\ \ldots\ldots & \ldots\ldots \quad \ldots\ldots \quad \ldots\ldots \end{vmatrix}$$

in which $a_{pq} = a_{qp}$.

294. If $\nabla(a, b, c)_{2n+1} \equiv \begin{vmatrix} a & b & 0 & 0 & \ldots & 0 & 0 & 0 \\ c & a & b & 0 & \ldots & 0 & 0 & 0 \\ 0 & c & a & b & \ldots & 0 & 0 & 0 \\ 0 & 0 & c & a & \ldots & 0 & 0 & 0 \\ \ldots & \ldots & \ldots & \ldots & \ldots & \ldots & \ldots & \ldots \\ 0 & 0 & 0 & 0 & \ldots & c & a & b \\ 0 & 0 & 0 & 0 & \ldots & 0 & c & a \end{vmatrix}$,

then will

$$\nabla(a, b, c)_{2n+1} = a\nabla(a^2 - 2bc, b^2, c^2)_n.$$

295. Also $\quad \nabla(x, 1, 1)_{2n} = \dot{\nabla}(x^2 - 2, 1, 1)_n + \nabla(x^2 - 2, 1, 1)_{n-1}.$

296. If $f(m, n, p) \equiv \begin{vmatrix} 1 & 1 & 1 & 1 \\ a^m & \beta^m & \gamma^m & \delta^m \\ a^n & \beta^n & \gamma^n & \delta^n \\ a^p & \beta^p & \gamma^p & \delta^p \end{vmatrix}$, prove that

$$(a+\beta+\gamma+\delta)f(m,n,p) - a\beta\gamma\delta f(m-1, n-1, p-1)$$
$$= f(m+1, n, p) + f(m, n+1, p) + f(m, n, p+1).$$

297. If $\Delta \equiv |a_1, b_2, c_3, \ldots k_n|$ and $\Delta\begin{pmatrix} \lambda, \mu, \ldots \\ l, m, \ldots \end{pmatrix}$ denote the result of replacing the lth, mth, $\ldots$ columns of Δ by $a_l, \beta_l, \ldots, a_m, \beta_m, \ldots$, respectively, show that

$$\begin{vmatrix} \Delta\begin{pmatrix} \lambda \\ l \end{pmatrix}, & \Delta\begin{pmatrix} \lambda \\ m \end{pmatrix}, & \ldots \\ \Delta\begin{pmatrix} \mu \\ l \end{pmatrix}, & \Delta\begin{pmatrix} \mu \\ m \end{pmatrix}, & \ldots \\ \ldots & \ldots & \ldots \end{vmatrix} = \Delta^{r-1} \times \Delta\begin{pmatrix} \lambda, \mu, \ldots \\ l, m, \ldots \end{pmatrix},$$

the determinant on the left of "=" being of the order r.

298. $\begin{vmatrix} 1 & 1 & 1 & 1 & 1 & 0 & 0 & 0 \\ 0 & 1 & 1 & 1 & 1 & 1 & 0 & 0 \\ 0 & 0 & 1 & 1 & 1 & 1 & 1 & 0 \\ 0 & 0 & 0 & 1 & 1 & 1 & 1 & 1 \\ 1 & 0 & 0 & 0 & 1 & 1 & 1 & 1 \\ 1 & 1 & 0 & 0 & 0 & 1 & 1 & 1 \\ 1 & 1 & 1 & 0 & 0 & 0 & 1 & 1 \\ 1 & 1 & 1 & 1 & 0 & 0 & 0 & 1 \end{vmatrix} = 5.$

Show that the value of a determinant formed like the above, but with m units and n zeros in each line, is m if m be prime to n, but is zero if m be not prime to n.

299. If $\quad (a, b, c, f, g, h)(\theta\phi, \theta+\phi, 1)^2 = 0,$

and $(a, b, c, f, g, h)(\phi\chi, \phi+\chi, 1)^2 = 0,$

find $(\alpha, \beta, \gamma, \kappa, \lambda, \mu)(\theta\chi, \theta+\chi, 1)^2 = 0,$

and show that if

$$\alpha = a, \ \beta = b, \ \gamma = c, \ \kappa = f, \ \lambda = g, \ \mu = h,$$

then will

$$ac + b^2 + 2bg - 4fh = 0.$$

300. The minors of order $2n - 1$ of a skew symmetric determinant of order $2n$ are divisible by the square root of the determinant.